Biological Treatment of
Hazardous Wastes

Biological Treatment of Hazardous Wastes

Gordon A. Lewandowski
Distinguished Professor and Chairperson
Department of Chemical Engineering, Chemistry, and Environmental Science
New Jersey Institute of Technology
Newark, New Jersey

Louis J. DeFilippi
Independent Consultant, Palatine, Illinois

A Wiley-Interscience Publication

JOHN WILEY & SONS, INC.

New York • Chichester • Weinheim • Brisbane • Singapore • Toronto

CHEM

Copyright © 1998 by John Wiley & Sons, Inc.

All rights reserved. Published simultaneously in Canada.

Library of Congress Cataloging in Publication Data

Lewandowski, Gordon A.
    Biological treatment of hazardous wastes/Gordon A. Lewandowski,
  Louis J. DeFilippi.
       p. cm.
    Includes index.
    ISBN 0–471–04861–5 (cloth: alk. paper)
      1. Hazardous wastes—Biodegradation.   I. DeFilippi, Louis J.
  II. Title.
    TD1061.L48   1998
    628.4′2–dc21   97–10384

Printed in the United States of America

10 9 8 7 6 5 4 3 2 1

CONTENTS

v

▬▬▬ PREFACE

Our purpose in putting this book together is to provide a combination of both fundamental principles and practical applications for the biological treatment of hazardous wastes that goes beyond a formularized exposition of specific solutions for particular pollutants. We hope to accomplish this by including hydrogeological, engineering, and microbiological fundamentals in the context of biotreatment applications. Some of the chapters will appeal to readers interested in the application of mathematics to engineering design considerations, while other readers will be more interested in chapters dealing with microbiology and the like.

We are inclined in favor of the use of biological systems for many aspects of waste treatment. Unlike physical approaches, biotreatment has the potential to transform organic pollutants into innocuous products rather than merely transferring the pollutant to another medium. Furthermore, by comparison to other chemical transformation techniques, such as incineration, biotreatment is generally cheaper and enjoys a greater degree of public acceptance.

Microorganisms can transform virtually any organic compound, whether man-made or naturally occurring. It is up to engineers and scientists using biotreatment techniques to manipulate, whenever possible, environmental conditions (oxygen content, chemical composition, temperature, etc.) in order to effect complete transformation to acceptable products in the most cost-effective manner.

The media in which pollutants occur can be aqueous, gaseous, or associated with sediments and soils, with the possibility of pollutant transfer between these media. In fact, soil-bound pollutants must often be transferred to an aqueous or gaseous phase in order to effect their treatment.

Pollutants can be treated ex situ (which requires design of an engineered reactor in which the biological reactions take place) or in situ (which requires manipulation of the conditions in a naturally occurring subsurface 'reactor'). In situ methodologies are particularly complex, dealing as they do with a hydrogeology that is preexisting and that often presents severe constraints on the treatment process. Both ex situ and in situ reactors can involve microorganisms in a suspended or attached (fixed-film) state. These states may be physiologically different and are treated differently in methods used for engineering analysis and design.

Another critical factor in biological treatment is the presence (or absence) of oxygen or other oxidizing agent (such as nitrate or sulfate). In many fixed-film systems, both aerobic (oxygen-rich) and anaerobic (oxygen-poor) regions can occur in close proximity, relative to the surface of the biofilm.

We hope that the reader will obtain an appreciation of these factors in designing biotreatment processes. All too often, the failure of such processes has been ascribed to the serendipity of living organisms, when in fact such failures are the product of a lack of basic understanding of the complex factors involved.

GORDON A. LEWANDOWSKI
LOUIS J. DEFILIPPI

ACKNOWLEDGMENTS

Dr. Lewandowski would particularly like to thank graduate student Dilip Mandal (and his forbearing wife, Aparna) for his invaluable assistance in correcting manuscripts and transferring them in a uniform manner to disk. Dr. Lewandowski would also like to thank the New Jersey Institute of Technology for granting him a sabbatical, during which (among other tasks) he was able to initiate this book.

Dr. DeFilippi would like to thank his wife, Dr. Irene DeFilippi, for all of her support and suggestions during the editing of this book, and his good friend, F. Stephen Lupton, who has immeasurably enhanced his understanding of microbial processes.

Both editors would like to thank the contributing authors for the high quality of their chapters, their diligence, and their patience.

CONTRIBUTORS

Daniel Abramowicz, General Electric Corporate Research and Development, Schenectady, New York 12301. Natural Restoration of PCB-Contaminated Hudson River Sediments.

Piero M. Armenante, New Jersey Institute of Technology, Department of Chemical Engineering, Chemistry, and Environmental Science, Newark, New Jersey 07102. Suspended-Biomass and Fixed-Film Reactors.

Basil C. Baltzis, Department of Chemical Engineering, Chemistry, and Environmental Science, New Jersey Institute of Technology, Newark, New Jersey 07102. Biofiltration of VOC Vapors; Impact of Biokinetics and Population Dynamics on Engineering Analysis of Biodegradation of Hazardous Wastes.

Kelton D. Barr, Delta Environmental Consultants Inc., 2770 Cleveland Avenue, Roseville, Minnesota 55113. Hydrogeologic Factors Affecting Biodegradation Processes.

Edward J. Bouwer, Department of Geography and Environmental Engineering, Johns Hopkins University, 3400 North Charles Street, Baltimore, Maryland 21218. Design Considerations for In Situ Bioremediation of Organic Contaminants.

Christos Christodoulatos, Center for Environmental Engineering, Stevens Institute of Technology, Hoboken, New Jersey 07030. Bioslurry Reactors.

Louis J. DeFilippi, Independent Consultant, 208 Edgewood Lane, Palatine, Illinois 60067. Introduction to Microbiological Degradation of Aqueous Waste and Its Application Using a Fixed-Film Reactor.

Neal D. Durant, Department of Geography and Environmental Engineering, Johns Hopkins University, 3400 North Charles Street, Baltimore, Maryland 21218. Design Considerations for In Situ Bioremediation of Organic Contaminants.

John A. Hogan, Department of Environmental Sciences, Cook College, Rutgers University, New Brunswick, New Jersey 08903. Composting.

Peter R. Jaffé, Department of Civil Engineering and Operations Research, Princeton University, Princeton, New Jersey 08544. Assessment of the Potential for Clogging and Its Mitigation During In Situ Bioremediation.

David Kafkewitz, Department of Biological Sciences, Rutgers University, Newark, New Jersey 07102. Microbes in the Muck: A Look into the Anaerobic World.

Matthias Kniebusch, Technische Universität Hamburg, Hamburg, Germany. Membrane Biofilm Reactors.

Frank R. Kolb, Wassergütewirtschaft, Technische Universität München, D-85748, Garching, Germany. Membrane Biofilm Reactors.

Agamemnon Koutsospyros, Department of Civil and Environmental Engineering, University of New Haven, New Haven, Connecticut, 06516. Bioslurry Reactors.

Gordon A. Lewandowski, Department of Chemical Engineering, Chemistry, and Environmental Science, New Jersey Institute of Technology, Newark, New Jersey 07102. Impact of Biokinetics and Population Dynamics on Engineering Analysis of Biodegradation of Hazardous Wastes.

Carol D. Litchfield, Department of Biology, George Mason University, Fairfax, Virginia 22030. Pentachlorophenol Biodegradation: Laboratory and Field Studies.

F. Stephen Lupton, AlliedSignal Environmental Systems and Services, Des Plaines, Illinois 60067. Introduction to Microbiological Degradation of Aqueous Waste and Its Application Using a Fixed-Film Reactor.

Frank J. Mondello, General Electric Corporate Research and Development, Schenectady, New York 12301. Natural Restoration of PCB-Contaminated Hudson River Sediments.

Madhu Rao, Department of Biology, George Mason University, Fairfax, Virginia 22030. Pentachlorophenol Biodegradation: Laboratory and Field Studies.

James R. Rhea, HydroQual Inc., 4914 West Genesee Street, Suite 119, Camillus, New York 13031. Natural Restoration of PCB-Contaminated Hudson River Sediments.

Stewart W. Taylor, Bechtel International, Inc., Oakridge, Tennessee. Assessment of the Potential for Clogging and Its Mitigation During In Situ Bioremediation.

Monica Togna, Center for Agricultural Molecular Biology, Cook College, Rutgers University, New Brunswick, New Jersey 08903. Microbes in the Muck: A Look into the Anaerobic World.

Peter A. Wilderer, Wassergütewirtschaft, Technische Universität München, D-85748, Garching, Germany. Membrane Biofilm Reactors.

Liza P. Wilson, National Center for Environmental Assessment, U.S. Environmental Protection Agency, 401M Street S. W. (8620), Washington, D.C. 20460. Design Considerations for In Situ Bioremediation of Organic Contaminants.

Wei-xian Zhang, Department of Civil and Environmental Engineering, Lehigh University, Bethlehem, Pennsylvania 18015. Design Considerations for In Situ Bioremediation of Organic Contaminants.

Biological Treatment of Hazardous Wastes

CHAPTER 1

Suspended-Biomass and Fixed-Film Reactors

Piero M. Armenante

New Jersey Institute of Technology, Department of Chemical Engineering, Chemistry, and Environmental Science, Newark, New Jersey 07102

CLASSIFICATION OF REACTORS FOR BIOLOGICAL TREATMENT

A large number of reactor configurations exist as a consequence of the many parameters involved in any biotreatment process and the possibility of optimizing different aspects of the process. The following is a review of bioreactor classification according to the way in which:

- Mechanical energy is delivered to the reactor's contents
- Gas is sparged and off-gases are collected
- The reactor is operated (continuously, batchwise, or sequencing batch)
- The desired degree of homogeneity is achieved
- High biomass concentration is maintained

Mechanical Energy Delivery Systems

A biological reactor must be able to satisfy a number of different and sometimes contrasting requirements in order to operate properly. Examples of such requirements include the maximization of the microbial concentration in the entire volume of the reactor, the achievement (as much as possible) of good internal homogenization to make the nutrients available to the entire biomass, the dispersion of a sparged gas phase (typically in aerated reactors) to generate a large gas–liquid interfacial area, and the enhancement of mass transfer from air bubbles to microorganisms through sufficiently high turbulence intensity.

Biological Treatment of Hazardous Wastes, Edited by Gordon A. Lewandowski and Louis J. DeFilippi
ISBN 0-471-04861-5 ©1998 John Wiley & Sons, Inc.

In order to accomplish all this, mechanical energy must be supplied to the reactor in one or more of the following ways:

- By mechanical agitators (e.g., stirred reactors)
- By a moving liquid (e.g., jet reactors with a recirculation pump)
- By an expanding gas (e.g., airlift reactors)

In general, mechanically agitated reactors are able to deliver the greatest amount of power per unit liquid mass in the reactor, typically resulting in high gas–liquid mass-transfer rates. The pumping action of many impellers can also provide a good level of homogeneity within the reactor. However, the efficiency of mechanical agitation systems (expressed as amount of oxygen transferred per unit energy delivered) is also much lower than the other two reactor types (especially the expanding gas system).

Gas Sparging and Off-Gas Collection Requirements

Depending on the type of microorganisms used, biotreatment processes can be classified into aerobic or anaerobic. Aerobic processes require that oxygen be supplied to the microorganisms in the bioreactor, typically by sparging air into the liquid waste. Unfortunately, the saturation concentration of oxygen in water is quite small (of the order of 8 mg/L at room temperature). Therefore, aerobic reactors usually have some provisions for dispersing air (or sometimes oxygen-enriched air) through the waste to form small bubbles with a large interfacial area. The transfer of oxygen to the biomass is also enhanced by the turbulence that results from the input of mechanical energy.

Anaerobic reactors do not typically require that any gas be dispersed into the reactor contents (although in some reactor configurations the off-gas is reinjected into the liquid to mix the reactor contents). Therefore the external mechanical energy input they require is typically quite small and limited to that necessary for the generation of a recirculation flow capable of maintaining uniformity of the reactor contents. Since anaerobic metabolism is slower than aerobic metabolism, anaerobic reactors typically require longer retention times (or bigger reactor volumes) than aerobic reactors. In addition, the growth rate of anaerobes is slower than aerobes, which makes continuous anaerobic reactors more susceptible to hydraulic overloading than aerobic reactors. One way in which this problem can be minimized is by immobilizing the microorganisms inside the reactor (Aivasidis and Wandrey, 1988).

Most anaerobic organisms are poisoned by oxygen. In addition, anaerobiosis typically results in the generation of compounds that present an odor problem. Therefore, whereas many aerobic reactors are often open to the atmosphere, anaerobic reactors are not. Closed reactors offer the further advantage of facilitating the collection of anaerobic off-gases that may be rich in methane and can be used as an energy source.

Continuous and Batch Reactor Operations

In *continuous operation* the reactor is continuously fed with the waste stream, while the treated stream is continuously removed from the reactor. Alternatively, a reactor can be operated in a *batch mode* in which the waste material is charged to the reactor, the degradation reaction is allowed to proceed until completion, and the treated waste is discharged.

In *semibatch processes*, the waste material is continuously fed to an otherwise batch-operated reactor. In some applications (e.g., in *sequencing batch reactors*, or SBRs) the reactor is sequentially operated in a batch or semibatch mode by loading the waste to be treated, allowing the biomass to grow and treat the waste, and finally discharging the treated effluent. All these phases of the process are carried out in single reactors, as shown in Figure 1.1. If several such reactors are operated in parallel with a staggered time sequence, the overall process is practically continuous, from a user point of view. SBRs have found applications in a number of treatment processes (Irvine and Busch, 1979; Baltzis et al., 1991).

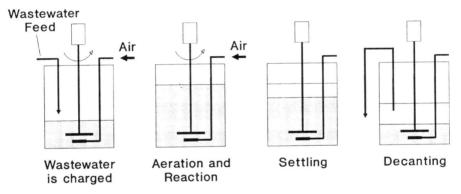

Figure 1.1 Operation of a sequencing batch reactor (after Armenante, 1993).

In continuous reactor operation, the *residence* or *detention time*, θ, defined as the ratio of the reactor volume, V, to the volumetric flow rate, Q:

$$\theta = \frac{V}{Q} \tag{1}$$

is a measure of the average amount of time spent in the reactor by the waste being processed (Levenspiel, 1972). Consequently, the longer the residence time, the larger the reactor volume will be for a given flow rate. Continuous operation is quite common in large-scale operation, especially if the waste is being produced at a uniform rate. However, the ability of a continuous flow reactor to accommodate fluctuations in flow or waste concentration is generally less than that of a sequencing batch reactor.

Batch operation is more common when the amount of waste is small or when the time required for the degradation process is too long for effective continuous treatment. Batch biotreatment is more flexible than continuous treatment since the treatment time can be shortened or lengthened depending on how fast the degradation process proceeds. A drawback of many batch processes is that they are labor intensive (although this limitation is being ameliorated with the advent of advanced microprocessors). In addition, storage of the waste between batches is necessary.

Degree of Homogeneity in Reactors: Well-Mixed vs. Plug-Flow Reactors

Although perfect mixing of reactor contents is only theoretically possible, continuous reactors or batch reactors in which a gas is continuously sparged are often designed as, and often resemble, *well-mixed* systems (Fig. 1.2*a*). In such

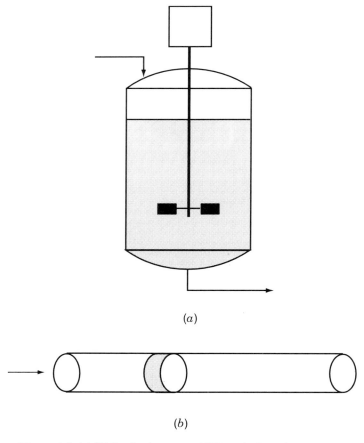

(*a*)

(*b*)

Figure 1.2 (*a*) Well-mixed reactor (CSTR). (*b*) Plug-flow reactor.

cases, the concentration inside the reactor is assumed to be homogeneous because of the presence of a mixing device (e.g., an impeller) or good internal recirculation. This assumption is often made in the design of bioreactors because most biodegradation reactions are typically much slower than the reactor's blending time, that is, the average time required for the homogenization of the reactor contents. This simplifies the modeling of well-mixed batch reactors. For example, if a pollutant does not inhibit growth of the microorganisms, and is furthermore the limiting growth nutrient (as opposed to oxygen, nitrogen, or phosphorous), the unsteady state mass balances for the biomass and the pollutant in a batch reactor can be written, respectively, as (Sundstrom and Klei, 1979; Horan, 1990; Metcalf & Eddy, 1991):

$$\frac{dX}{dt} = \frac{\mu_m C}{K_m + C} X - k_d X \tag{2}$$

$$-\frac{dC}{dt} = \frac{1}{Y} \frac{\mu_m C}{K_m + C} X \tag{3}$$

where X is the biomass concentration, C is the pollutant concentration, μ_m and K_m are the Monod kinetic parameters, k_d is the cell death/endogenous respiration constant, and Y is the yield coefficient (biomass formed per unit of substrate utilized). These equations can be easily integrated if the initial conditions are known.

By contrast, for a well-mixed continuous-flow reactor [also referred to as a *continuous stirred-tank reactor* (CSTR)], the corresponding steady-state mass-balance equations are (Bailey and Ollis; 1986; Horan, 1990; Metcalf & Eddy, 1991; Sundstrom and Klei, 1979):

$$\frac{X - X_{in}}{\theta} + \frac{\mu_m C}{K_m + C} X - k_d X = 0 \tag{4}$$

$$\frac{C_{in} - C}{\theta} - \frac{1}{Y} \frac{\mu_m C}{K_m + C} X = 0 \tag{5}$$

where the subscript '*in*' denotes the concentration in the incoming stream, θ is the residence time, and C is the pollutant concentration in the effluent (which is equal to the concentration in a perfectly mixed reactor). More complicated equations can be developed for the case in which a recycle is present (Metcalf & Eddy, 1991; Sundstrom and Klei, 1979).

At the other extreme of the operating spectrum a reactor can be operated as a *plug-flow* system in which the liquid moves through the reactor as it would ideally in a narrow pipe, that is, without any mixing fluid elements with those preceding or following it in the pipe (Fig. 1.2*b*). The mathematical representation of such a system can be obtained by performing a material balance in an infinitesimally thin section of the plug-flow reactor resulting in the following steady-state equations (Metcalf and Eddy, 1991; Sundstrom and Klei, 1979):

$$\frac{Q}{A} \frac{dX}{dZ} = \frac{\mu_m \, C}{K_m + C} X - k_d \, X \tag{6}$$

$$\frac{Q}{A} \frac{dC}{dZ} = -\frac{1}{Y} \frac{\mu_m \, C}{K_m + C} X \tag{7}$$

where Z is the axial distance along the reactor, and A is the cross-sectional area of the reactor. These equations can be integrated if the boundary conditions are known. In such a reactor, biomass must either be present in the incoming feed or immobilized on packing material within the reactor for degradation to occur.

Since the rate of degradation is typically proportional to the concentration of the pollutant, these equations would predict that the overall degradation process is more efficiently carried out (e.g., smaller reactor volume for a given flow) in plug-flow rather than well-mixed systems (assuming ideal behavior in both). In well-mixed systems the concentration in the reactor is the same as the exit concentration, which must be kept low for effective treatment of the waste. In practice, the choice between these two reactor configurations (well-mixed vs. plug-flow) depends on a number of other design and operating factors (Sundstrom and Klei, 1979; Metcalf and Eddy, 1991). Most reactors actually operate somewhere between these two ideal modes.

Suspended Biomass and Fixed Films

Microorganisms can grow in suspended culture and/or attached to a surface forming a microbial film. However, since the waste material to be treated in bioreactors is typically dissolved in an aqueous environment, it is advantageous to utilize microorganisms capable of attacking the target waste compounds where they are, that is, in solution. Suspended microbial cultures have the additional advantage of being more easily supplied with oxygen (if required) by sparging air in the liquid waste. Therefore, many bioreactor configurations utilize suspended biomass and are designed and operated accordingly. They differ primarily by the way mechanical energy for air sparging, mass transfer, and internal recirculation is supplied.

When continuous, open systems are used in which the treated waste stream is continuously removed from the reactor, the suspended biomass (i.e., the reactor "biocatalyst") is partially but continuously lost with that stream. Unlike other reactors, bioreactors have the advantage of internally regenerating their biocatalyst through microbial growth. Therefore the biocatalyst loss may not be as significant, provided that the rate of microbial growth is equal to the rate of loss at the outlet. Since this is not typically the case in industrial applications, continuous suspended-biomass reactors are often provided with a recycle system that separates the biomass from the treated waste stream and partially recycles it back into the reactor (this is the principle of operation of any conventional activated sludge process). As a consequence, the separating device, typically a clarifier, becomes an integral and essential part of the overall reactor design.

Another way of retaining the biomass inside the bioreactor is by immobilizing it on a solid support. This is not particularly difficult to achieve since most microorganisms show a preference for attaching themselves to solid surfaces. Many bioreactors are designed with this approach in mind, and many design configurations are possible. They differ primarily in the way the liquid waste is supplied to the biofilm (e.g., by percolating the liquid waste on the biofilm, as in trickle-bed reactors, or by feeding it from the bottom as in upflow anaerobic reactors), the way oxygen is supplied to the biofilm (e.g., by periodic exposure of the biofilm to air, as in rotating biological contactors, or by gas sparging), and the way in which the solid support is retained within the reactor (e.g., packed-bed reactors or fluidized bed reactors).

SUSPENDED-BIOMASS REACTORS

Stirred Reactors

Stirred reactors are vessels provided with a mechanical agitation system. The reactors can be open to the atmosphere, as in the case of large-scale aerobic biological treatment, or closed, as in the case of anaerobic digestors. The vessel construction material is typically steel or concrete, the latter application being especially common in activated sludge treatment plants in which the reactor is open to the atmosphere.

If the reactor is cylindrical in shape, baffles must be provided to prevent swirling and vortex formation. Baffles are vertical fins having a width typically within $\frac{1}{12}$ to $\frac{1}{10}$ of the vessel diameter, placed near the wall, and spaced 90° apart. Baffles may not be required if the tank has a square cross section (as in most aeration basins in the activated sludge process), or if the impeller is placed nonsymmetrically or nonvertically (Oldshue, 1983).

In stirred reactors, the agitation system typically consists of one or more externally driven impellers placed either in the fluid bulk or near the air–liquid interface. As these impellers rotate they produce:

- Homogenization of the reactor's content to attain uniform concentration and temperature distribution
- Dispersion of a gas (typically air) in the liquid
- Suspension of solids in the liquid
- Dispersion of immiscible liquids
- Heat exchange, if necessary

Hence, a single piece of equipment—the impeller—must be designed and operated to satisfy most if not all the requirements specified above. Different types of impeller exist for this purpose. Figure 1.3 shows some common types of submerged impellers.

Radial Flow Impellers

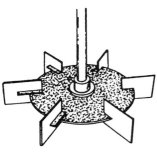

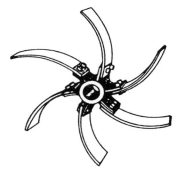

Disk Style Flat Blade Turbine
Commonly Referred to as
the Rushton Impeller

Sweptback or Curved Blade Turbine
(a Spiral Turbine)

Axial Flow Impellers

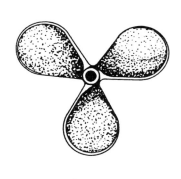

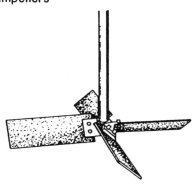

Propeller

45° Pitched Blade Turbine

Figure 1.3 Common types of submerged impellers (after Tatterson, 1991).

A simple impeller classification is based on the types of predominant flow pattern generated in baffled vessels, which, in turn, results in different values of the impeller power consumption. Impellers are classified as radial, axial, and mixed flow. Radial impellers produce a strongly radial discharge flow near the impeller, such as that produced by disk turbines (having horizontal blades sweeping through the fluid). As this radial jet impinges on the vessel wall, it is diverted upward and downward forming the typical "double eight" circulation pattern shown in Figure 1.4a. Axial impellers, such as marine propellers or many of the recently developed fluid foil impellers, pump the liquid downward (or upward, depending on the rotation direction) because of their inclined blades, thus creating a strong top-to-bottom

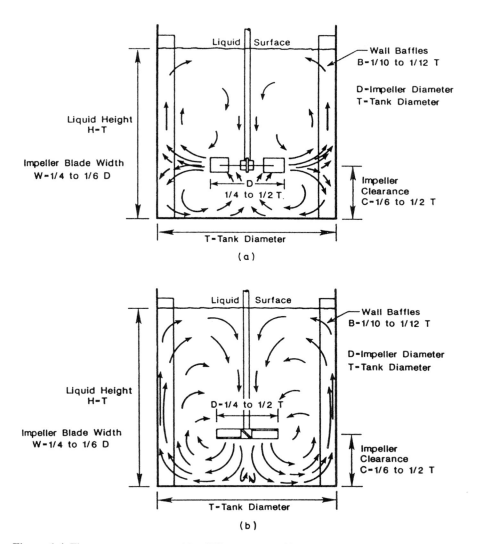

Figure 1.4 Flow patterns generated by different types of impellers: (*a*) radial-flow impeller; (*b*) axial-flow impeller (after Tatterson, 1991).

recirculation pattern (Fig. 1.4*b*). Finally mixed-flow impellers, such as the 45° pitched-blade turbine, produce an intermediate type of flow pattern (Harnby et al., 1985). However, because of the strong axial component they generate, pitched-blade turbines are often also classified as axial-flow impellers.

One of the most important characteristics of impellers is their power consumption, that is, the mechanical power that must be supplied by the motor to overcome the resistance of the impeller as it rotates in a fluid. Impeller power consumption is a function of a large number of parameters, primarily the agitation speed, the type, size, and position of the impeller, the type and size of the vessel, and the characteristics of the fluid. In nondimensional terms, this dependence can be expressed as (Bates et al., 1966):

$$\frac{P}{\rho N^3 D^5} = N_p = f\left(\frac{D^2 N \rho}{\mu}, \frac{DN^2}{g}, \frac{D}{T}, \frac{D}{H}, \frac{D}{C}, \frac{D}{W}\right) \tag{8}$$

in which the term on the left-hand side is the nondimensional power consumption, also referred to as the power number, N_p, and the terms in parenthesis are, respectively, the impeller Reynolds number, Re ($= D^2\rho/\mu$), the Froude number, Fr ($= DN^2/g$), and a number of geometric ratios including the impeller diameter, D, the vessel diameter, T, the vessel height, H, the impeller clearance off the vessel bottom, C, and the impeller blade width, W. For the case in which geometrically similar systems are examined, all the geometric ratios in this equation are constant. In baffled systems in which no vortexing occurs (and hence the gravitational effects are unimportant), the Froude number, (Fr $= DN^2/g$) is unimportant. The power equation then becomes (Bates et al., 1966):

$$\frac{P}{\rho N^3 D^5} = N_p = f\left(\frac{D^2 N \rho}{\mu}\right) = f(\text{Re}) \tag{9}$$

This equation indicates that the power number is only a function of the impeller Reynolds number in geometrically similar, baffled agitation systems. The shape of the function $N_p = f(\text{Re})$ depends not only on the type of impeller but also on the agitation regime (turbulent, transitional, or laminar). Figure 1.5 shows a plot of N_p vs. Re for different impeller configurations in which $D = T/3$, $H/T = 1$, and $C/D = 1$. If the flow is turbulent (by far the most common occurrence in the biotreatment of wastes), then N_p becomes a constant independent of Re and the power consumption, and can be calculated from:

$$P = N_p \rho N^3 D^5 \tag{10}$$

Power numbers in turbulent regime range from about 0.3 for marine propellers or fluid foil impellers to about 5 for six-blade disk turbines (Bates et al., 1966; Oldshue, 1983).

A single impeller is typically used in cylindrical vessels for which $H/T < 1.5$. If the H/T ratio is greater, multiple impellers arranged on the same shaft are used. The power consumed by multiple-impeller assemblies depends on the types of impeller and the spacing between them, S. For example, when two disk turbines are placed relatively close to each other ($S/D < 1$), they consume only a fraction of the sum of the powers that would be required by each impeller if mounted alone. However, if the spacing between them is significant ($S/D > 2$), the total power consumed is nearly twice the power dissipated by a single impeller mounted alone. On the other hand, the power consumed by two pitched-blade turbines mounted on the same shaft

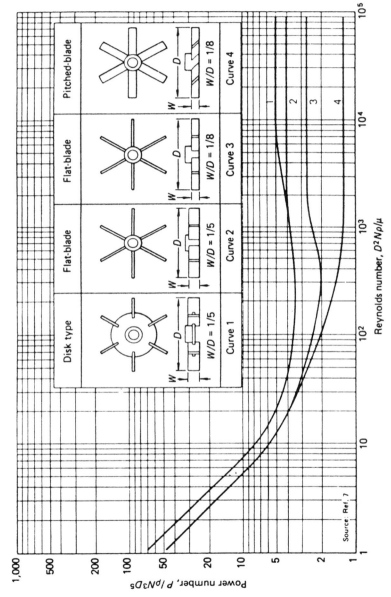

Figure 1.5 Power number–Reynolds number correlation for various types of impeller designs (after Bates et al., 1963).

is always smaller than twice the power of a single impeller alone, even when the spacing is very large (S/D up to 4) (Bates et al., 1963, 1966; Oldshue, 1983).

The above discussion applies to impellers rotating in homogenous fluids. However, in most aerobic bioreactors, air is sparged just beneath the impeller since one of the most important functions of the impeller is to disperse air across the reactor (Fig. 1.6). The power drawn by the impeller under aerated conditions is typically a fraction of that drawn in the absence of the gas phase. The ratio P_g/P (where P_g is the power drawn under gassed conditions) is a complex function of the impeller type and size, the agitation speed, the aeration rate, Q_g, as well as the characteristics of the gas and the liquid involved (since these may have a significant impact on the size and coalescing property of the gas bubbles). A plot of P_g/P as a function of the gas flow number, Q_g/ND^3, typically shows a rapid drop of P_g/P from 1 to between 0.3 and 0.6 as the gas flow number approaches 0.035–0.05. In most practical cases the ratio P/P_g is in the range 0.4–0.7 (Middleton, 1985; Tatterson, 1991). Different equations have been proposed to calculate the aerated power consumption. Of these, the most widely used is the Michel–Miller equation (Michel and Miller, 1962; Bailey and Ollis, 1986):

$$P_g = m\left(\frac{P^2ND^3}{Q_g^{0.56}}\right)^{0.45} \tag{11}$$

where m is an experimentally determined constant. Despite its common use, this equation has the disadvantage of being dimensional, and of blowing up when Q_g goes to zero.

Power dissipation is not only important to determine the energy requirement of a process but also because the performance of a reactor, especially as far as mass transfer is concerned, depends on it. In general, the external mass transfer between the bulk of the fluid and individual microorganisms or microbial flocs is not the limiting step in the oxygen transfer process. Much more common is the case in which the mass transfer of oxygen from the air bubbles to the liquid is limiting. The reason for this is indicated by the mass transfer equation:

$$M_{O_2} = k_L a\,(C_{O_2s} - C_{O_2}) \tag{12}$$

in which the rate of oxygen transfer per unit liquid volume, M_{O_2}, is a function of the oxygen saturation concentration in water, C_{O_2s}, the actual oxygen concentration in water, C_{O_2} (assuming that the liquid side mass transfer is limiting, as in most cases), the mass-transfer coefficient, k_L, and the interfacial area per unit volume, a. The driving force ($C_{O_2s} - C_{O_2}$) is typically very small due to the low solubility of oxygen in water. Even if the water is saturated with oxygen (saturation concentration: 10.15 mg/L at 15°C) and the value of C_{O_2} is as low as the critical value for most aerobic organisms (typically about 0.3–0.7 mg/L, i.e., $3\,K_{mO_2}$) (Bailey and Ollis, 1986), the driving force can be only as large as 9.5–9.8 mg/L. Such low value of the driving force can result in low oxygen transfer rates. To partially compensate for the reduced oxygen, solubility mechanical energy is used to increase the values of k_L

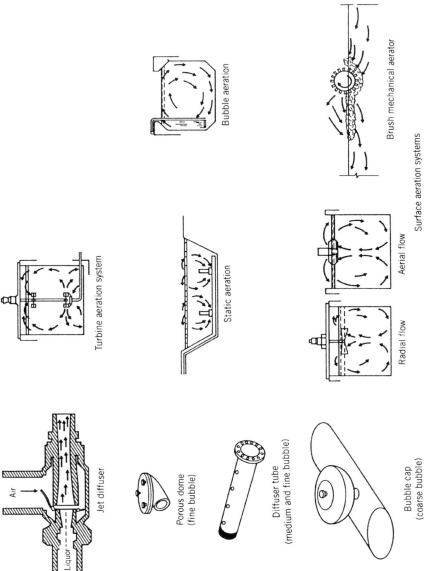

Figure 1.6 Different types of aeration systems (after Eckenfelder, 1989).

and a, which are both functions of power consumption. Since it is quite difficult to measure these two parameters independently, it is common practice to produce correlations for their product, $k_L a$. A typical correlation for $k_L a$ is:

$$k_L a = \beta \left(\frac{P_g}{V} \right)^{0.7} v_g^{0.6} \tag{13}$$

where v_g is the superficial gas velocity and β is a constant of the order of 1.2–2.3 (in SI units; Middleton, 1985). Typical oxygen transfer efficiencies in stirred reactors are of the order of 1–1.8 kg O_2 transferred per kWh consumed (Armenante, 1993; Eckenfelder, 1989). This equation shows that increasing the impeller power dissipation does not produce a linear increase in $k_L a$. Equations have also been developed to calculate gas holdup. Although no correlation of general applicability has been proposed, different equations exist for gas–water systems (in which the gas bubbles have a strong tendency to coalesce) as well as gas–electrolyte solution systems (in which coalescence is significantly reduced) (Tatterson, 1991).

Another important characteristic of agitated reactors is the degree of homogeneity that can be achieved as a result of mechanical agitation. This depends not only on the flow distribution inside the reactor, but also on the way the reactor is operated (e.g., CSTR, plug-flow, batch). In general, a measure of the homogeneity achievable in a reactor is provided by the mixing time, θ_m, defined as the time required to achieve a predefined spread of concentrations of a tracer injected in the reactor.

In nondimensional terms, the typical equation for mixing time in baffled, nonaerated cylindrical vessels is (Gray, 1966):

$$\theta_m N = \gamma \left(\frac{N^2 D}{g} \right)^{1/6} \left(\frac{H^{0.5} T^{1.5}}{D^2} \right) \tag{14}$$

where γ is a constant having a value typically in the range 5–9. General correlations for mixing time in aerated vessels have also been produced and are reviewed elsewhere (Tatterson, 1991). Equations such as Eq. (14) show that to keep the mixing time constant in geometrically similar systems the term $\theta_m N$ must approximately be kept constant during scale-up, which implies maintaining nearly constant agitation speed upon scale-up. This may be extremely difficult, if not impossible, to achieve since power consumption increases dramatically with both scale ($P \propto D^5$) and agitation speed ($P \propto N^3$). For example, if a reactor is to be scaled up by a factor of 20, keeping constant impeller speed would mean increasing the power consumption per unit liquid volume in the reactor, P/V, by a factor of 400, a nearly impossible feat in most cases. To keep P/V constant in the same scale-up process would require a *decrease* in the agitation speed by a factor of 7. On the other hand, mixing time becomes important only if the process of interest has a characteristic time of the same order of magnitude of, or smaller than, the mixing time. For example, if an extremely fast reaction is occurring in the reactor upon the addition of a reactant, it is intuitive that even a "short" mixing time of a few tens of seconds may lead to inhomogeneities in the vicinity of the injection point and result in an incomplete reaction or by-product formation. On the other hand, if the same reaction takes many hours to complete, even a "long" mixing time of the order of several minutes may have no impact. Most biological reactions are very slow. Hence, long-term homogenization of the bioreactor may well be achieved even when mild agitation is present. On the contrary, oxygen depletion in the presence of active biomass is a relatively fast process (even starting with an oxygen-saturated medium), unless oxygen is continuously supplied and distributed throughout the bioreactor. This is an additional reason why anaerobic bioreactor typically require low mechanical energy input (just to keep the reactor contents homogeneous, and the biomass suspended), whereas aerobic reactors need to be vigorously agitated, not only to disperse air bubbles and create large gas–liquid interfacial area, but also to quickly distribute the rapidly depleting dissolved oxygen throughout the entire reactor, thus preventing the formation of anoxic regions.

Suspended solids in bioreactor are typically microbial flocs that are relatively easily maintained in suspension. However, solid suspension is also a design requirement in a bioreactor. Several correlations have been proposed to determine the minimum agitation speed, N_m, to achieve off-bottom solid suspension. The most widely used is the Zwietering correlation (Zwietering, 1958; Gray and Oldshue, 1986):

$$N_m = s \frac{d_p^{0.2} \mu^{0.1} (g \Delta \rho)^{0.45} B^{0.13}}{\rho^{0.55} D^{0.85}} \tag{15}$$

which relates N_m to a number of geometric or physical parameters. The constant s is a function of the impeller type and geometric parameters such as the impeller-to-tank diameter ratio or the impeller off-bottom clearance.

Mechanical agitation is one of the most common features of suspended-growth reactors and is found in both aerobic systems (typically coupled with air sparging) and anaerobic digestors (e.g., the anaerobic contact process) in which intimate mixing between the nutrient-rich liquor and the suspended microorganisms is vital to proper operation.

Surface Aerated Reactors

In conventional aerated stirred reactors, compressed air is sparged within the liquid bulk, well under the liquid-free surface, so that the impeller can disperse it. An alternative approach is to place the impeller close enough to the surface to produce air entrainment directly from the atmosphere without the need for compressed air. This is the approach used in surface-aerated reactors.

Most impellers tend to entrain atmospheric air if their liquid coverage is below $0.5D$. In surface aeration, the impeller is typically placed at or near the free surface (Fig. 1.6). Oxygenation of the liquid occurs as a result of the splashing and spraying of the liquid around the impeller, resulting in a large interfacial area between the liquid droplets and the air. Surface aerators are classified according to the direction of the liquid flow they generate (radial, axial, or mixed), and their rotational speed (high or low). Two basic configurations are common: (1) a low-speed type in which a radial impeller placed at the liquid surface produces a radial liquid flow responsible for entraining the air, and (2) a high-speed type, typically mounting an axial impeller placed in a draft tube below the surface, and producing an upward, axial flow, then diverted radially, spraying liquid above the liquid surface.

The oxygen transfer efficiency of surface aerators is quite high, typically in the range of 1.5–2.5 kg O_2 transferred per kWh consumed (Zlokarnik, 1979; Schügerl, 1987). Correlations have been proposed in which the oxygen transfer rate per unit power consumed was shown to be proportional to the power consumption per unit surface area (Eckenfelder, 1989). Other investigators have established a weak dependence of the oxygen transfer rate per unit power consumed and the power consumption per unit liquid volume, P/V (Schügerl, 1987).

Surface aeration is especially common in large basins such as lagoons in which the surface aerator assembly (including the motor) floats on the liquid, thus providing aeration even when significant variations in the basin's liquid level occur. Alternatively, the aerator is suspended from a fixed rig above the reactor.

Surface aerators also produce homogenization of the liquid bulk and suspension of solids off the reactor bottom. However, because of their location the suspension activity of surface aerators can only be effective if the basin's depth is typically within 3–5 m (Eckenfelder, 1989).

A different type of surface aerator is the brush mechanical aerator, consisting of a horizontal cylinder provided with metal fins and rotating at high speed at the liquid–air interface. The oxygen transfer rate of these aerators is proportional to the power consumed and is typically in the range of 1.5–2 kg O_2/kWh (Schügerl, 1987; Eckenfelder, 1989; Metcalf & Eddy, 1991).

Reactors Using Diffused Aeration or Gas Dispersion

In a number of aerobic treatment facilities oxygenation is achieved by sparging air more or less homogeneously throughout the reactor. Diffusers placed at the bottom of the reactor are used for this purpose. Two kinds of diffuser are typically used (Fig. 1.6). One type consists of porous material devices through which air is forced, producing fine or medium size bubbles with large interfacial area and long residence time in the reactor. These diffusers require the use of filtered air to prevent clogging. The second type of diffuser produces coarser bubbles using a number of simpler devices such as inverted bubble caps or perforated pipes or sieve plates. These diffusers are easier to maintain and more flexible to operate than the previous type.

Air sparging without mechanical agitation is used in bubble columns (i.e., tall columns in which the gas is sparged at the base of the column), as well as in large aeration basins in activated sludge treatment plants. Since mechanical agitation is eliminated, the power consumption in diffused aeration systems is quite modest, typically of the order of 0.1–0.5 kW/m^3. However, the oxygen transfer per unit power consumed is quite high and typically in excess of 3 kg O_2/kWh. An estimation of the mass–transfer coefficient in bubble columns is given by the equation (in SI units; Schügerl, 1987, p. 161):

$$k_L a = 0.0023 \left(\frac{v_g}{d_b} \right)^{1.58} \tag{16}$$

One of the drawbacks of diffused aeration is that the liquid circulation pattern is complex, the recirculation time possibly long, and the turbulence intensity at the bottom of the reactor quite low. Hence solid suspension can be a problem. Different arrangements can be made to increase the level of recirculation. The first is to locate the air-sparging diffusers only in a section of the reactor. For example, if the diffusers are staggered near one of the vertical walls of an aeration basin, the presence of the bubble swarm will produce a gas-induced liquid rising flow near that wall. This, in turn, creates a recirculation pattern throughout the reactor (Fig. 1.6).

Another common alternative is that used in airlift (or gaslift) reactors, containing a draft tube within the reactor (Fig. 1.7). Air is sparged at the bottom of the draft tube. This results in an upward gas–liquid flow within the tube at the end of which most of the bubbles disengage while the liquid flows downward in the annulus external to the draft tube. These types of columns are also quite efficient, with power consumptions comparable to those of bubble columns and specific oxygen transfer rates above 4 kg O_2/kWh (Hines et al., 1975; Schügerl, 1987).

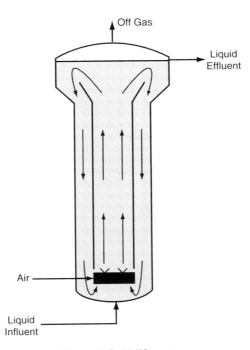

Figure 1.7 Airlift reactor.

It should not be assumed that gas-induced liquid recirculation is limited to aerobic reactors. In anaerobic digestion, mixing is essential to distribute the nutrients throughout the suspended anaerobic biomass. Since there is no need for intense mixing (no high gas–liquid interfacial area must be generated, and no oxygen must be distributed), the less turbulent recirculation pattern produced by gas sparging is often more than adequate to provide a sufficient level of homogenization. In this case the fermentation gas collected at the top of the reactor is compressed and sparged at the bottom of the reactor using one of the methods described above, such as gas diffusion, gaslift (as opposed to airlift) via draft tubes, or gas injection in a section of the reactor.

Jet Reactors

In jet reactors, internal recirculation of the reactor's content is provided by an external pump that is fed through an external loop with liquid taken from the reactor. The pump discharges back into the reactor, typically through a nozzle, so that a liquid jet is formed within the reactor maximizing recirculation effectiveness. By properly locating the intake and discharge points of the external loop and by using internals (such as draft tubes) within the reactor, an effective circulation pattern can be obtained. If the reactor is aerated or gas sparged, the gas may be introduced with the nozzle using a venturi orifice (Jackson, 1964; Prokop et al., 1982; Tojo and Miyanami, 1982). The shearing action of the forming jet disperses the gas promoting the formation of small bubbles. In addition, the liquid jet may produce a suspension of solids or microbial flocs similar to what occurs in fluidized beds. Liquid jets are encountered in anaerobic digesters to promote internal liquid recirculation as well as in large aerobic bioreactors where they are also employed as air diffusers (Fig. 1.6). Typical oxygen transfer rate efficiencies are of the order of 1.5 kg O_2/kWh but can be as high as 3.8 kg O_2/kWh (Schügerl, 1987).

The shearing action generated by the pump or in the nozzle (where the gas can be introduced) can also have, at least in principle, an impact on the microorganisms. In general, individual microorganisms are too small to be affected by such shearing action (since even the smallest turbulent eddy may be many times larger than them) unless they are especially sensitive to shear. This is not typically the case in wastewater treatment. Microbial flocs or filamentous organisms can, however, be disrupted by this mechanism and may become impaired at shear stress levels even lower that those produced by a rapidly rotating pump vane. For example, Armenante et al. (1992b) argued that the efficiency in the biodegradation of 2-chlorophenol by the filamentous organisms *Phanerochaete chrysosporium* decreased when the fungus was not protected against the shear stress generated by impellers.

Liquid jets are also utilized in a different type of reactor design, namely impinging-jet reactors. In these reactors, the liquid is recirculated through the external loop via a pump and then forced through a high-speed jet nozzle placed inside the reactor above the liquid level. As the jet emerges from the nozzle, it entrains some of the gas in the reactor headspace, carrying it inside the liquid bulk in the form of small bubbles. As the jet plunges into the reactor liquid, it sprays liquid droplets in the gas, further promoting gas transfer. The k_La in this type of reactor is directly proportional to the power supplied to the liquid jet. An equation of the type

$$k_La = 9 \times 10^{-2}\, \frac{P}{V} \tag{17}$$

(in SI units) can be used to predict k_La (Schügerl, 1987). The oxygen transfer rate efficiencies are in the range of 1–8 kg O_2/kWh (Tojo and Miyanami, 1982; Schügerl, 1987).

Biomass Separators

In continuous, suspended-growth treatment processes, such as the activated sludge process or the anaerobic contact process, the rate of biomass growth due to waste degradation is often smaller than the rate of biomass lost with the effluent. If left uncorrected, this would lead to washout of the reactor. Therefore, the suspended biomass leaving the reactor is often separated from the reactor effluent and partially recycled back to the reactor (Fig. 1.8). Biomass separation also serves the purpose of removing the suspended solids in the treated wastewater prior to its discharge. Although biomass separation equipment is not, strictly speaking, an element of most reactors, it is nevertheless an integral part of the overall system, since without it most continuous bioreactors could not operate.

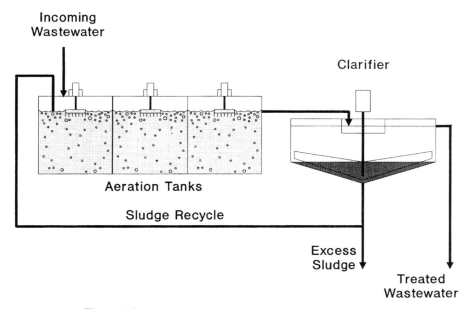

Figure 1.8 Activated sludge process (after Armenante, 1993).

Some reactors are designed in such a way as to have a zone where biomass settling occurs prior to the effluent being discharged from the reactor, as in the case of Hoechst's Biohoch Reactors (Mudrack et al., 1987). However, in the vast majority of cases biomass separation is conducted in a separate device—the clarifier. Although different types of clarifiers exist (primarily rectangular or circular), their operating principle is always the same and is based on biomass sedimentation.

Rectangular clarifiers consist of elongated basins through which the effluent from the reactor travels at a sufficiently low velocity. As the treated wastewater moves along the clarifier, the biomass settles. By the time the clarified, treated wastewater leaves the clarifier, the biomass has settled to the bottom of the clarifier where it is collected and partially recycled to the reactor. Circular clarifiers work in a similar fashion except that the feed is typically introduced in the center and the clarified effluent is collected at the rim (Fig. 1.9). Design equations for clarifiers can be found in a number of references (Talmage and Finch, 1955; Sundstrom and Klei, 1979; Corbitt, 1990; Metcalf & Eddy, 1991; Christian, 1994).

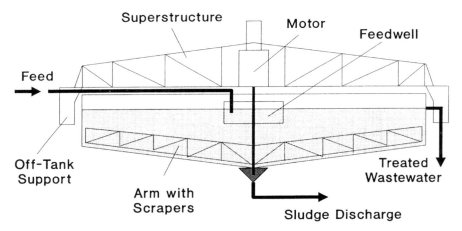

Figure 1.9 Side view of a circular clarifier (after Armenante, 1993).

The operation of the clarifier is vital to most suspended treatment processes. The vast majority of failures of activated sludge processes occur precisely because the biomass does not settle rapidly enough. This condition, known as "bulking," occurs when filament organisms create bridges between the microbial flocs, preventing them from agglomerating and settling. Failure of the biomass to settle results in loss of biomass with the overflow of the clarifier and the failure of the entire treatment process. For this reason, suspended-growth bioreactors for continuous treatment should always be designed together with their biomass separation units.

FIXED-FILM REACTORS

Packed-Bed Reactors

Packed-bed reactors consist of vessels filled with granular material, or with other packing materials that provide a high internal surface area per unit reactor volume. Packed-bed reactors can be designed and operated for two different types of service: with continuous gas phase or with continuous liquid phase. The former type, also referred to as trickle-bed reactor in the wastewater industry, is examined in the next section. In the latter type of packed-bed reactor, the liquid is fed from the bottom of the reactor and completely fills the reactor. If a gas phase is present, it is dispersed by supplying it from the bottom of the reactor. The presence of the packing enhances gas dispersion by breaking up the gas bubbles and preventing them from coalescing. It also enhances contact between the gas and liquid phases, and the biofilm attached to the packing surfaces.

Different types of packing materials are used in packed-bed reactors, such as gravel, ceramic materials (e.g., Rashig rings), or plastic materials in a variety of forms. The microorganisms colonize the surface of the packing by forming a biofilm that constitutes the biocatalyst of the reactor. The immobilization of the microorganisms is especially important in the case of anaerobic reactors because it

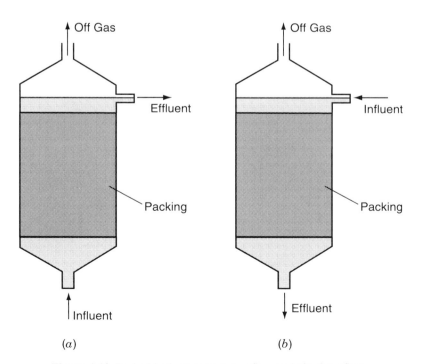

Figure 1.10 Packed-bed reactors: (*a*) upflow and (*b*) downflow.

minimizes the possibility of washout, a problem especially significant in anaerobic systems because of the slow growth rate of anaerobes.

Packed-bed reactors are used in both anaerobic and aerobic applications. Anaerobic packed-bed reactors (also referred to as anaerobic filters, although no filtration takes place in them) are fed from either the top or the bottom (Fig. 1.10). The packing material is typically coarse (20–60 mm) because of the tendency of anaerobic filters to plug as a result of microbial growth in the interstitial space of the packing. Anaerobic packed-bed reactors are typically operated as plug-flow reactors, with or without a recycle. The methane gas produced during the anaerobic process is collected at the top of the reactor. Anaerobic reactors of this kind have been successfully used to treat high chemical oxygen demand (COD) wastewaters (incoming concentration: 10,000–20,000 mg/L; removal efficiency: 75–85%) with residence times between 24 and 48 h and organic loadings of 1–5 kg COD/m^3 day (Metcalf and Eddy, 1991). Processes associated with higher removal rates (4–10 kg COD/m^3 day) and shorter residence times (4–18 h) have also been reported (Armenante, 1993; Schügerl, 1987).

Figure 1.11 shows an example of the application of a packed-bed reactor to the removal of 2,4,6-trichlorophenol (2,4,6-TCP) from a wastewater. In this two-step biotreatment system anaerobic microorganisms were immobilized in a downflow packed-bed reactor in which reductive dechlorination of 2,4,6-TCP to

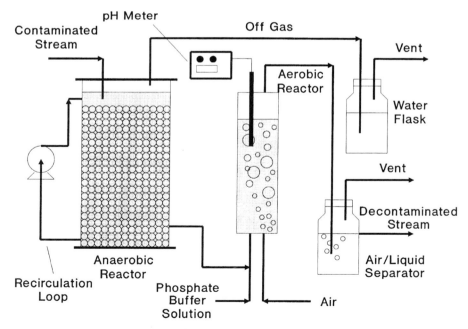

Figure 1.11 Sequential anaerobic–aerobic bioprocess for the treatment of 2, 4, 6–trichlorophenol (after Armenante et al., 1992a).

2,4-dichlorophenol and eventually to 4-chlorophenol (4-CP) took place. The effluent from the anaerobic reactor was continuously fed to a suspended growth aerobic reactor in which the 4-CP produced in the anaerobic reactor was mineralized (Armenante et al., 1992a). This system clearly shows the advantage of using an immobilized-film reactor, such as the packed-bed reactor, to retain the slow-growing anaerobic microorganisms within the reactor. In this system there was no need to immobilize the aerobic microorganisms in the aerobic reactor since the aerobes grew more rapidly.

Aerobic packed-bed reactors require air to be sparged in the reactor or to be fed with wastewater already saturated with air. Reactors of this type have been used in a variety of treatment applications, such as nitrification, removal of carbonaceous biological oxygen demand (BOD) (Metcalf and Eddy, 1991), mineralization of trichlorophenol using a white-rot fungus (Armenante et. al., 1989, 1992b; Lewandowski et al., 1990), and treatment of groundwater contaminated with ethylene dichloride (Friday and Portier, 1989; LaGrega et al., 1994). A recent example of the use of such a reactor is given by Pal et al. (1995) who used a packed-bed reactor sparged with air, in which the white rot fungus *P. chrysosporium* was immobilized, to mineralize 2,4,6-trichlorophenol and 2,4,5-trichlorophenol (Fig. 1.12). The packed-bed reactor was used not only to retain the biomass, but also to minimize shear stress effects. Previous work has

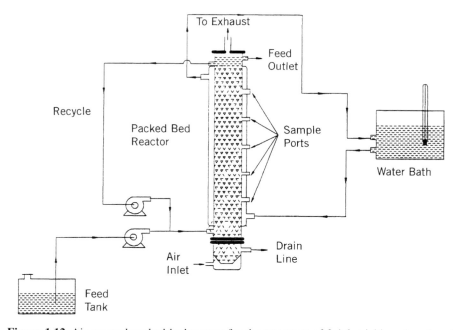

Figure 1.12 Air-sparged packed-bed reactor for the treatment of 2,4,6–trichlorophenol and 2,5,6-trichlorophenol using *P. chrysosporium* immobilized on polyethylene terephthalate packing (after Pal et al., 1995).

shown that this type of filamentous organism can be quite sensitive to high shear, as in the case in which it is grown in a stirred vessel (Lewandowski et al., 1990; Armenante et al., 1992b).

Please refer to Chapter 2 for further discussion on fixed-film reactors, especially those containing mixed support media.

Trickle-Bed Reactors

Trickle-bed reactors are packed-bed reactors operated so that the gas phase is the continuous phase. As in any packed-bed system, trickle-bed reactors consist of vessels containing a packing material having a high surface area per unit volume. The liquid phase, typically a wastewater, is introduced from the top of the reactor and trickles over the wetted packing. The space between the packing is also occupied by the gas phase that flows either upward or downward. The packing serves the dual purpose of immobilizing the biomass growing on its surface and producing a high gas–liquid contact area that enhances gas–liquid mass transfer. The latter aspect is especially important in the aerobic utilization of this type of reactor.

The most important application of trickle-bed reactors is in the aerobic treatment of wastewater, where they are typically referred to as "trickling filters," an obvious misnomer since their mode of operation does not rely at all on filtration but on biodegradation. Most trickling filters are concrete vessels having a circular shape and an open top (Fig. 1.13). The reactors can be packed with a variety of packing materials such as rocks, slag, wood, or plastic media. Rocks typically have a diameter of 25–100 mm, a void space of 40–60%, and are packed to heights varying between 0.9 and 2.5 m (commonly 1.8 m). Higher packing heights are difficult to

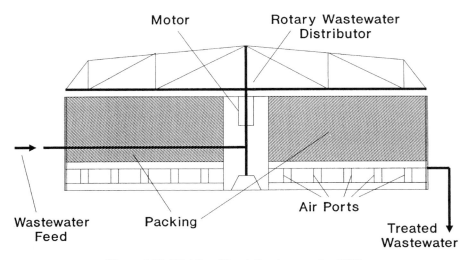

Figure 1.13 Trickling filter (after Armenante, 1993).

achieve because of the packing weight. Plastic media come in a variety of structured shapes (e.g., vertical-flow packing, cross-flow packing) all having a high surface area-to-volume ratio and a void fraction of about 94–97%. Plastic random packing materials are also used. The depth of these types of packing (4 to 12 m) is typically higher than in rock-filled reactors because of the reduced weight. Because of the higher surface area per unit volume and greater height, trickle-bed reactors with plastic packing are typically operated at much higher wastewater loading rates than rock-filled reactors (Sundstrom and Klei, 1979; Grady and Lim, 1980; Eckenfelder, 1989; Horan, 1990; Metcalf and Eddy, 1991).

In trickle-bed reactors, the wastewater is dispersed at the top of the reactor by a distribution mechanism typically consisting of two or more distributor arms rotating just above the packing and provided with nozzles to discharge the wastewater. The distributor sweeps the entire surface of the reactor, which can be as large as 60 m in diameter. After traveling the entire depth of the bed, the effluent water is collected by an underdrain system and pumped away.

Air is circulated through the bed by either natural convection—by far the most common case—or forced convection. The driving force for the air movement is the temperature difference between the wastewater and the air. The air circulates upward or downward depending on whether it is warmer or colder than the wastewater. Wind may also produce a draft helping natural convection. Natural convection is also promoted by ventilation ports located along the bottom of the reactor, between the packing and the underdrain system. Typically, the area of the ventilation ports should be no less than 15% of the cross-sectional area of the reactor. Forced convection is used in trickling filters in which natural convection is difficult to achieve, and in tall trickle-bed reactors (Eckenfelder, 1989; Corbitt, 1990).

The flow of the nutrient-rich wastewater on the surface of the packing promotes the growth of a biofilm, typically 0.1–0.2 mm thick (Wentz, 1995). As more waste is degraded, the biofilm thickness increases until the point where no oxygen reaches the biomass layer in contact with the solid support. When this happens, the microorganisms enter a phase of endogenous respiration and lose their ability to remain attached to the packing. This produces the periodic detachment of portions of the biofilm—the so-called sloughing process—requiring the presence of a clarifier downstream to remove the suspended solids from the effluent. The clarifier in trickle filter operation does not serve the same purpose as in suspended-growth processes where biomass recycle is necessary for the viability of the process. However, effluent recycle can also be used in the operation of trickling filters to dilute the concentration of the pollutants in the incoming stream, and to minimize odor problems (Metcalf and Eddy, 1991).

Because of their mode of operation, trickle-bed reactors are typically modeled as plug-flow reactors. In addition to kinetic expressions to describe the rate of biodegradation, these models also employ transport equations to describe the diffusion of nutrients (particularly oxygen) through the water film enveloping the biomass, and through the biofilm itself. In conventional wastewater treatment, oxygen transfer can become rate limiting if the concentration of organic material is above 400–500 mg/L (expressed as BOD). Empirical equations have also been proposed to model the degradation process. One of the most commonly used equations of this type is (Sundstrom and Klei, 1979; Eckenfelder, 1989):

$$\frac{C_{\text{out}}}{C_{\text{in}}} = \exp\left[-\frac{KH_p}{(Q/A)^n}\right] \tag{18}$$

where C_{in} and C_{out} are the concentration of the pollutant removed in the process, H_p is the packing depth, Q/A is the flow rate per unit cross-sectional area, and K and n are empirical constants.

Rotating Disk Reactors

Rotating disk reactors are exclusively used in the aerobic treatment of wastewater. A typical reactor configuration of this kind is the rotating biological contactor (RBC), consisting of a series of polystyrene or polyvinyl chloride disks mounted on a horizontal shaft slowly rotating around its axis, as shown in Figure 1.14. The disks are covered with a biofilm that is partially submerged in the wastewater to be treated. As the disks rotate, the microorganisms in the biofilm are alternatively immersed in the wastewater (which contains organic and inorganic nutrients) and exposed to the air (which provides the oxygen for the aerobic process).

The disks are typically tightly spaced in order to maximize the surface area per unit shaft length. Standard surface areas are in the range of 100–120 m²/m³ of reactor (Rich, 1973; Eckenfelder, 1989). Typical disk diameters are in the range of 1.5–3.6 m. Rotational speeds are within 2–5 rpm. In a conventional RBC unit the disks are about 40% immersed. In other types of RBCs (provided with air capture cups), air is also sparged just below the disks that are about 90% immersed.

The biomass on the disk is contained in a biofilm 0.3–4 mm thick that is typically formed over a period of about 1–2 weeks (Wentz, 1995). The degradation activity of the biofilm is carried out not only when the biofilm is immersed but also when it is exposed to the air since a film of wastewater is always present around the

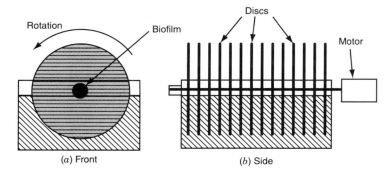

Figure 1.14 Rotating disk contactor.

biofilm. As in other continuous fixed-film processes, the excess biomass in RBCs is periodically sloughed off as a result of the growth of the biofilm and the difficulty of the microbial layer closest to the disk surface to receive nutrients. When the nutrients do not penetrate deep enough in the biofilm, the microbial layer loses cohesiveness with the solid support and biomass separates from the disk. A clarifier is typically used to recover and separate this biomass from the treated effluent.

For the most typical RBC application, namely BOD removal, the following design equation has been developed (Wu et al., 1980):

$$\frac{C_{out}}{C_{in}} = 14.2 \left[\frac{(Q/A)^{0.558}}{\exp(0.32\ N_s)\ C_{in}^{0.684}\ T_w^{0.248}} \right] \tag{19}$$

where C_{out} and C_{in} are the concentration of the pollutants in grams of BOD per cubic meter, Q is the BOD mass rate in grams per, A is the disk area in feet squared, N_s is the number of stages, and T_W is the wastewater temperature in degrees Celsius. Other design equations based on a controlling external mass-transfer resistance model can be found in the literature (LaMotta, 1976; Sundstrom and Klei, 1979)

Continuous RBCs can be arranged in a number of different ways with the wastewater flowing either parallel to the disks or perpendicular to them. The energy consumption of RBCs is typically quite low (Opatken et al., 1988).

Fluidized-Bed Reactors

In fluidized-bed reactors, the microorganisms are immobilized on the surface of particles suspended in the reactor by the velocity of the incoming liquid entering through a bottom distribution plate. The immobilized biomass can be retained within the reactor, since the top part of the vessel provides a calming zone that allows the particles to settle back to the fluidized portion of the reactor.

The materials on which the biofilm is formed are typically sand, activated carbon, or resin particles. The velocity that the liquid must have in order to achieve minimum fluidization conditions can be calculated from a momentum balance around the bed. For very small particles settling according to Stokes law, it can be shown that the minimum fluidization velocity is given by (Geankoplis, 1993):

$$\frac{v_p}{v_{fl}} \cong 90 \tag{20}$$

where v_{fl} is the superficial fluidization velocity in the reactor and v_p is the settling velocity of the particle in a quiescent liquid. For large particles settling in turbulent flow, the corresponding relationship is:

$$\frac{v_p}{v_{fl}} \cong 9 \tag{21}$$

Other fluidization equations are also available (Ngain et al., 1980; Schügerl, 1987). Fluidized-bed reactors can contain biomass in very high concentration (up to

40,000 mg/L). This concentration is much larger (about 10 times) than that typically found in suspended-growth reactors (Armenante, 1993; Cheremisinoff, 1990).

Fluidized-bed reactors have been used in nitrification/denitrification processes (Schügerl, 1987) as well as the anaerobic treatment of municipal and industrial wastewaters (Kobayashi et al., 1988). One of the advantages of fluidized-bed reactors in wastewater treatment is that the biomass concentration can be extremely high, and relatively short detention times can be achieved. High COD removal rates (20–27 kg/m^3/day) have been obtained in large-scale fluidized-bed reactors (diameter: 4.6 m; height: 21 m; packing height; 13 m) used in the treatment of industrial effluents such as those from brewer's yeast plants (Schügerl, 1987).

Bed expansion—a very mild type of fluidization in which the particles in the bed become just separated from each other as a result of an upward moving liquid stream—is also used advantageously in some bioreactors (Jewell et al., 1981). This principle is used in a particular type or anaerobic reactor: the upflow anaerobic sludge blanket reactor (UASB), shown in Figure 1.15. In this reactor the liquid injected through the bottom of the reactor encounters a layer of microbial pellets forming a sludge blanket. The upward direction of the flow maintains the pellets suspended and promotes the transfer of nutrients from the liquid to the microorganisms in the pellets. The biomass concentration in this blanket layer is extremely high—from 60,000 up to 150,000 mg/L (Eckenfelder, 1989). As a result, very effective biodegradation takes place (Lettinga et al., 1980; Collivignarelli et al., 1990).

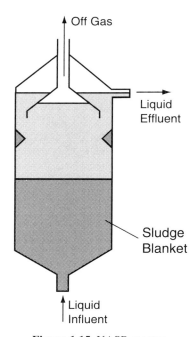

Figure 1.15 UASB reactor.

The density of the microbial pellets is low enough for bed expansion to take place at fluidization velocities in the range of 0.6–0.9 m/h (Metcalf and Eddy, 1991), but high enough to prevent most of the pellets from escaping from the blanket zone, and hence from the reactor. Above the blanket area the biomass concentration is still high (15,000–25,000 mg/L) but much lower than in the blanket. The methane gas produced in the process contributes toward mixing the liquid phase, as well as maintaining the biomass pellets in suspension. The pellets rise due to bubbles becoming attached to them, eventually reaching a series of degassing baffles that produce a disengagement of the gas bubbles, which coalesce and are collected as off-gas. The pellets then resettle back into the blanket. Very high organic loadings can be achieved in UASB reactors (4–12 kg COD/m^3 day) with removal efficiencies comparable to those of other anaerobic systems (75–85%) (Metcalf and Eddy, 1991).

NOTATION

a	Surface area per unit volume (m^2/m^3)
A	Cross-sectional area or surface area (m^2)
B	Solid fraction in liquid in Equation (15) (percent by weight); or baffle width (m)
C	Concentration of pollutant (g/L, mg/L, or mol/L); or impeller clearance off the reactor bottom in Figure 1.4 and Equation (8) (m)
C_{in}	Inlet concentration of pollutant (g/L, mg/L, or mol/L)
C_{out}	Outlet concentration of pollutant (g/L, mg/L, or mol/L)
C_{O_2}	Dissolved oxygen concentration (g/L, mg/L, or mol/L)
C_{O_2s}	Dissolved oxygen concentration at saturation (g/L, mg/L, or mol/L)
d_b	Bubble diameter (m)
d_p	Particle diameter (m)
D	Impeller diameter (m)
g	Acceleration of gravity (m/s^2)
Fr	Froude number $(= DN^2/g$; nondimensional)
H	Liquid height in reactor (m)
H_p	Height of packing (m)
k_d	Cell death/endogenous respiration constant (s^{-1})
k_L	Liquid-side mass-transfer coefficient (m/s)
K	Constant in Eq. (18) (appropriate units)
K_m	Substrate saturation constant (g/L, mg/L, or mol/L)
m	Experimentally obtained constant in Eq. (11) (appropriate units)
M_o	Rate of oxygen transfer per unit liquid volume (mg O_2/m^3/s or moles O_2/m^3/s)
n	Constant in Eq. (18) (appropriate units)
N	Agitation speed (rev/s)
N_m	Minimum agitation speed for complete solid suspension (rev/s)
N_p	Power number (non dimensional)

N_s Number of stages (nondimensional)
P Power dissipation (Watts)
P_g Power dissipation under gassed conditions (W)
Q Liquid flow rate (m³/s)
Q_g Gas or air flow rate (m³/s)
Re   Impeller Reynolds number (= $\rho\, D^2\, N/\mu$; nondimensional)
s Constant in Eq. (15) (nondimensional)
S Spacing between impellers (m)
t Time (s)
T Reactor diameter (m)
T_w Wastewater temperature (°C)
v_{fl} Fluidization superficial velocity (m/s)
v_g Superficial gas velocity (m/s)
V_p Settling velocity of particle (m/s)
V Volume (m³)
W Impeller blade width (m)
X Biomass concentration (g/m³ or mg/L)
X_{in} Inlet biomass concentration (g/m³ or mg/L)
Y Yield coefficient (g biomass produced/g nutrient consumed)
Z Axial distance along the reactor (m)

Greek Symbols

β Constant in Eq. (13) (appropriate units)
γ Constant in Eq. (14) (nondimensional)
μ Viscosity (cp or kg/m·s)
μ_m Maximum specific growth rate constant (s⁻¹)
ρ Liquid density (kg/m³)
$\Delta\rho$ Solid–liquid density difference (kg/m³)
θ Residence time (s)
θ_m Mixing time (s)

REFERENCES

Aivasidis, A. and C. Wandrey (1988). Recent Developments in Process and Reactor Design for Anaerobic Wastewater Treatment, *Water Sci. Tech.* **20**:211–218.

Armenante, P. M. (1993). Bioreactors, in *Bioremediation: Principles and Practice*, M. A. Levine and M. A. Gealt, Eds., McGraw-Hill, New York.

Armenante, P. M., G. Lewandowski, and D. Pak. (1989). Reactor Design Employing a White Rot Fungus for Hazardous Waste Treatment, in *Biotechnology Applications in Hazardous Waste Treatment*, G. Lewandowski, P. M. Armenante, and B. Baltzis Eds., Engineering Foundation, New York, pp. 185–198.

Armenante, P. M., D. Kafkewitz, G. Lewandowski, and C. M. Kung (1992a). Integrated Anaerobic-Aerobic Process for the Biodegradation of Chlorinated Aromatic Compounds, *Environ. Progr.* **11**:113–122.

Armenante, P. M., G. Lewandowski, and I. U. Haq (1992b). Mineralization of 2-Chlorophenol by *P. Chrysosporium* Using Different Reactor Configurations, *Haz. Waste Haz. Mat.* **9**(3):213–229.

Bailey, J. E. and D. F. Ollis (1986). *Biochemical Engineering Fundamentals*, 2nd ed., McGraw-Hill, New York.

Baltzis, B., G. Lewandowski, and S. Sanyal (1991). Sequencing Batch Reactor Design in a Denitrifying Application, in *Emerging Technologies in Hazardous Waste Management II*, D. W. Tedder and F. Pohland, Eds., ACS Symposium Series. American Chemical Society, Washington, DC.

Bates, R. L., P. L. Fondy, and R. R. Corpstein (1963). An Examination of Some Geometric Parameters of Impeller Power, *Ind. Eng. Chem. Proc. Des. Dev.* **2**(4):310–314.

Bates, R. L., P. L. Fondy, and J. G. Fenic (1966). Impeller Characteristics and Power, in *Mixing. Volume 1*, V. W. Uhl and J. B. Gray Eds., Academic, New York.

Cheremisinoff, P. N. (1990). Biological Treatment of Hazardous Waste, Sludges and Wastewater, *Pollut. Eng.*, **22**(5):87–94.

Christian, J. B. (1994). Improve Clarifier and Thickener Design and Operation, *Chem. Eng. Progr.*, **89**(7):50–56.

Collivignarelli, C., G. Urbini, A. Farneti, A. Bassetti, and U. Barbaresi (1990). Anaerobic-Aerobic Treatment of Municipal Wastewaters with Full-Scale Upflow Anaerobic Sludge Blanket and Attached Biofilm Reactors, *Wat. Sci. Tech.* **22**:475–482.

Corbitt, R. A. (1990). *The Standard Handbook of Environmental Engineering*, McGraw-Hill, New York.

Eckenfelder, W. W., Jr. (1989). *Industrial Water Pollution Control*, McGraw-Hill, New York.

Friday D. D. and R. J. Portier (1989). Modular Reactor Approaches for Remediation of Ground Water: A Case Study with Volatile Chlorinated Aliphatics, *Proc. 10th Nation. Conf.—Superfund '89*, Hazardous Material Control Research Institute, Washington, DC, Nov. 27–29.

Geankoplis, C. J. (1993). *Transport Processes and Unit Operations*, 3rd edn., Allyn and Bacon, Boston.

Gray, J. B. (1966). Flow Patterns, Fluid Velocities, and Mixing in Agitated Vessels, in *Mixing Volume 1*, V. W. Uhl and J. B. Gray, Eds., Academic, New York.

Gray J. B. and J. Y. Oldshue (1986). Agitation of Particulate Solid-Liquid Mixtures, in *Mixing. Volume 3*, V. W. Uhl and J. B. Gray, Eds., Academic, New York.

Grady, C. P. L., Jr., and H. C. Lim (1980). *Biological Wastewater Treatment*, Marcel Dekker, New York.

Harnby, N., M. F. Edwards, and A. W. Nienow (1985). *Mixing in the Process Industries*, Butterworths, London.

Hines, D. A., M. Bailey, J. C. Oursby, and F. C. Roesler (1975). Novel Aeration System Bows, *Water Wastes Eng.*, **12**(12):59–64.

Horan, N. J. (1990). *Biological Wastewater Treatment Systems, Treatment and Operation*, Wiley, New York.

Irvine R. L. and A. W. Busch (1979). Sequencing Batch Reactors: An Overview. *J. Water Pollut. Contr. Feder.*, **51**:235–243.

Jackson, M. L. (1964). Aeration in Bernoulli Types of Devices, *AIChE J.*, **10**:836–842.

Jewell, W. J., M. S. Switzenbaum, and J. M. Morris (1981). Municipal Wastewater Treatment with the Anaerobic Attached Microbial Film Expanded Bed Process. *J. Water Pollut. Contr. Feder.*, **53**:482–490.

Kobayashi, H. A., E. Conway de Macario, R. S. Williams, and A. J. L. Macario (1988). Direct Characterization of Methanogens in Two High-Rate Anaerobic Biological Reactors. *Appl. Environ. Microb.*, **54**:693–698.

LaGrega, M. D., P. L. Buckingham, and J. C. Evans (1994). *Hazardous Waste Management*, McGraw-Hill, New York.

LaMotta, E. J. (1976). External Mass Transfer in a Biological Film Reactor, *Biotech. Bioeng.*, **18**:1359–1365.

Lettinga, G., A. F. M. van Velsen, S. W. Hobma, W. De Leeuw, and A. Klapwijk (1980). Use of the Upflow Sludge Blanket (USB) Reactor Concept for Geological Wastewater Treatment, Especially for Anaerobic Treatment, *Biotech. Bioeng.*, **22**:699–734.

Levenspiel, O. (1972). *Chemical Reaction Engineering*, 2nd ed., Wiley, New York.

Lewandowski, G., P. M. Armenante, and D. Pak (1990). Reactor Design for Hazardous Waste Treatment Using a White Rot Fungus, *Water Res.*, **24**:75–82.

Metcalf & Eddy, Inc. (1991). *Wastewater Engineering: Treatment, Disposal, and Reuse*, 3rd ed., McGraw-Hill, New York.

Michel, B. J. and S. A. Miller (1962). Power Requirements of Gas-Liquid Agitated Systems, *AIChE J.* **8**:262–266.

Middleton, J. C. (1985). Gas-Liquid Dispersion and Mixing, in *Mixing in the Process Industries*, N. Harnby, M. F. Edwards, and A. W. Nienow, Eds., Butterworths, London, pp. 322–355.

Mudrack, K., H. Sahm, and W. Sittig (1987). Environmental Biotechnology, in *Fundamentals of Biotechnology–1987*; P. Praeve, U. Faust, W. Sittig, and D. A. Sukatasch, Eds., VCH, Weinheim, Germany, pp. 623–660.

Ngian, K. F. and W. R. B. Martin (1980). Bed Expansion Characteristics of Liquid Fluidized Particles with Attached Microbial Growth, *Biotech. Bioeng.*, **22**:1843–1856.

Oldshue, J. Y. (1983). *Fluid Mixing Technology*, McGraw-Hill, New York.

Opatken, E. J., H. J. Howard, and J. J. Bond (1988). Biological Treatment of Leachate from a Superfund Site, *Environ. Progr.* **7**:12–18.

Pal, N., G. Lewandowski, and P. M. Armenante (1995). Process Optimization and Degradation Modeling of Chlorophenol by *Phanerochaete chrgsosporium*, *Biotech. Bioeng.*, **46**(6):599–609.

Prokop, A., P. Janik, M. Sobotka, and V. Krumphanzl (1982). Hydrodynamics, Mass Transfer and Yeast Culture Performance of a Column Bioreactor with Ejector, *Biotech. Bioeng.* **25**:1147–1160.

Rich, L. G. (1973). *Environmental Systems Engineering*, McGraw-Hill, New York.

Schügerl, K. (1987). *Bioreaction Engineering*, Vol. 2, Wiley, New York.

Sundstrom, D. W. and H. E. Klei (1979). *Wastewater Treatment*, Prentice–Hall, Englewood Cliffs, NJ.

Talmage, W. P. and E. B. Finch (1955). Determining Thickener Unit Area, *Ind. Eng. Chem.*, **47**(1):38–41.

Tatterson, G. B. (1991). *Fluid Mixing and Gas Dispersion in Agitated Tanks*, McGraw-Hill, New York.

Tojo, K. and K. Miyanami (1982). Oxygen Transfer in Jet Mixers, *Chem. Eng. J.*, **24**:89–97.

Wentz, C. (1995). *Hazardous Waste Management*, 2nd ed., McGraw-Hill, New York.

Wu, A. C., E. D. Smith, and Y. T. Hung (1980). Modeling of Rotating Biological Contactor Systems, *Biotech. Bioeng.*, **22**:2055–2064.

Zlokarnik, M. (1979). Scale-up of Surface Aerators for Waste Water Treatment, in *Advances in Biochemical Engineering*, Vol. 11, T. K. Ghose, A. Fichter, and N. Blackebrough, Eds., Springer Verlag, Berlin, pp. 157–180.

Zwietering, T. N. (1958). Suspending of Solid Particles in Liquids by Agitators, *Chem. Eng. Sci.*, **8**:244–253.

CHAPTER 2

Introduction to Microbiological Degradation of Aqueous Waste and Its Application Using a Fixed-Film Reactor

Louis J. DeFilippi and F. Stephen Lupton

Independent Consultant, 208 Edgewood Lane Palatine, Illinois 60067 (L.J.D.) and
AlliedSignal Environmental Systems and Services, Des Plaines, Illinois 60017 (F.S.L.)

Biological treatment of industrial and municipal wastewaters is well established. Biological treatment of contaminated soils and groundwater is also becoming widely accepted as a remediation option, especially because of its ability to reduce costs over the lifetime of the project. Recent advances in immobilized (fixed-film) biological systems show great promise to increase the efficiency of biological treatment systems by:

1. Increasing process stability and resistance to shock loadings
2. Allowing much greater organic loads to be processed per given reactor volume
3. Producing considerably less biological solids (biosolids or sludge) as by-product.

A fixed-film bioreactor system has been developed that utilizes a highly porous polyurethane foam coated with activated carbon. This support is blended with cylindrical plastic rings to form a packed bed of 'mixed-media' supports that allows a high surface area-to-volume ratio but that also reduces potential problems of plugging, distribution of air and water, and mass-transfer limitation. Performance data for fixed-film bioreactors treating a number of industrial and municipal wastewaters as well as groundwater remediation systems showed organic removal

Biological Treatment of Hazardous Wastes, Edited by Gordon A. Lewandowski and Louis J. DeFilippi
ISBN 0-471-04861-5   ©1998 John Wiley & Sons, Inc.

rates in the range of $2-8$ kg COD/(m^3 day) or $16-67$ lb COD/(kgal day) with biological solids formation in the range of $0.08-0.15$ kg solids/kg BOD_5.

In addition to bacteria and other microorganisms there are immobilized cell reactors that employ mammalian cells, insect cells, or plant cells for various chemical transformations. An inspection of a single issue of *Chemical Abstracts Service Biotech Updates* (one week's worth of literature) in the area of biochemical immobilization and biocatalytic reactors (Issue 16, August 7, 1995) reveals 29 references on the subject, 9 of which discuss systems involved in wastewater treatment. In this chapter we will concentrate on microorganisms (especially bacteria) and the treatment of aqueous waste. Furthermore, we will employ the AlliedSignal Immobilized Cell Bioreactor (ICB^{TM}) as an example of an engineered system that utilizes an advanced fixed-film biological treatment process for wastewater treatment. The ICB was developed by AlliedSignal Research and Technology to meet the demand for a high-performance biological treatment system. It is an advanced fixed-film reactor that has a high rate of organics removal and that reduces these components to very low levels. It has the advantage of producing considerably less biological sludge than conventional biological treatment systems. The reader is referred to Chapter 1 for comparative data on related technologies as well as reactor design.

USE OF MICROORGANISMS IN WASTE TREATMENT

Microorganisms have been employed in a systematic fashion for the destruction of aqueous waste for many years (McCarty, 1982). Both aerobic (Rittmann, 1987) and anaerobic (McCarty and Smith, 1986; Jewell, 1987) biological processes have been reviewed along with chemical treatment processes for removal of particles and dissolved organic and inorganic contaminants (Lawler, 1986; Weber and Smith, 1986; Clifford et al., 1986). Also see Belhateche (p. 32, 1995) for "guidance on when to apply what type of treatment to which waste streams" and a summary of various chemical, thermal, and biological industrial wastewater treatment processes. Biological processes have the advantage over certain competing technologies by being relatively inexpensive and able to almost quantitatively convert or mineralize organic pollutants to innocuous products. Mineralization in the microbial degradation sense is defined as the conversion of the pollutant to inorganic constituents: carbon dioxide, salts (when applicable; e.g., when organic halides or organic amines are present), and water. It should not be confused with the hydrogeological definition of mineralization (see Chapter 7).

Bacteria in the wild "reproduce under scientifically scandalous circumstances, growing in mixed cultures on a miscellany of sources of carbon, nitrogen, phosphorous and sulphur" (Paigen and Williams, p. 252, 1970). Attempts have been made to bring logic to the extremely important decision as to whether the components of a particular waste stream may be economically treated employing biological processes. Indices have been compiled to help guide this decision. However, these guides should be consulted with circumspection. One should not be confused by inspection of refractory index tables such as appears on page 376 in Fan and Tafuri (1994) to determine if a particular compound may be efficiently degraded by biological means. A table such as this gives a ratio of BOD_5 (biological oxygen demand) to COD (chemical oxygen demand). It is a purported indication of the difficulty of biological degradation. A low BOD_5/COD ratio is associated with a compound that is relatively hard to degrade biologically and a number approaching one relatively easy to biodegrade. Many of these investigations utilize a standard microbial inoculum and a standard incubation time (often 5 days, the subscript in BOD_5) and reaction conditions. Although useful for certain comparisons, it does not reflect the actual conditions under which a reactor (or in situ bioremediation) with selected or adapted microorganisms will optimally function. Thus, the potential for mineralization of a particular component may be incorrectly judged.

In support of this contention, we have found that the use of classically enriched common consortia of wild-type microorganisms, what we sometimes euphemistically term "street bugs," perform better in a fixed-film biological reactor than do pure strains of specialized laboratory microorganisms. For example, although Fan and Tafuri (1994) report o-xylene, m-xylene, and ethylbenzene have a refractory index of <0.009 (very hard to degrade), we have found that enriched soil bacteria (as opposed to the specialized laboratory strains employed in the BOD_5 test) devour these gasoline components quite voraciously in a fixed-film reactor. The ready utilization of hydrocarbons by microorganisms has been known for many years (Greig-Smith, 1914; also see Bushnell and Haas, 1941, and references contained therein dating to end of the nineteenth century).

Similarly, it has been found that bioaugmentation (addition of microorganisms) plus addition of nutrients to crude oil-contaminated land yielded microbial mineralization of the oil no faster than if nutrients alone were added without bioaugmentation (Venosa, 1995). One concludes from this that indigenous populations are sufficient to adapt to and bioremediate the oil, as long as sufficient nutrients were present. Similarly, Otte et al., (1994) has reported that laboratory studies with a fed-batch bioreactor indicated that no addition of a pentachlorophenol (PCP) degrading consortium was required for PCP mineralization at PCP concentrations up to 700 ppm. Indigenous microorganisms were sufficient. The observed initial lag period of 2.5 days may be decreased somewhat by augmentation with enriched inocula, but dramatic effects were not demonstrated. It should be recognized that naturally occurring microorganisms are extremely effective at adapting themselves to utilizing a wide variety of carbon sources or substrates to grow and extract energy (Ottengraf et al., 1986).

Pollutants are sometimes degraded in a sequential, rather than simultaneous, fashion. From the point of view of the microorganisms, "survival may depend on utilization of the optimal substrate [carbon source or pollutant], for the cell that saves the best carbon source for last may well find that its neighbor has not been so forbearing" (Paigen and Williams, p. 252 1970). That is, when multiple carbon sources are present, not only may different substrates be degraded at different rates, some may be ignored almost entirely by the microorganism until certain preferential substrates are depleted below a critical repressive or inhibitory level. Certain substrates, such as glucose, will exert their repressive effects on a genetic level (catabolite repression) that will inhibit the formation of enzymes required for certain metabolic processes or on enzyme activity in a process termed catabolite inhibition.

Processes other than mineralization occur during biological treatment of wastewater and impacted groundwater. Under particular conditions (often where there is an excess of carbon source) certain organic compounds such as phenols are polymerized rather than oxidized to CO_2 and the like. Chloroaromatics form polymers through mechanisms similar to the formation of humic substances (Knackmuss, 1983). The end result is the formation of polymers that resist further microbial breakdown. Under certain anaerobic conditions methane is produced. This major by-product of anaerobic metabolism is a valuable source of energy (McCarty, 1982). Furthermore, biological processes also have beneficial effects by their action on inorganic constituents. These benefits are realized by the release of halides from toxic halogenated organic compounds to yield less toxic products (a partial or complete mineralization process depending upon conditions, time frame, and the chemical nature of the halocarbon; see Chapters 10 and 11 and Hinchee et al., 1994), the fixing of phosphate (McClintock and Frazier, 1994), and metals (Van Voorhis, 1990) in biomass, the release of ammonia from organic amines, the conversion of ammonia to nitrate and molecular nitrogen (Tsuno et al., 1992; Wijffels et al., 1995), and by the precipitation of many unwanted metals as their sulfides or oxides (Lester et al., 1983; Revis et al., 1991; DeFilippi, 1994). However, the microorganisms do leave behind their living and dead remains. Along with the above precipitates, these are more-or-less readily dealt with as sludge (Cheremisinoff, 1993, 1995).

Soils and sludges may also be treated using biological means (see Chapters 3 and 13). This observation notwithstanding, one must remember all microbial reactions must have an aqueous component. Thus, the more tightly a pollutant adheres to soil or to a nonaqueous phase liquid, relative to its propensity to dissolve in the aqueous phase, the less it is available to microorganisms as a food source. Surfactants, whether natural or xenobiotic, aid the transfer to the aqueous phase. Hence the use of surfactants to augment the biodegradative process (Lupton and Marshall, 1979).

One can facilitate the action of native microorganisms by altering the environment (either in situ or ex situ) to take advantage of the natural propensity of the microorganisms to degrade pollutants. For example, taking a microbial ecology approach to bioremediation can often be quite effective since "what you are really doing in *in situ* bioremediation is creating environments that are conducive to the right organisms doing our work" [Bruce Rittmann, quoted in Hart p. 398A, (1996) and references therein]. It is now being accepted that under the right circumstances the least invasive approach, allowing "intrinsic bioremediation" or "natural attenuation" to perform the tasks of bioremediation can sometimes be the method of choice, or at least one component of an approach, for eliminating environmental pollutants.

Potentially detrimental reactions may also occur as a result of the action of microorganisms on chemicals. Certain chemicals that are either innocuous or of low toxicity may be transiently converted to compounds of greater toxicity or to compounds that are more easily transported. Examples of this are N-nitrosation of secondary amines to yield carcinogenic daughter compounds, partial dehalogenation of pentachlorobenzyl alcohol to yield the more phytotoxic tri- and tetrachloro cogeners, and the methylation of mercury to yield a more toxic and volatile product. See Alexander (1980) for a further discussion of this.

FIXED FILM IN THE MICROBIAL WORLD

A fixed-film biological reactor is arguably the most "natural" of reactors. Why is that? First we must define a fixed-film reactor. In the simplest terms it is a reactor where microorganisms occupy a surface and perform a metabolic process. This occupation is in the form of a layer of cells of varying thicknesses. The presence of this film is made possible by adhesion mechanisms, whether natural or artificial.

Fixed-film systems are part of a larger class of immobilized cell systems. Immobilized cells are attached or immobilized on substrata. The substrata are composed of a variety of materials and formed into a number of different configurations. A variety of mechanisms, including entrapment, are employed for cell immobilization. Gels are often the medium chosen for this technique.

By definition fixed films are not composed of free swimming microorganisms. The microbes are not milling about hunting for food, but rather food (and essential nutrients) must be bought, or transported, to them. This transportation occurs by a number of different mechanisms. The mechanisms include, on the most gross scale, having dissolved, gaseous or vaporous nutrients and the target compound(s), in our case pollutants, flow past the microorganisms that are bound to a particular surface of a substratum. On a smaller scale, the pollutants must diffuse from the aqueous phase to the surface of the microorganism. The substratum and a large portion of the microorganisms stay in the reactor while the purified water, sloughed and free swimming cells, and often vapors, exit the reactor.

A question that a wastewater engineer may ask is: In what ways may an understanding of native microbial ecology aid in the design of a fixed-film reactor? That is, how does fixed-film reactor design benefit from an understanding of microbial processes in the "natural" world? The answer is that, in the simplest of terms, many (most?) wild-type microorganisms have an innate propensity to adhere to surfaces. Wetlands are one form of fixed-film reactor, be they constructed (perhaps in the most "natural" of settings) or naturally occurring. In this system microorganisms are fixed on the surface of sediment particles and the roots and stems of the wetland flora (Bhamidimarri et al., 1991). Pollutants are removed as they pass though.

Microorganisms often occur naturally on surfaces in greater numbers than suspended in liquid (Zobel, 1943). Furthermore, although much of our under-standing of microorganisms has advanced with the use of pure cultures in suspended- (not fixed-film) growth studies, wild-type microorganisms almost invariably live, grow, and divide as part of a mixed culture of genetically different microorganisms known as consortia, and these consortia often live in fixed films. Characklis (1981) and Marshall (1992) present summaries of the formation and development of biofilm consortia. They recognize that in virtually all situations where natural and artificial surfaces are exposed to a natural aqueous environment, microorganisms and a source of carbon and nutrients will develop a biofilm. This biofilm consists of a sticky polymer (the exopolysaccharide produced by the microorganisms) as well as the microorganisms themselves (Fletcher and Floodgate, 1976). The bulk of the metabolic processes that occur in the natural aqueous environment occur at these biofilms.

Scientists and engineers have long known microorganisms tend to attach to and proliferate on surfaces in contact with aqueous systems (Whipple, 1901). Proliferation is the greatest where the rate of movement of the aqueous environment past the substratum, that is, rate of fluid flow, is the greatest. The reasons for this are well summarized by Marshall (1992) and are based on selective survival advantages. The sequences involved in the attachment of microorganisms to and detachment of microorganisms from surfaces have been characterized (see, e.g., Characklis, 1981, and Tijhuis, 1995). In essence, an initially clean surface binds dissolved chemicals (both organic and inorganic) found in the bulk fluid, followed by attachment of microorganisms. The microorganisms proliferate and simultaneously produce the

Figure 2.1 Electron micrograph of bacteria and their adhesive polymers that have attached to a nylon rod inserted into an ICB fixed-film reactor. Note the long rod-shaped bacilli as well as the smaller, football-shaped bacteria. The white reference bar in the lower right-hand corner is 10 μm in length.

extracellular polymers that help them to remain attached. Highly subcultured laboratory strains must often be attached or entrapped artificially since they have frequently lost the ability to produce the appropriate extracellular polymers. As the biofilm becomes thicker, portions are lost due to fluid sheer stress at the biofilm surface. Figure 2.1 is an electron micrograph of the bacteria that were found to be denizens of an ICB. Note their adhesive polymers binding them to the surface of the fixed film. These polymers are not observed in highly subcultured laboratory strains.

FIXED-FILM REACTOR IN INDUSTRIAL APPLICATION

Comparisons

Fixed-film biological reactors (also known as biofilm or immobilized cell reactors) possess certain performance properties superior to suspended-growth systems (Criddle et al., 1991). A fixed-film reactor has the ability (or at least potential) to mimic, and hopefully be employed to optimize in a controlled fashion, the environment in which microorganisms have existed, competed, and evolved for countless millennia. It has been a major objective of ours to create an engineered fixed-film reactor that can best take advantage of the positive aspects of a microbial ecology approach to waste treatment. At the same time the system needs to be designed to best mitigate process limitations inherent in a fixed-film reactor. The classic fixed-film reactor is the trickling bed or filter (Crine et al., 1990). Unlike the ICB, which is a submerged fixed-film reactor, the bulk of the volume in a trickle-bed reactor is occupied by air. One of the main drawbacks of a trickling bed is that maldisribution of liquid that is directed along preferential flow paths results in non-wetted zones, which are therefore incapable of contributing to the degradation of the pollutants, thereby decreasing the reactor efficiency. Sand beds have also been used with reasonable success. For example, Chin et al. (1993) have used such a system for the treatment of widely varying volumes of fish farm pond water and stored rainwater for use in recycling back to the ponds, with little buildup of metals and total solids.

Advanced Fixed-Film Reactors

Biological reactors that retain their active biomass are termed "advanced" (Guiot and van den Berg, 1985). However, as one might expect, some are more advanced than others. We have found that the most often published method for biomass retention involves immobilization by entrapment in a hydrogel such as calcium-hardened alginate. Inspection of the literature indicates that this approach is most important when the study of a specialized laboratory strain (one that is often incapable of effective adherence to a surface by natural means) is desired. In contrast, it appears that at present self-attaching strains (almost invariably wild-type) are the only ones of use in full-scale application. There are, of course, exceptions.

The use of advanced fixed-film bioreactors for treatment of industrial wastewaters is gaining acceptance (Bernard, 1990; Gassen, 1993). Fixed-film reactors are also employed for the destruction of vaporous pollutants and odors in an airstream (Ottengraf, 1986). This process is often referred to as biological air filtration or biofiltration, meaning the biological removal of impurities, although filtration as known to chemists, that is, use of porous substance for removal of particles, is clearly not involved in the process. Classically biofiltration technology employs passive approaches, such as peat beds, through which air to be cleaned is passed. The AlliedSignal countercurrent, trickle-bed approach has been termed biological air treatment (DeFilippi, 1993; DeFilippi et al., 1993a, 1995, 1996). In almost any system where microbiologically effected degradation is involved, one may substitute "fixed–film (reactor)" for "filter" without problem.

Advanced fixed-film bioreactors can be formatted in a number of configurations including: rotating biological contactors, or RBCs (Galil and Rebhun, 1990; Tyagi et al., 1992), fluidized beds (Huppe et al., 1990), airlift suspension reactors (Tijhuis et al., 1995), packed beds (Morsen and Rehm, 1990; Sanyal et al., 1993), rotating fiber disks (Clyde, 1983), microporous membranes (Lakhwala et al., 1990), and cross-flow and vertical-flow high-rate trickling filters (Battistoni et al., 1992). The different forms of fixed-film reactors typically demonstrate high rates of organic removal and lower sludge yields than activated-sludge systems, but there are substantial capital, operating, and maintenance cost differences between these biofilm systems as well as different levels of ease of operation.

A number of fixed-film reactor designs developed by AlliedSignal and based on the commercial ICB are being considered for space station application, including trickling bed and submerged bed reactors (Petrie and Lupton, 1991; Petrie and Nacheff-Benedict 1991; Nacheff-Benedict et al., 1994). Mass is tremendously important on a space mission. As compared to chemically based systems, the microbial purification of water for long-duration space missions yields the greatest mass savings of life support resources.

Where a fixed-film reactor is employed in industrial application, it is almost invariably one component of a treatment train consisting of primary, secondary, and often tertiary treatment components. This is true for both air and water applications. Please refer to Mashayekhi (1993) for an excellent discussion of this topic as it applies to creosote wood treating and tar distillation plants.

Whatever the configuration, all fixed-film reactors appear to exhibit characteristics that give them advantages (and certain disadvantages!) over more conventional suspended growth reactors such as activated-sludge systems and the lower performance fixed-film systems such as trickling filters. These *advantages* include:

1. *Higher Hydraulic Loading Rates.* Higher loading rates [shorter hydraulic residence times, (HRTs)] are possible due to greater mineralization rates in a given reactor volume. This is due to the greater concentration of biomass within the reactor. This can in turn translate to lower capital outlay and less land usage.

2. *Retention of High Biomass Concentration within the Reactor.* Biomass retention yields an effective biomass concentration in the fixed-film reactor significantly greater than activated-sludge systems. Furthermore, this retention eliminates the tedious chore of sludge recycling required by activated-sludge systems. Furthermore, one major drawback of activated-sludge reactors, that of washout of the biomass during periods of high flow (e.g., during heavy rains) is very much less of a problem since the biomass is retained in a fixed-film system.

3. *Greater Stability.* The high concentration of biomass, especially when immobilized, increases the process stability and resistance to environmental shock. For example, Galil and Rebhun (1990) compared RBCs to activated sludge for the biological effluent quality and resistance to upset by phenolic spike loading and found the fixed-film RBC to be superior. Inclusion of activated carbon can further enhance this upset resistance (Lupton and Zupancic, 1991; DeFilippi, 1996).

4. *Growth of Microbes Not Required for Maintaining High Biocatalyst Concentrations within the Reactor.* When growth is not required, this can result in less sludge. As sludge disposal costs can often be a major factor in processing all but the most dilute waste streams, this can contribute considerable savings in operation and maintenance costs.

5. *Promotion and Maintenance of the Activities of Slow Growing Microbes.* Microorganisms such as nitrifying bacteria that are prone to washout due to long dividing times (slow rate of growth) in a suspended-growth system such as an activated-sludge or continuously stirred tank reactor (CSTR) can be maintained in a reactor by immobilization even at short HRTs (Wijffels et al. 1995).

6. *Development of Ecological Niches or Zones.* Unlike essentially homogeneous reactors such as single-stage CSTRs, fixed-film reactors develop zones where different classes of microorganisms predominate. This zonation is dependent on the nature of the:

   a. Food source (pollutant)
   b. Partial decomposition and intermediate metabolism products
   c. Terminal electron acceptor(s)
   d. Reactor design

   The zonation occurs at three levels.

   a. First, as depth into a *film* increases, oxygen is depleted and the metabolic processes become progressively more anaerobic. The film depth where little aerobic process takes place is usually on the order of 0.1 mm with moderate to high BOD concentrations. At progressively lower BOD concentrations the zone of predominance of aerobic processes increases in thickness. In the anaerobic zones microbial utilization of terminal electron acceptors other than oxygen predominate. These include nitrate and sulfate. Other electron acceptors may also function under these conditions (Laverman et al., 1995).

b. Second, as depth into a *porous substratum* [such as the 2 in. (5 cm) foam cubes used in the ICB] increases, oxygen tension decreases. This is the case even in a well-aerated system with only modest occlusion of the pores. For the porous foam used in the ICB and other fixed-film reactors, this may be on the order of centimeters. See Figure 2.2 for a representation of this. Again, BOD levels will determine the approximate boundaries of the zones and therefore the proportions of the metabolic processes for any given substratum and set of operating conditions.

c. Third, *zonal development* will also occur along the path taken by the wastewater through the reactor if back mixing is retarded by employing a multiple-chambered or baffled reactor design. Chambered, semi-plug-flow, along with relatively level loading rates, encourages these stable ecological niches. In this case the proportion of the metabolic processes that are aerobic, relative to those that are anaerobic, increases as BOD is depleted in the latter chambers.

7. *Decreased Sensitivity to Low Temperatures.* Slow growing microorganisms, as typified by nitrifying bacteria, display relative insensitivity to temperature drops when immobilized. For example, dropping the temperature from 30° to 12°C yields a 90% reduction in nitrification by suspended *Nitrobacter agilis* but only a 20% reduction when immobilized. Part of the explanation for this lies in an increased affinity for the food source and part from greater diffusion limitation resulting from immobilization (Wijffels et al., 1995).

Fixed-film reactors sometimes suffer from *disadvantages*. One of the biggest potential problems is one of high influent concentrations of total suspended solids (TSS) (> 100 ppm). Particulate material can tend to accumulate in the reactor. Fortunately this problem can be readily dealt with through the use of coagulating polymers and a primary settling (clarification) system prior to entry to the fixed-film reactor.

A problem common to all biodegradative systems is the presence of dense nonaqueous phase liquids (DNAPL; these liquids are more dense than water and thus tend to sink when suspended in an aqueous environment) and light nonaqueous phase liquids (LNAPL; these liquids tend to float). These highly hydrophobic organic oils are not readily biodegradable per se (remember, materials must be in an aqueous phase to be degraded). They can be even more troublesome when polyurethane foams are involved as the microbial support since a high enough concentration of certain of them (e.g., creosote oils) can swell the foam, adversely affecting the foams' structural properties. As with the TSS, this problem is readily dealt with through a pretreatment step. This entails the installation of an oil–water separator (Mashayekhi, 1993) that removes the oils prior to the entry of the wastewater to the biological reactor.

Activated-sludge systems (Rittmann, 1987) can be operated at high sludge recycle ratios to achieve high biomass concentrations within the reactor and to minimize biological solids formation. However, the efficiency of sludge recycle is completely dependent on the efficiency of the clarification step (Cheremisinoff,

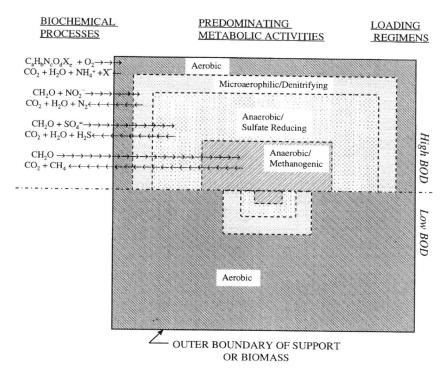

Figure 2.2 Schematic representation of the major metabolic processes occurring in a porous support. Depicted as an example is the cross section of a porous cubic support with an approximate delineation of the biochemical process (chemical equations at left) and predominating metabolic processes (word descriptions at center) that occur under two BOD loading rate regimens (right). Imaginary boundaries of these processes are delineated by the dashed (– – –) lines. The boundaries of process occurring at relatively *high* BOD loading rates are depicted above the – · – · – line and *low* below the – · – · – line. Note the preponderance of aerobic processes at low BOD loadings (bottom) compared to high BOD loadings (top). A generalized chemical pollutant molecule is represented by the molecular formula $C_aH_bN_cO_dX_e$. The capital letters have the usual elemental definitions, with X representing a halogen (Cl, Br, I). The lower case subscripts denote the actual number of atoms of each type appearing in a molecule, some of which may be zero. There may be any number of pollutant types in a given waste stream. The linked arrows at the left of the figure indicate that the individual molecules must diffuse into, and products and partial decomposition products out of, the fixed-film and porous substratum. The molecules must get at least as far as the final arrow in the foam in order for the various described metabolic processes to occur at that zone. Nutrient molecules (including the appropriate electron acceptors) must also diffuse into (and sometimes out of) the film and substratum. When atoms such as N, which also function as nutrients, are present in the pollutants in proportions greater than the needs of the microbial population, the amount in excess of the microorganisms' catabolic and anabolic needs is excreted in some form (such as ammonium, NH_4^+) along with metabolic by-products. No attempt has been made to balance the chemical equations or charges. For a discussion of diffusional effects and scaling procedures the reader is referred to Manem and Rittmann (1990).

1995). Without clarification the sludge age is equal to hydraulic retention age due to the completely mixed nature of the activated-sludge reaction vessel. Difficulties are often encountered in activated-sludge plants that attempt to lower sludge production by increasing sludge age. Two of the major problems encountered with this approach are:

- Poor sludge settleability
- Scum and foam formation

High sludge recycle ratios often lead to the growth of filamentous bacteria such as the actinomycete *Norcardia* (Soddell and Seviour, 1990), which promotes sludge bulking, scum formation, and increased sludge wasting. When the biomass is fixed in a submerged immobilized system, the reactor does not suffer this disadvantage. In contrast, random pack media trickling filter reactors such as the fixed-growth reactor described by Orr and Lawty (1990) have serious problems with this "troublesome" bacterium.

Description of ICB Reactor Media   The ICB system developed by Allied Signal has a number of proprietary and distinguishing features compared to other biofilm reactor systems. The ICB media or packing consists of two major components: (1) relatively stiff reticulated polyurethane foam and (2) rigid

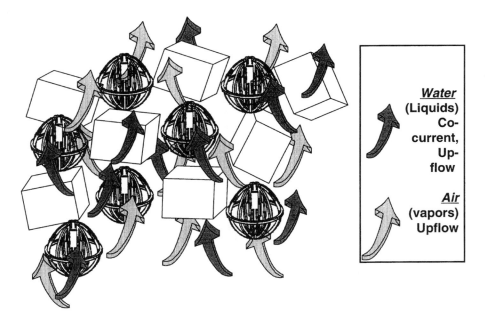

Figure 2.3 Schematic representation of the packing distribution and the movement of water and air through an ICB bed. The cubes represent the polyurethane foam blocks. Here Pall rings are substituted for Munipak rings.

polypropylene open-structure spacers (DeFilippi et al., 1993b) such as Rauschert Munipak rings. Since one component is blended with the other, we call this a mixed-media system. Figure 2.3 schematically depicts the co-current nature of the movement of water and air through the packed mixed-media bed. Note how the foam blocks are bathed on all sides by the water and air passing through the spacers. The ICB mixed media contains an evenly distributed mixture of reticulated polyurethane foam and rings in a 1:1 ratio by count. Specific dimensions and characteristics of the foam blocks and spacers are given in Table 2.1. The use of polyurethane foam as a substratum is gaining acceptance in a number of fixed-film reactor applications such as a fluidized bed (Tsuno et al., 1992) or RBC (Tyagi et al., 1992).

One cubic foot of commercial mixed media contains about 75 ± 5 pieces of foam blocks and 75 ± 5 pieces of spacers. These two packings are placed in the bed in a random configuration so that pockets of dead zones are not formed.

The polypropylene rings of the mixed media have multiple functions: They act as spacers in the mixed media system and provide a rigid open structure around the foam block. Because of the open nature of the spacers, the gas–liquid distribution rate is substantially greater than if foam alone is used. This is because foam-block-to-foam-block contact is minimized. The spacers keep the blocks of foam apart, allowing them to be bathed in liquids and gases on all sides. Therefore, an improper distribution of spacers in the foam bed would only cause channeling and give poor BOD removal rates. Ideally, one would like to have a packing as a spacer that will have a minimum surface area and a maximum void space so that filamentous biological growth cannot plug the spacer and cause channeling and poor BOD removal.

The second property that inclusion of rigid spacers imparts to the bed is that it becomes considerably less compressible. This tends to decrease the compaction of the bed such that channeling of liquids and gas is minimized.

Table 2.1   Characteristics of ICB Support Media

Characteristic	Product Name	
	Carbon-Coated Foam	Munipak^{TM} (Spacers)[a]
Chemical of fabrication	Reticulated polyether polyurethane foam	Polypropylene rings
Shape	Cube	Cylinder
Dimensions	$2 \times 2 \times 2$ in. ($5.1 \times 5.1 \times 5.1$ cm)	2×2 in., $L \times D$ (5.1×5.1 cm)
Void	93%	95%
Density	3.6 ± 0.1 lb/ft^3 (58 kg/m^3)	2.56 ± 0.1 lb/ft^3 (41 kg/m^3)
Pore Sizes[b]	15 ± 5 ppi (5.9 ppcm)	1.0 ppi (0.39 ppcm)
Surface area, (macro)	250 ft^2/ft^3 (820 m^2/m^3)	25.9 ft^2/ft^3 (85 m^2/m^3)

[a] Munipak is a packing commonly used in scrubbing towers and is distributed by Rauschert.
[b] As pores per inch (ppi) or per centimeter (ppcm), measured linearly for the foam and diagonally for the spacers.

Laboratory-Scale Studies

Mass Transfer   One of the crucial factors that must be addressed in biological reactor design is the ability to deliver oxygen to the microorganisms. In a fixed-film reactor mass transfer of dissolved oxygen from the bulk liquid to the individual microorganisms involves greater barriers than are encountered in mixed systems such as a CSTR. This is because films of microorganisms are usually considerably thicker than microorganisms grown in suspension. Furthermore, fixed films will have just one of two surfaces exposed to bulk liquids and gases whereas suspended microorganisms are exposed on all sides. An additional potential problem when foam blocks are used is one of penetration of air into the block and entrapment of air. Lack of air penetration into inappropriately sized foam pores results in all but the very outermost surface of the foam blocks being anaerobic. Air entrapment results in unacceptable buoyancy of the blocks that then tend to float and place pressure on uppermost hold-down grids.

The experimental setup (Sanyal et al., 1993; Defilippi et al., 1993b) for bulk liquid oxygenation consisted of a 55 gal (208 L) Nalgene® tank, 36 × 22 in. (91 × 56 cm) height (h) × inner diameter (ID), filled with Tri-Packs® test packing (spacers) and water to a volume of 45 gal (170 L) with h = 26 in. (66 cm). Spacers and foam blocks were retained between two perforated Plexiglas® disks with coarse bubble diffusers placed below the bottom disk. In Table 2.2 PUF-1 (polyurethane foam) refers to a small pore foam of ~60 pores per inch, ppi (23.6 pores per centimeter, ppcm), PUF-2 to a large pore foam of 15–20 ppi (5.9–7.9 ppcm), and mixed PUF to a 1:1 random mix of spacers and the appropriate foam. The system was first deaerated with nitrogen, then aerated and the time taken to attain a particular % dissolved oxygen (DO) recorded. YSI Inc. DO probes were employed for the oxygen determinations.

Inspection of the table indicates that the unique combination of a high surface area packing (e.g., polyurethane foam) with a relatively open volume packing (the spacers) yields oxygenation rates equivalent to when only the spacers alone are used, and a rate only approximately 50% longer than when water only is present in the tanks. In comparison, when foam is employed without spacers, oxygenation times are about 3–5 times longer than when water only is present.

Table 2.2   Time Taken to Reach 80% DO Level with Different Air Flow Rates and Bed Compositions

Bed Composition	Time (min)	
	2.0 scfm (0.06 m³/min)	3.6 scfm (0.1 m³/min)
Water only	9.2	6.3
Tri-Packs only	15.0	9.0
PUF-1 only	42.0	24.0
Mixed PUF-1	15.0	6.0
PUF-2 only	30.0	12.0
Mixed PUF-2	15.0	9.0

The small-pore foam exhibited process problems in that it had a strong tendency to trap air. This made the foam float, resulting in unacceptable pressure against the upper hold-down grate. Furthermore very little air appeared to penetrate and pass through this foam. In comparison, the larger-pore foam exhibited very little air entrapment. Furthermore, air was observed to readily pass through the foam blocks. The conclusion from these observations is that mixed media is quite significantly superior to foam-only systems as a biological support in a packed-bed fixed-film reactor and that large-pore foam is superior to small-pore foam.

Since the above experiments indicated that the ICB mixed media functions about as well as a spacer-only packing system when oxygenation is the criterion, why use any foam at all? This question is answered by the following set of experiments.

In this set of experiments the pollutant-degrading performance of fixed-film reactors using different packings is compared. One cylindrical 7.5-gal (28 L) ICB is packed with 6 gal (23 L) of a 50/50 combination of $2 \times 2 \times 2$ in. ($5.1 \times 5.1 \times 5.1$ cm) blocks of PUF and 2-in. Jaeger Tri-Packs® spacers, while a second reactor is packed with the Tri-Packs alone. The PUF is reticulated and possessed 15–20 ppi (5.9–7.9 ppcm). The reactors are operated in a up-flow mode, that is, both air and wastewater flowing from the bottom to the top of the reactor. Compressed air (40 psig) is used to aerate the reactor through three sintered glass diffusers located at the bottom of the tank. The reactors are seeded with a phenol-degrading consortium. Phenol at a nominal concentration of 1000 ppm in a nutrient-containing solution (0.1 g diammonium phosphate, 0.2 g diammonium sulfate, 0.1 g magnesium sulfate, 0.05 g calcium chloride, and 0.01 g yeast extract, all per liter) was used as a model pollutant wastewater. This wastewater is pumped through the ICBs with hydraulic residence time varying from 11 to 15 h. The reactors are sparged with between 8 and 13 L air/min. Phenol is quantitated using the 4-aminoantipyrene method. All readings below 1 ppm are reported as 1 ppm. Representative influent and effluent data are depicted versus sample date in Figure 2.4. As can be readily observed by inspection of this figure, the phenol concentration in the effluent of the reactor containing mixed media is consistently lower than the effluent from the all-Tri-Pack reactor. In other experiments (not shown) where high-ppi PUF was substituted for the low-ppi PUF in the mixed-bed reactor, effluent phenol concentrations were many times (up to 100) greater. We feel the superior performance observed when the low-ppi PUF is used with the mixed media is due to a number of factors: (1) The *macroporosity contributed by the low-ppi PUF*; this provides high surface area for biological growth [approximately 250 ft²/ft³ (820 m²/m³)] yet does not have pores so small as to restrict movement of air and nutrients and that would thus render the internal portions of the support unavailable to biocatalysis; (2) the *openness and rigidity of the spacers*; this increases the gas–liquid distribution rate. The spacers have the dual function of contributing strength to the reactor internals by virtually eliminating the propensity of the packing to collapse while maintaining sufficient open space to function as continuous redistributes of both nutrients/pollutants and air.

The foam may also be optionally coated with powdered activated carbon. It is well known that the presence of activated carbon has advantageous effects on

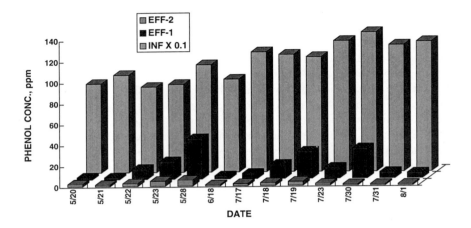

Figure 2.4 Bulk removal of phenol in laboratory: performance comparison of mixed-media bed and bed consisting only of spacers. Plot of phenol concentration vs. sampling date. The HRT averaged 13.2 h; reactor bed volume 45 gal. EFF-1 is the phenol concentration in the effluent from the reactor packed with spacers only. EFF-2 is the phenol concentration in the effluent from the reactor packed with spacers plus polyurethane blocks (mixed media). Note the values for phenol concentration in the influent are 10 times greater than depicted.

biological degradation of organic pollutants in wastewaters, even if the mechanisms by which this occurs are not yet well documented (Criddle et al., 1991). In this process (Sanyal et al., 1993; DeFilippi, 1996), the carbon is slurried in a noninhibiting polymer composition and coated and cured on the polyurethane foam. The carbon functions as a molecular buffer, adsorbing pollutants under conditions of high loading or spike-load conditions and releasing them under conditions of lower BOD loadings or normal operating conditions. Under conditions where there is very little adsorption of the target compounds, and especially when these compounds have little or no microbial toxicity, as say with sugars in food-derived wastewater, we expect there would be little advantage to the use of carbon.

Bench-Top Bioreactor   A laboratory-scale fixed-film reactor study is often the first step in determining operational variables and scale-up parameters. One must determine performance vs. HRT and aeration rate as well as the degree of stripping of volatiles and semivolatiles. Various measures of performance are described below.

For the purpose of determining approximate residence times and other variables, a bench–top ICB system is employed. It is set up as a vertical glass column with a length to diameter ratio (L/D) of 8 in./2.5 in. (20 cm/3.4 cm) = 3.2. Reticulated foam of approximately 15–20 ppi (5.9–7.9 ppcm) is cut into $20 \times 20 \times 20$ mm cubes. Small plastic 1 in. (2.54 cm) Pall rings are added as spacers at the ratio of 1:1, foam:ring. The unit is horizontally separated in half by a Teflon redistribution plate containing five concentric 6 mm holes; there is a ground glass frit at the bottom through which air passes at an aeration rate of between 100 and 300 mL/min; air and hydraulic flows are co-current. The approximate volume for the empty reactor is 750 mL, while its void volume is approximately 650 mL. Peristaltic pumps are used to pump sample wastewater through the system; house air is supplied through a gassing manifold to the reactor. Aeration flow rates are measured by flow meters; liquid flow rates are measured by collecting and quantitating an effluent volume in a known time period. Air stripping of volatile organics [e.g., volatile aromatic constituents of gasoline: benzene, toluene, ethylbenzene, and xylene isomers (BTEX); acrylonitrile] in the bench-top ICB can be determined by collecting vent gas organics onto a charcoal tube from a known gas volume, eluting with the appropriate solvent, and analyzing for the component(s) on a gas chromatograph fitted with a flame ionization detection (GC/FID). Occupational Safety and Health Administration (OSHA) Method 12 for BTEX in air samples is the general protocol followed (*Fed. Reg.* 52, No. 176, p. 34572).

Field-Scale Bioreactor Studies

Field Pilot-Scale Reactors   For development of accurate scale-up variables for the treatment of groundwater or process water in the field, skid-mounted ICB systems [reactor volume = 1500 gal (5680 L); bed volume = 1000 gal (3790 L)] are employed. These are self-supported pilot systems having a packed fixed-bed reactor, influent pump, pH adjustment tank, nutrient addition tank, and air blower. This system possesses an *L/D* ratio of 2:1. Please refer to Gromicko et al. (1994) for a schematic of such a pilot reactor.

Industrial Wastewater Treatment and Groundwater Remediation

For industrial wastewater treatment, a horizontal reactor design is utilized that contains four compartmentalized sections as described in Lupton et al. (1994). This compartmentalization is essential in effecting a semi-plug-flow of the process water. Compartmentalization therefore retards back-mixing allowing rather effective reduction of pollutant levels. Both air and water flow are co-current, although countercurrent designs are now being considered. In the co-current design water flow from one compartment to the other is via down-flow weir boxes. Individual reactors range in bed volume capacity between 3000 and 30,000 gal (11.4 and 114 m^3) per unit. Larger reactor capacities are achieved by linking reactors in either series or parallel or by the use of cylindrical units.

Organic Loading Rates   The performance of a reactor is ultimately determined by its ability to mineralize pollutants. The organic mineralization rates can be approximated by the differential change in COD (or sometimes BOD) between the reactor influent and effluent. The rates of both laboratory-scale and commercial-scale ICB systems treating a wide range of industrial wastewaters are shown in Table 2.3. Please recall that mg/L is equivalent to ppm. At high organic loading rates of 3.5–8.0 kg COD/(m^3 day), or 29–67 lb/(kgal day), the COD removal averaged

Table 2.3   COD Removal Efficiencies for Various Wastewaters Using the ICB

Wastewater	COD (mg/L)		COD Mineralization Rate		
	Influent	Effluent	kg/ (m^3 day)	lb/ (kgal day)	% Removal
Cutting oil emulsion[a]	16,000	2,500	6.7	55.9	84%
Specialty chemicals production[c]	13,000	9,000	8.0	66.8	69%
Textile mill[a]	1,100	260	1.7	14.2	76%
Cosmetics (fats & oils)[a]	4,400	1,000	3.4	28.4	77%
Flavors and fragrances[b]	6,430	5,050	15.6	130	22%
Shower and wash water (gray water)[b]	650	40	1.6	13.3	94%
Coal tar wastewater[c]	3,500	330	1.8	15	91%
Creosote wood treatment[c]	19,500	650	3.4	28	97%
Creosote wood treatment[c]	5,000	300	2.2	18.4	94%
Glycol wastewater[a]	2,000	100	1.5	12.5	95%

[a] Laboratory-scale system.
[b] Field pilot system.
[c] Commercial.

82±12% whereas at low organic loading rates of 1.5–3.5 kg COD/(m^3 day), or 13–29 lb/(kgal day), the COD removal averaged 90±8%. Refer to Sheridan (1993) for a good description of practical empirical methods to determine scale-up parameters.

BOD$_5$ removal rates in the ICB system range from 1.0 to 2.5 kg BOD$_5$/(m^3 day), or 8.3–21 lb/(kgal day), with values being dependent on such factors as microbiological considerations (V_{max}, K_M), choice of tester strain (s), diffusional effects, temperature, pressure, degree of salinity, desired percentage of COD reduction, and others. BOD$_5$ removal rates in the ICB system are in the same range as that for certain activated-sludge systems and are higher than conventional fixed-film bioreactors such as high rate trickling filters and rotating biological contactors (Table 2.4).

Table 2.4   Comparative Removal Rates by Biological Treatment Systems

Treatment System	Removal Rate	
	kg BOD$_5$/(m^3 day)	lb BOD$_5$/(kgal day)
Conventional activated sludge[a]	0.3–0.6	2.5–5.0
High aeration activated sludge[a]	0.6–2.0	5.0–16.7
Pure O$_2$ activated sludge[a]	1.6–3.0	13.3–25.0
High rate trickling filter[a]	0.3–1.0	2.5–8.3
Rotating biological contactor[a]	0.4–1.6	3.3–13.3
ICB[b]	1.0–2.5	8.3–20.8

[a] Data from *Wastewater Engineering: Treatment, Disposal and Reuse*, Metcalf & Eddy, Inc., edited by G. Tchobanoglous, McGraw-Hill Series in Water Resources and Environmental Engineering.
[b] Data from laboratory, field pilot, and commercial ICB systems.

Economics Efficient removal of BOD can have a substantial effect on lowering the costs associated with other components in a complete bioremediation system. For example, biological treatment of a moderately impacted groundwater (concentrations of COD $\gtrsim$ 418 ppm, phenols $\gtrsim$ 14 ppm, PAH $\gtrsim$ 9.8 ppm, and BTEX $\gtrsim$ 0.23 ppm) prior to an activated-carbon adsorption step greatly decreased the consumption of carbon and thus the estimated cost of groundwater remediation. Calculations are based on the results of two parallel field 1000-gal (3784–L) pilot studies (Gromicko et al., 1994). In one test the groundwater was passed through, in sequence, equalization tanks, a pH control and polymer addition system, flocculation tank (for solids removal), API separator (for removal of NAPL and DNAPL), ICB, sand filters, and finally granular activated-carbon absorbers. In the second test the ICB was omitted. HRTs of 16.6–2.8 h were employed. It was found that polycyclic aromatic hydrocarbon (PAH) removal ranged from 98.8 to 88%, phenolics from >99.9 to 72%, and BTEX from 98.4 to 91.5%. It was calculated that in a 60-gpm (227-L/min) bioremediation system, the use of the ICB for a 30-year project would yield a total project net present worth (cost to the customer) of $1,900,000 in 1994 U.S. dollars, whereas if the ICB is not included, activated-carbon change-out costs would drive this value to $5,000,000. Furthermore, where the impacting constituents are not activated carbon adsorbable, a biological treatment step may be indispensable.

Biological Solids Formation The biological solids present in the ICB system have been determined in a number of field applications. The concentration of solids in a commercial ICB system that is treating wastewater from a coal tar distillation plant is shown in Table 2.5. The average amount of biomass per piece of foam is 480 mg biomass/in.^{3} (29 mg/cm^3). The average concentration of biomass in the cylindrical rings (spacers) is 61 mg biomass/in^3. (3.7 mg/cm^3). This difference in biomass concentration is due to the inherent differences in physical shape and structure between the reticulated foam and the rings. The foam blocks are capable of retaining a higher concentration of biomass than the rings due to their much smaller pore size. The pores of the foam are mostly filled with biomass and act by protecting against shear forces and physical attrition due to aeration of the packed bed. The spacers have very large pores that are poorly protective against sheer forces. Thus, since the surfaces of the Hiflow are directly exposed to air agitation, they are not capable of retaining high concentrations of biomass on their exposed surfaces.

The equivalent mixed-liquor solids (EMLS) in a fixed-film reactor is determined as the average concentration of biomass within the system. That is, it is the

Table 2.5 Typical Distribution of Total Solids Present in the ICB System

Parameter	Reticulated Foam	Hiflow Cylinders
Biological solids, mg/packing piece	3840	380
Packing density, pieces/ft^3 (pieces/m^3)	72(2550)	72(2550)
Equivalent mixed liquor solids, mg/L	9780	970

concentration of biomass suspended in the void liquid as well as the concentration immobilized. In the case of the ICB the location of the latter is the foam and Hiflow rings. The EMLS of the ICB system is usually in the range of 8000–10,000 mg/L.

The solids retention time (SRT) for a fixed-film reactor is calculated as:

$$SRT = \frac{EMLS \times V}{TSS_{eff} \times Q}$$

For a typical full-scale ICB system, SRT is calculated to be 133 days with V = reactor volume [120,000 gal (454 m^3)], TSS_{eff} = total suspended solids or biomass concentration in effluent (150 mg/L), and Q = wastewater flow rate [60,000 gpd (227m^3/day)]. In comparison, the SRT of typical activated-sludge systems is 2–4 days. The long SRT of the ICB system is a major factor in promoting the low sludge yield of the system. The sludge yield can be calculated as follows:

$$Y_{obs} = \frac{(TSS_{eff} - TSS_{inf}) \times Q}{(BOD_{inf} - BOD_{eff}) \times Q}$$

where, using the above ICB treatment scenario as an example, Y_{obs} (observed sludge yield) is is calculated to be 0.06 kg of TSS/kg BOD$_5$ when TSS_{inf} (TSS in the influent wastewater) is 36 mg/L, TSS_{eff} (TSS in the reactor effluent) is 150 mg/L, BOD_{inf} (BOD in the reactor influent) is 2144 mg/L, BOD_{eff} (BOD in the reactor effluent) is 160 mg/L and Q represents the wastewater flow rate (which cancels out).

A comparison of the sludge yields between the ICB system and alternative biological wastewater treatment systems are shown in Table 2.6. The ICB system produces less than 20% of the amount of sludge generated by a typical activated-sludge system and less than 30% of the sludge generated by alternative fixed-film systems such as high-rate trickling filters.

The view expressed by Lawler (p. 856, 1986) that "biological treatment can be viewed as the conversion of primarily soluble organic pollutants into micro-organisms (particles) for subsequent removal" as the objective of biological treatment is essentially true (the soluble and vaporous inorganic products notwithstanding). Since there is a cost associated with disposal of the microbial sludge, wastewater treatment is most beneficial when the yield of biomass relative

Table 2.6 Comparative Sludge Yields for Waste Treatment Systems

Biological Treatment System	Sludge Yield[a]
Activated sludge	0.45–0.56
Sybron up-flow biotower	0.21–0.30
ICB	0.07–0.15

[a] Sludge yield defined as dry weight solids produced/dry weight BOD consumed by waste treatment process.

to the consumption of pollutants is minimized. That is, lower sludge yields result in lower disposal costs for the sludge, a clearly advantageous position as landfill disposal costs continue to increase.

Sludge generated in a fixed-film reactor appears in the effluent by a number of processes. A portion of the biomass remains in the reactor, while a portion detaches from the support, combines with biomass formed in suspension (i.e., never attached), exits the reactor, and appears in the effluent in the form of TSS which, when removed in a separation step (by, e.g., a clarifier), is designated "sludge." Four mechanisms by which biofilm detachment occurs have been summarized by (Rittmann, 1989): (1) grazing, (2) sloughing, (3) erosion, and (4) abrasion. An additional mechanism, (5) viral lysis, should also be considered as a potential contributing factor (Mathias et al., 1995). The first and fifth (subsets of both bioparticulate turnover and predation) result in a reduction of effluent sludge. The other three contribute to its increase.

A number of factors determine cell yield, which ultimately controls total biomass (and thus sludge) production. They include certain metabolic properties, such as energy yield, as well as process parameters such as growth temperature and SRT. Only the latter will be discussed here. SRT is an important variable in activated-sludge systems. It is a measure of how long sludge remains in the reactor. It is important in relation to sludge production since it has been found that biomass yield declines hyperbolically with SRT (Sykes, 1991). SRT also appears to have a substantial effect on the amount of sludge formed in fixed-film reactors. We do not completely understand how the long SRTs of fixed film biological reactors in general, and the ICB system in particular, translate into lower sludge yields. Yet, a number of microbiological processes can be evoked to explain this phenomenon:

1. *Bioparticulate Turnover.* The biomass yield can be defined as the ratio of unit dry weight of microbial biomass produced divided by unit dry weight of wastewater organic substrates utilized. It is equivalent to the cell yield expression in classical microbiology. In aerobic processes, biomass results from the utilization of part of the energy produced during biooxidation of a fraction of the organic carbon in the wastewater to convert a further fraction of the organic carbon into particulate biomass. In the short SRT usually present in suspended-growth (activated-sludge) systems, this is the major process occurring. However, when SRTs are increased, further reactions can occur such as the reutilization of the previously formed particulate cellular biomass. This will produce new cellular mass with a further concomitant production of carbon dioxide. The longer the solids retention time, the more 'particulate turnovers' can occur and the more the fraction of original wastewater organic is transformed into carbon dioxide (Hamer, 1990).

2. *Uncoupling of Metabolism and Growth.* If growth can be uncoupled from respiration, then biomass yield can be greatly minimized (Chudoba et al., 1992). That is, at the metabolic level anabolic processes (the energy-requiring microbial/enzymatic synthesis of relatively large molecular components of cells) are uncoupled from catabolic processes (the energy-producing microbial/enzymatic degradation of food/pollutant molecules). This uncoupling is the metabolic

equivalent of an automobile with its engine running but with its belts slipping and gear stripped. It burns a lot of fuel but gets nowhere! Transient environmental conditions such as cycles of oxic and anoxic environments can lead to metabolic uncoupling and cause bacteria to wastefully utilize organic substrates. Immobilization of biomass in a porous substratum such as reticulated polyurethane foam can induce apparent effects that promote this phenomenon (Bryers and Mason, 1987).

3. *Anaerobic Processes.* Even when a fixed-film reactor such as the ICB is operated in a fully aerobic mode, anaerobic processes occur (see Fig. 2.2). From the point of view of efficiency as measured by production of cell mass (reproduction) relative to mass of organic foods metabolized, anaerobic processes are much less efficient than aerobic processes. However, inefficient microbial usage of organics for the production of more biomass is advantageous for minimization of the operation and management costs associated with a waste treatment system since it means less mass of sludge is produced per mass of organic pollutant consumed. This can result in significant savings in sludge disposal costs.

4. *Predation.* Predatory microorganisms will ingest and consume the pollutant-degrading microorganisms (usually bacteria). They will predate the bacteria that are attached to the surface as well as those that are suspended or free-swimming. When the object of this predation is attached microorganisms, the process is termed

Figure 2.5 Electron micrograph of a rotifer (sack like organism to the right) that has attached to a nylon rod inserted into an ICB fixed-film reactor. This sessile predator is attached by means of a stalk (center, top) to the surface. The rings of cilia that occupy the mouth (lower right) and act to draw food particles into the organism are not visible due to an artifact of the sample preparation. The white reference bar in the lower right-hand corner is $10\,\mu m$ in length.

grazing. We have observed large numbers of predatory organisms inhabiting the ICB including paramecia, rotifers (Fig. 2.5) and even flat worms.

5. *Viral Lysis.* There is some data that indicates that viral (bacteriophage-induced) lysis of bacteria in marine systems may be on the order of 20% of the bacterial population (Mathias et al., 1995). The relevance of viral activity to biomass turnover in fixed-film reactor systems is yet to be fully determined.

The reader is referred to the section "Effect of Mass Transfer on Microbial Processes," this chapter, for metabolic effects on sludge yield.

Pollutant Removal Efficiency   The ICB system has the ability to reduce organic pollutants to very low concentrations at relatively short hydraulic retention times. Table 2.7 shows the removal efficiency of priority pollutants from a coal tar distillation wastewater at a hydraulic retention time of 16 h and an organic loading of 1.0 kg $BOD_5/(m^3$ day), or 8.3 lb/(kgal day). Removal of aromatic and polynuclear aromatic hydrocarbons ranged from 98.8 to 99.99% and BOD_5 removal was of the order of 98.1%

Table 2.7   ICB Removal Efficiency of Priority Pollutants from Coal Tar Distillation Wastewater

	Concentration (mg/L)		
Pollutant	Influent	Effluent	% Removal
Phenol	322–368	0.005–0.020	99.99
2,4-Dimethylphenol	4–6	BDL[a]–0.010	99.8
Benzene	4.1–6.8	BDL–0.003	99.94
Toluene	2.2–3.3	BDL–0.001	99.97
Ethylbenzene	0.6–0.8	BDL–0.001	99.86
Naphthalene	20–26	0.001–0.005	99.98
Acenaphthene	1.8–3.0	0.001–0.005	99.97
Fluorene	0.8–1.2	0.001–0.005	99.5
Phenanthrene	1.2–2.2	BDL–0.003	99.8
Anthracene	0.4–1.0	BDL–0.004	99.4
Fluoranthene	0.7–1.3	0.005–0.010	99.0
Pyrene	0.5–0.9	0.002–0.008	98.9

[a] BDL = below detection limit of 0.001 ppm.

The presence of activated carbon on the surface of the polyurethane foam blocks (DeFilippi, 1996) appears to enhance the removal of organic pollutants from wastewaters. A comparison of the pollutant removal by two 6000-L (1600-gal) field pilot ICB systems, one containing mixed media with carbon-coated foam and one containing mixed media with uncoated foam is shown in Table 2.8. The addition of

Table 2.8   Comparison of Carbon-Coated and Uncoated Foam as Supports During Biological Treatment of Coal Tar Distillation Wastewater in a Pilot-Scale ICB Reactor

Pollutant	Influent (mg/L)	% Time Effluent Limit Achieved		Effluent Discharge Limit (mg/L)
		Carbon	No Carbon	
Phenol	437–637	72	18	0.019
2,4-Dimethylphenol	3.5–7.5	100	59	0.019
Benzene	1.0–4.5	100	100	0.037
Ethylbenzene	0.1–0.98	100	100	0.032
Naphthalene	17.9–36.6	80	82	0.022
Acenaphthylene	0.22–0.43	100	100	0.022
Acenaphthene	1.8–7.8	100	18	0.022
Fluorene	1.1–2.0	100	82	0.022
Phenanthrene	1.4–2.6	100	82	0.022
Anthracene	0.35–0.86	100	82	0.022
Fluoranthene	0.55–0.75	100	12	0.022
Pyrene	0.35–0.5	100	12	0.022

activated carbon greatly improves the ability of the ICB system to meet very low effluent criteria for aromatic and polynuclear aromatic removal from coal tar distillation wastewater.

Process Stability   The high concentration of fixed biomass within the foam blocks enables the ICB system to tolerate widely fluctuating organic loads and still maintain high removal efficiencies. A 60,000-gal (227-m^3) ICB system was installed at a specialty chemical manufacturing facility that produces dye intermediates, surfactants, and other specialty organic chemicals (Sheridan, 1993). This plant uses batch production modes involving reactions with a variety of phenolic derivatives, aniline derivatives, and aromatic organic amines. The plant produces over 20,000 gal (76 m^3) of wastewater daily that is highly variable in both strength and composition (see Fig. 2.6 legend). The performance of the ICB system under these conditions is shown in Figure 2.6. The ICB system was able to maintain an effluent BOD$_5$ concentration of less than the POTW mandated 300-mg/L limit (average 70 mg/L) even when the influent BOD$_5$ concentration varied rapidly and repeatedly between levels of 1000 and 15,000 mg/L. The term POTW is an abbreviation of publicly owned treatment works, itself a euphemism for sewage treatment plant.

The presence of activated carbon also appears to give the ICB system enhanced stability in the face of highly fluctuating concentrations of toxic aromatic pollutants present in coal tar processing wastewater. The performance of a 350-L (92-gal) field pilot ICB system (L/D of 10) containing carbon-coated foam was compared with a

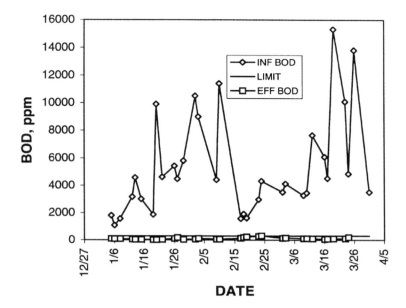

Figure 2.6 Performance of a 60,000-gal (227-m^3) bed ICB system at a specialty chemical plant. Plot of BOD concentration vs. sampling date. The bulk of the BOD present in the wastewater is in the form of acrylonitrile with phenolics, anilines, and aromatic amines or their derivatives being present at various times. The permissible discharge limit is 300 ppm. Influent and effluent BODs are denoted as inf BOD and eff BOD, respectively. Loadings varied from 112 to 2401 (av. 484 ± 422 SD) lb BOD$_5$/day [51 to 1089 (av. 220 ± 191 SD) kg BOD$_5$/day]. Flow rates varied from 3.4 to 14.5 gpm (12.9 to 54.9 L/min). Only a single point exceeded the discharge limit.

160,000-L (42,300-gal) full-scale fixed-film bioreactor system that utilized only Hiflow rings as the support matrix under equivalent conditions of hydraulic loading rate, organic load, and wastewater composition (see Fig. 2.7). The ability of the ICB system containing the carbon-coated foam to handle shock loads of phenolics was significantly greater than that of the bioreactor containing Hiflow rings only.

Reduced Air Stripping of Volatile Organic Compounds (VOCs) A common concern in wastewater and groundwater treatment processes involves the potential for stripping (volatilization) of VOCs. This is particularly important during biological treatment of groundwaters that are often contaminated with low levels of highly volatile solvent- or fuel-derived hydrocarbons. Table 2.9 shows the biodegradation of halogenated (1,2-dichlorobenzene) and unhalogenated (BTEX) aromatic hydrocarbon compounds during a laboratory-scale ICB treatment of contaminated groundwater at three different retention times. The results show that chlorobenzenes at a concentration of about 5 mg/L can be lowered to concentrations of between 0.1 and 1 mg/L at a 6-h HRT. Off-gas analysis of the bench-scale ICB

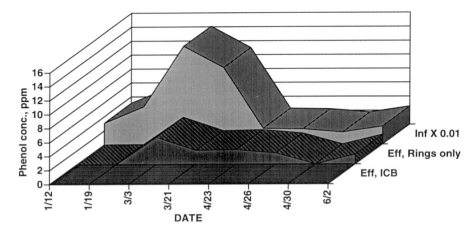

Figure 2.7 Effluent phenol concentrations after biological treatment of coal tar wastewater in field operation performance comparison. Plot of phenol concentration vs. sampling date. The influent to both reactors is identical and is labeled Inf X 0.01. Eff ICB depicts the concentration of phenol in the effluent from a mixed-media bed pilot ICB whereas Eff rings only depicts the concentration of phenol in the effluent from an installed commercial reactor with a bed consisting only of Hiflow packing rings. The wastewater is similar to that tabulated in Tables 2.5 and 2.6, although during this test the phenol concentrations are lower. Note the values for phenol concentration in the influent are 100 times greater than depicted. See text for further details.

system as determined by OSHA method 12 (*Fed. Reg.* 52, No 176, p. 34572) demonstrate that volatilization of the VOCs is less than 0.1%. Spiking studies were also performed by the addition of chlorobenzene to the groundwater sample. At a concentration of approximately 30 mg/L of chlorobenzene, the ICB was able to achieve 99.5% removal of the chlorobenzene at a 6-h HRT. However, the concentration of chlorobenzene remaining in solution is in the order of 150 µg/L. This

Table 2.9 Biodegradation of Aromatic Compounds in a Bench-Top ICB at Three Different Hydraulic Retention Times

		Concentration (µg/L)		
		Effluent		
Compound	Influent	HRT (h) 12	HRT (h) 6	HRT (h) 3
Benzene	894	<0.1	0.1	0.1
Toluene	122	<0.1	<0.1	<0.1
Ethylbenzene	26	<0.1	<0.1	<0.1
Chlorobenzene	4576	0.9	<0.2	4.0
1,2-Dichlorobenzene	174	0.5	1.3	0.6

indicates that HRTs longer than 6 h are required when lower discharge limits are required. The presence of activated carbon in the ICB system may contribute to reduction of volatile organic compound stripping during aeration of the bioreactor.

Similarly, the degradation and volatilization of chlorinated benzenes from groundwater was investigated in a 6000-L field pilot ICB system (data not shown). The off-gas from this reactor was analyzed by Environmental Protection Agency (EPA) reference method 18 (Title 40, CFR Part 60, Appendix A) to determine volatilization of organics. The volatilization of chlorobenzenes is also less than 0.1%.

Versatility

Industrial Wastewater.   Fixed-film reactors have proved to be quite versatile at degrading pollutants. As an example of a modern fixed-film reactor, the ICB has been successfully applied to effectively and economically remediate a diverse class of waste- and groundwaters in bench-top and field demonstrations as well as full-scale commercial units. Industrial wastewaters that have been treated in the ICB system include coal tar processing wastewater, wood treatment (creosote) wastewater, specialty chemical manufacturing wastewater containing acrylonitrile and other organics, and textile and food processing wastewaters. These applications demonstrate the ability of the ICB system to remove BOD and COD at high rates but at the same time generate only a reduced quantity of wasted biological sludge. The ICB removes OCPSF (organic chemicals, plastics, and synthetic fibers pretreatment standards) mandated pollutants to below their regulatory limits at effective hydraulic retention times within the bioreactor.

Groundwater.   Groundwaters that have been successfully remediated by the ICB system include: BTEX and methylene chloride in landfill groundwater, BTEX and chlorobenzenes in groundwater at an abandoned chemical plant, PAHs in groundwater from an abandoned wood preserving site, as well as TCE (trichloro-ethylene) at an abandoned mechanical engineering lab site. These applications demonstrate the low stripping of volatile pollutants from the bioreactor system with greater than 99% of the volatiles being biodegraded rather than volatilized.

Dependence of Performance on Microbiology and Process Control

HRT and SRT   The two fundamental parameters of biological remediation of wastewater are the hydraulic retention time (HRT) and the microbial solids retention time (SRT). The HRT required to achieve a given amount of BOD removal with a particular waste stream is a measure of bioreactor efficiency in that a short HRT translates into a relatively small reactor volume required to treat a given flow of wastewater. That is, the mass loading rate of BOD per unit volume of reactor is greatest at lowest retention times. The SRT is a factor that determines the concentration and thus efficiency of biocatalysis within the reactor. Long SRT translates into a high concentration of biomass within the reactor and promotes high rates of organics destruction and stability and shock resistance to variations in the organic mass loading to the bioreactor. Furthermore, as mentioned above, high SRTs also lead to lower sludge (biomass) yields per volume of wastewater treated and thus lower sludge disposal costs.

In CSTRs with the biological mass (biomass) existing in the suspended or unattached mode, the SRT is approximately equal to the HRT as the biomass is evenly distributed throughout the reactor and is being carried out of the reactor by the hydraulic flow. In this situation, short HRTs are not obtainable due to the fact that if the HRT exceeds the fastest growth rate of the microbial biomass, the microbes are washed out of the reactor at a faster rate than which they can replenish themselves by growth. The relatively short SRT also makes the CSTR a high sludge-producing system as growth and thus biomass yield are maximized.

The activated-sludge process is an attempt to overcome these difficulties by using a clarification step that allows the biomass washed out of the bioreactor to be collected by natural flocculation and settling aided by polymer addition and then be recycled back into the reactor. This is a considerable improvement upon the CSTR, and activated-sludge systems are by far the dominant wastewater system currently being utilized for the treatment of municipal and industrial wastewater. However, the activated-sludge system does suffer from some real limitations with regards to maintaining high SRTs. The efficiency of sludge recycling is dependent on the settling step. For high efficiency of settling, large clarification basins are required as well as the addition of expensive clarification polymers. Also, high sludge recycle ratios lead to low food-to-microorganism (F/M) ratios that lead to the development of filamentous bacteria that promote scum formation and low settling rate sludge (Soddell and Seviour, 1990). This not only leads to removal of biomass from the reactor and a subsequent drop in organic removal efficiency but also leads to enhanced sludge formation. This sludge is also often difficult to dewater and its disposal can be problematic.

In recent years, especially in Europe (Gassen, 1993) and Japan, biofilm reactors have been gaining in popularity due to their inherent advantages over the activated-sludge process. There is a major difference between the biofilm reactors and the CSTR and activated-sludge suspended-growth systems: The immobilization of the biomass on the reactor substratum leads to very much greater solids retention times as the bacteria are prevented from being washed away by the flow of wastewater. This leads to a high concentration of active biomass within the biofilm reactors. This high concentration of catalyst is reflected in a high performance and thus reduced reactor size, as well as resistance to toxic shocks and organic overloads (Santiago and Grady, 1990).

Effect of Mass Transfer on Microbial Processes One of the major problems encountered by biofilm reactor systems is that of mass transfer and distribution of air and water through the packed bed as well as the related problem of clogging. When immobilizing substrata (packings) are completely colonized and an extensive biofilm begins to bridge the gaps between this packing, clogging can occur that reduces the effective volume of the reactor by creating dead zones with minimal mass transfer and biocatalytic activity. This situation is exacerbated by small-sized reactor packing. This problem is addressed in the ICB by selecting relatively large blocks of porous packing with very high internal surface area (the use of mixed media). Once the bioreactor is fully colonized, these highly reticulated polyurethane foam blocks are completely filled with biomass. This gives the foam blocks a very high biomass density [>8000 mg/L mixed liquor volatile suspended solids (MLVSS), particulate matter that is lost on ignition, including micro-organisms and organic matter] and due to the highly immobilized nature of this biomass, a very long solids retention time of 150 days. Relatively large spacing between the packing is effected by evenly mixing a very open plastic spacer with the reticulated foam blocks. The biomass concentration on the spacers is much lower than the foam due to elutriation of excess biomass by hydraulic flow and aeration shear forces. This maintenance of open space between the foam blocks ensures good distribution of both gas and water throughout the packed bed of the ICB. In effect, the spacers perform the function of continuous redistributers of the fluids, ensuring delivery of O_2, pollutants, and nutrients to the microorganisms and removal of metabolic products.

Based upon observations on laboratory-scale ICB systems and in field pilot and commercial systems, a basic model of operating microbiological processes within the ICB system can be drawn (see Fig. 2.2). The fully colonized foam blocks contain an outer oxic region where there is a flux of oxygen and soluble organics into the biofilm with a subsequent degradation of the organics and formation of particulate biomass material. Products of partial and complete mineralization diffuse both deeper into the foam blocks and out into the fluid bathing the blocks. In the inner regions of the reticulated foam block anoxic conditions are increasingly prevalent.

Alternating anoxic/oxic environments may be present in a well-aerated ICB system. These conditions appear to promote the energetic uncoupling of microbial biomass and can led to reduced biomass yields (Chudoba et al., 1992). Within the anoxic zones of the porous polyurethane foam, a significant proportion of particulate microbial material formed in the oxic zones goes through a process of a "cryptic" growth involving death, cell lysis and hydrolysis, and fermentation of microbial polymers to generate fermentation products such as volatile fatty acids (VFAs). Dissection of the reticulated foam blocks removed from an operating ICB reveals the presence of VFAs. Some of these VFAs migrate back to the oxic zone where they are aerobically metabolized to CO_2 and water whereas some are apparently consumed in the anoxic zone when nitrate and sulfate are available as electron acceptors. Hydrogen sulfide, an easily detected end product of biological sulfate reduction (DeFilippi, 1994) is also found in the anoxic zones of the dissected foam blocks but not in the effluent gases. We have not looked for the presence of methanogenesis (methane formation) in the aerobic foam-supported biomass blocks. Still, we have set up anaerobic bench-top ICBs that readily generate methane from fermentable carbon sources. Aerobic methanotrophic reactors have also been run with success for the co-metabolic degradation of halogenated hydrocarbons.

SUMMARY

Biological reactors are effective and efficient at treating a wide variety of waste streams and impacted groundwaters. Using the ICB as an example of an advanced fixed-film reactor, it is clear that many process changes have led to an improved biological reactor. The net result of the combination of two packings with specific purposes along with activated carbon has been the creation of a fixed-film system that has a very high level of immobilized biomass and high biocatalytic activity but which also has minimal problems of mass transfer and clogging that afflict biofilm reactors aiming to reach similar levels of biomass immobilization. The fruits of this development are a bioreactor system with a high efficiency for removing wastewater organics but that also reduces dramatically the amount of biological sludge generated as a by-product of the process.

Acknowledgments

The authors would like to recognize the late Robert Detroy, Director of Biotechnology, for his support and efforts in the early stages of the development of the Environmental Biotechnology Program at AlliedSignal and G. Howard Collingwood for his recognition, leadership, and tireless encouragement during the development and commercial stages. We would also like to recognize the valuable technical contributions of Cathy E. Cornwall, Patricia Gillenwater, Walter Hribik, Tyronna Johnson, Mark Kaspar, Mark B. Koch, Tim P. Love, Mansour Mashayekhi, Glenn E. Petrie, Sugata Sanyal, William G. Sheridan, Constance M. Voellinger, Daniel R. Winstead, Edward F. Zinger, and Denise M. Zupancic to the development of the AlliedSignal ICB and BAT, without whose help these reactors would not have been developed or commercialized. We would also like to thank our respective spouses, Dr. Irene C. Gangl DeFilippi and Ms. Jane Lupton for their help and support.

REFERENCES

Alexander, M. (1980). Biodegradation of Chemicals of Environmental Concern, *Science*, **211**(9):132–138.

Bhamidimarri, R., A. Shilton, I. Armstrong, P. Jacobson, and D. Scarlet (1991). Constructed Wetlands for Wastewater Treatment: The New Zealand Experience, *Wat. Sci. Tech.*, **24**(5):247–253.

Battistoni, P., G. Fava, and A. Gatto (1992). Fish Processing Wastewater: Emission Factors and High Load Trickling Filters Evaluation, *Wat. Sci. Tech.*, **25**:1–8.

Belhateche, D. H. (1995). Choose Appropriate Wastewater Treatment Technologies, *Chem. Eng. Prog.*, **91**(8):32–51.

Bernard, J., Ed. (1990). Technical Advances in Biofilm Reactors, Proceedings of the IAWPRC Conference, Nice, Fr., April, 1989. *Wat. Sci. Tech.*, **22**(1/2),

Bryers, J. D. and C. A. Mason (1987). Biopolymer Particulate Turnover in Biological Waste Treatment Systems: A Review, *Biopro. Eng.*, **2**(3):95–109.

Bushnell, L. D. and H. F. Hass (1941). The Utilization of Certain Hydrocarbons by Microorganisms, *J. Bact.*, **41**:653–673.

Characklis, W. G. (1981). Fouling Biofilm Development: A Process Analysis, *Biotech. Bioeng.*, **XXIII**:1923–1960.

Cheremisinoff, P. N. (1993). Sludge: What Is It, Where Does It Come From? *Natl. Environ. J.*, **Nov./Dec.**:46–50.

Cheremisinoff, P. N. (1995). Gravity Separation for Efficient Solids Removal, *Natl. Environ. J.*, **Nov./Dec.**:29–32.

Chin, K. K., S. L. Ong, and S. C. Foo (1993). A Water Treatment and Recycling System for Intensive Fish Farming, *Wat. Sci. Tech.*, **27**(1):141–148.

Chudoba, P., J. Chudoba, and B. Capdeville (1992). The Aspect of Energetic Uncoupling of Microbial Growth in the Activated Sludge System, *Wat. Sci. Tech.*, **26**(9–11):2477–2480.

Clifford, D., S. Subramonian, and T. J. Sorg (1986). Removing Dissolved Inorganic Contaminants from Water, *Environ. Sci. Tech.*, **20**(11):1072–1080.

Clyde, R. A. (1983). Fiber Fermenter, U. S. Pat. 4,407,954 (Oct. 4).

Criddle, C. S., L. A. Alvarez, and P. L. McCarty (1991). Microbial Processes in Porous Media, in *Transport Processes in Porous Media*, J. Bear and M. Y. Corapcioglu, Eds., Klewer Academic, p. 639–691.

Crine, M., M. Schlitz, and L. Vandevenne (1990). Evaluation of the Performances of Random Plastic Media in Aerobic Trickling Filters, *Wat. Sci. Tech.*, **22**(1/2):227–238.

DeFilippi, L. J. (1993). Vapor Phase Biological Treatment Using Carbon Biomass Support, *Applied Bioremediation 93*, Intertech Conferences, Fairfield, NJ (October 25–26).

DeFilippi, L. J. (1994). Bioremediation of Hexavalent Chromium in Water, Soil and Slag Using Sulfate-Reducing Bacteria, in *Remediation of Hazardous Waste Contaminated Soils*, D. L. Wise and D. J. Trantolo, Eds., Marcel Dekker, pp. 437–457.

DeFilippi, L. J. (1996). Support Containing Particulate Adsorbent and Microorganisms for Removal of Pollutants, U. S. Pat. 5,580,770 (Dec. 3).

DeFilippi, L. J., M. B. Koch, C. M. Voellinger, D. R. Winstead, and F. S. Lupton (1993a). A Biological Air Treatment System Based upon the Use of a Structured Carbon Biomass Support, *IGT Symposium on Gas, Oil and Environmental Biotechnology*, Colorado Springs, CO (November 29–December 1).

DeFilippi, L. J., S. Sanyal, and T. P. Love (1993b). Performance Improvement of a Fixed-Film Biological Reactor by the Use of Mixed Packing Media, *I&EC Special Symposium*, American Chemical Society, Atlanta GA (September 27–29).

DeFilippi. L. J., F. S. Lupton and M. Mashayekhi (1995). Process for Biological Remediation of Vaporous Pollutants, U. S. Pat. 5,413, 714 (May 9).

DeFilippi. L. J., F. S. Lupton, and M. Mashayekhi (1996). Apparatus for Biological Remediation of Vaporous Pollutants, U. S. Pat. 5,503,738 (April 2).

Fan, Chi-Yuan and A. N. Tafuri (1994). Engineering Application of Biooxidation Processes for Treating Petroleum-contaminated Soil, in *Remediation of Hazardous Waste Contaminated Soils*, D. L. Wise and D. J. Trantolo, Eds., Marcel Dekker, pp. 373–401.

Fletcher, M and G. D. Floodgate (1976). The Adhesion of Bacteria to Solid Surfaces, *Soc. Appl. Bacteriol. Tech. Ser.*, **10**:101–107.

Gassen, M. (1993). Stand und Entwicklung der Festbettechnologie für die kommunale Abwasserreinigung (Status and Development of Fixed-Bed Processes for Municipal Wastewater Treatment), *Gewaesserschutz, Wasser, Abwasser*, **139**, (Neue Ansaetze im Integrierten Umweltschutz), 20/1–20/23.

Galil, N. and M. Rebhun (1990). A Comparative Study of RBC and Activated Sludge in Biotreatment of Wastewater From an Integrated Oil Refinery, *44th Purdue Ind. Waste Conf. Proc*, Lewis Publishers, Chelsea, MI. 711–717.

Greig-Smith, R.(1914). The Destruction of Paraffin by Bacillus Prodigiosus and Soil Organisms, *Proc. Linnean Soc. N. S. Wales*, **39**:538–541.

Gromicko, G. J., M. Smock, A. D. Wong, and B. Sheridan (1994). Pilot Study of a Fixed-Film Bioreactor for Biodegradation of Organic Wood Treating Compounds in Groundwater, *Seventh International IGT Symposium on Gas, Oil and Environmental Biotechnology*, Colorado Springs, CO (Dec. 12–14).

Guiot, S. R. and L. van den Berg (1985). Performance of an Upflow Anaerobic Reactor Combining a Sludge Blanket and a Filter Treating Sugar Waste, *Biotech. Bioeng.*, **XXVII**:800–806.

Hamer, G. (1990). Aerobic Biotreatment: The Performance Limits of Microbes and the Potential for Exploitation, *Trans. IChemE.*, **68(B)**:133–139, May.

Hart, S. (1996). In Situ Bioremediation: Defining the Limits, *Environ. Sci. Tech.*, **30**(9):398A–401A.

Hinchee, R. E., A. Leeson, L. Semprini, and S. K. Ong, Eds. (1994). *Bioremediation of Chlorinated and Polycyclic Aromatic Hydrocarbon Compounds*, Lewis, Boca Raton,

Huppe, P., H. Hoke, and D. C. Hempel (1990). Biological Treatment of Effluents from a Coal Tar Refinery Using Immobilized Biomass, *Chem. Eng. Technol.*, **13**:73–79.

Jewell, W. J. (1987). Anaerobic Sewage Treatment, *Environ. Sci. Tech.*, **21**(1):14–20.

Knackmuss, H. -J. (1983). Xenobiotic Degradation in Industrial Sewage: Haloaromatics as Target Substances, *Biochem. Soc. Symp.*, **48**:173–190.

Lakhwala, F., V. Sinkar, S. Sofer, and B. Goldberg (1990). A Polymeric Membrane Reactor for Biodegradation of Phenol in Wastewater, *J. Bioactive Compatible Polym.*, **5**:439–452.

Laverman, A. M., J. S. Blum, J. K. Schaefer, E. J. P. Phillips, D. R. Lovley, and R. S. Oremland (1995). Growth of Strain SES-3 with Arsenate and Other Diverse Electron Acceptors, *Appl. Environ. Microbiol.*, **61**(10):3556–3561.

Lawler, D. F. (1986). Removing Particles in Water and Wastewater, *Environ. Sci. Tech.*, **20**(9): 856–861.

Lester, J. N., R. M. Sterritt, and P. W. W. Kirk (1983). Significance and Behavior of Heavy Metals in Wastewater Treatment Processes, II. Sludge Treatment and Disposal, *Sci. Total Environ.*, **30**:45–83.

Lupton, F. S. and K. C. Marshall (1979). Effectiveness of Surfactants in the Microbial Degradation of Oil, *Geomicrobiol. J.*, **1**(3):235–247.

Lupton, F. S., and D. M. Zupancic (1991). Removal of Phenols From Waste Water by a Fixed Bed Reactor, U. S. Pat., 4,983,299 (Jan. 8).

Lupton, F. S., W. G. Sheridan, and M. R. Surgi (1994). Process for Removal of Organic Pollutants from Waste Water, European Patent, 0,467,969 B1 (June 15).

Marshall, K. C. (1992). Biofilms: An Overview of Bacterial Adhesion, Activity, and Control at Surfaces, *Am. Soc. Microbiol. News*, **58**(4):202–207.

Manem, J. A. and B. E. Rittmann (1990). Scaling Procedure for Biofilm Processes, *Wat. Sci. Tech.*, **22**(1/2):329–346.

Mashayekhi, M. (1993). Full-Scale Experiences with the Fixed-Film Immobilized Cell Bioreactor, *Applied Bioremediation 93*, Intertech Conferences, Fairfield, NJ (October 25–26).

Mathias, C. B., A. K. T. Kirschner, and B. Velimirov (1995). Seasonal Variations of Virus Abundance and Viral Control of the Bacterial Production in a Backwater System of the Danube River, *Appl. Environ. Microbiol.*, **61**(10):3734–3740.

McCarty, P. L. (1982). One Hundred Years of Anaerobic Treatment, in *Anaerobic Digestion 1981*, Hughes et al., Eds., Elsevier Biomedical Press, pp. 3–21.

McCarty, P. L. and D. P Smith (1986). Anaerobic Wastewater Treatment, *Environ. Sci. Tech.*, **20**(12):1200–1206.

McClintock, S. A. and J. A. Frazier (1994). Optimization of Biological Nutrient Removal in Sequencing Batch Reactors at the Dauphin Borough Wastewater Treatment Plant, *Proceedings, WEFTEC '94, 67ᵀᴴ Annual Conference and Exposition*, pp. 585–593.

Morsen, A. and H. J. Rehm (1990). Degradation of Phenol by a Defined Mixed Culture Immobilized by Adsorption on Activated Carbon and Sintered Glass, *Appl. Microbiol. Biotech.*, **33**:206–212.

Nacheff-Benedict, M. S., G. H. Kumagai, G. E. Petri, R. W. Schweickart, C. D. McFadden, and M. A. Edeen (1994). An Integrated Approach to Bioreactor Technology Development for a Regenerative Life Support Primary Water Processor, SAE Paper 941397, 24th Internat. Conf. Environ. Systems, Friedrichshafen, Germany, June 20–23.

Orr, P. and R. Lawty (1990). Operating Experience with Large Random Packed Biofilm Reactors, *Wat. Sci. Tech.*, **22**(1/2):203–214.

Otte, M. -P., J. Gagnon, Y. Comeau, N. Matte, C. W. Greer, and R. Samson (1994). Activation of an Indigenous Microbial Consortium for Bioaugmentation of Pentachlorophenol/ Cresote Contaminated Soils, *Appl. Microbiol. Biotech.*, **40**:926–932.

Ottengraf, S. P. P., J. J. P. Meesters, A. H. C. van den Oeve, and H. R Rozema (1986). Biological Elimination of Volatile Xenobiotic Compounds in Biofilters, *Biopro. Eng.*, **1**:61–69.

Paigen, K. and B. Williams (1970). Catabolite Repressions and Other Control Mechanisms in Carbohydrate Utilization, in *Advances in Microbial Physiology*, A. H. Rose and J. F. Wilkinson Eds., Academic, pp. 251–324.

Petrie, G. E. and F. S. Lupton (1991). Performance Characteristics of a Low Sludge Bioreactor for Wastewater Treatment, *Waste Manage. Res.*, **9**:471–476 (1991).

Petrie, G. E. and M. S. Nacheff-Benedict (1991). Development of Immobilized Cell Bioreactor Technology for Water Reclamation in a Regenerative Life Support System, SAE Paper 911503, 21st Internat. Conf. Environ. Systems, San Francisco, California, June 15–18.

Revis, N. W., J. Elmore, H. Edenborn, T. Osborne, G. Holdsworth, C. Hadden, and A. King (1991). Immobilization of Mercury and Other Heavy Metals in Soil Sediment, Sludge and Water by Sulfate-Reducing Bacteria, *Innovative Hazard. Waste Treat. Technol.*, Ser. 1991, **3**(Bio. Proc.), 97–105.

Rittmann, B. E. (1987). Aerobic Biological Treatment, *Environ. Sci Tech.*, **21**(2):128–136.

Rittmann, B. E. (1989). Detachment from Biofilms, in *Structure and Function of Biofilms*, W. G. Characklis and P. A. Wilder, Eds., Wiley, pp. 49–58.

Santiago, I. and C. P. Leslie Grady, Jr. (1990). Simulation Studies of the Transient Response of Activated Sludge Systems to Biodegradable Inhibitory Shock Loads, *44th Purdue Ind. Waste Conf. Proc.*, pp. 191–198.

Sanyal, S., T. P. Love, and L. J. DeFilippi (1993). Process and Apparatus for Removal of Organic Pollutants From Waste Water, U. S. Pat. 5,217,616 (Jun. 8).

Sheridan, W. G. (1993). System Performance of the AlliedSignal Immobilized Cell Bioreactor (ICB) from the Bench-Top and Pilot Studies to the Full-Scale ICB Installation at a Specialty Chemical Dye Intermediate Plant, *Applied Bioremediation 93, Intertech Conferences, Fairfield, NJ (October 25–26)*.

Soddell, J. A. and R. J. Seviour (1990). Microbiology of Foaming in Activated Sludge Plants, *J. Appl. Bacteriol.*, **69**:145–176.

Sykes, R. M. (1991). The Product-Maintenance Theory of the Activated Sludge Process, *J. Environ. Sci. Health*, **A26**(6):855–881.

Tsuno, H., I. Somiya, N. Matsumoto, and S. Sasai (1992). Attached Growth Reactor for BOD Removal and Nitrification with Polyurethane Foam Medium, *Wat. Sci. Tech.*, **26**:2035–2038.

Tijhuis, L., M. C. M. van Loosdrecht, and J. J. Heijnen (1995). Dynamics of Biofilm Detachment in Biofilm Airlift Suspension Reactors, *Biotech. Bioeng.*, **45**:481–487.

Tyagi, R. D., F. T. Tran, and A. K. M. M. Chowdhury (1992). Performance of RBC Coupled to a Polyurethane Foam to Biodegrade Petroleum Refinery Wastewater, *Environ. Pollution*, **76**:61–70.

Van Voorhis, D. B. (1990). Factors Affecting Chromium and Copper Removal in a Fixed Film Biological System, Thesis. Order Number 9100891, University Microfilms Inc., Ann Arbor, MI,

Venosa, A. D. (1995). Delaware Oil Spill Bioremediation Field Study, *Tech. Trends*, **21**:1.

Weber Jr., W. J. and E. H. Smith (1986). Removing Dissolved Organic Contaminants from Water, *Environ. Sci. Tech.*, **20**(10):970–979.

Wijffels, R. H., G. Englund, J. H. Hunik, E. J. T. M. Leenen, Å. Bakketun, A. Günther, J. M. Obón de Castro, and J. Tramper (1995). Effects of Diffusion Limitation of Immobilized Nitrifying Microorganisms at Low Temperatures, *Biotech. Bioeng.*, **45**:1–9.

Whipple, S. A. (1901). Changes That Take Place in the Bacterial Contents of Waters During Transportation, *Tech. Quart.*, **14**:21–29.

Zobel, C. E. (1943). The Effect of Solid Surfaces upon Bacterial Activity, *J. Bacteriol.*, **46**:39–56.

CHAPTER 3

Bioslurry Reactors

Christos Christodoulatos and Agamemnon Koutsospyros

Center for Environmental Engineering, Stevens Institute of Technology, Hoboken, New Jersey 07030 (C.C.) and Department of Civil and Environmental Engineering, University of New Haven, West Haven, Connecticut 06516 (A.K.)

INTRODUCTION

Bioremediation involves stimulation of microbially mediated processes applied exclusively or in combination with other physicochemical processes for the cleanup of diverse waste matrices including vapor, liquid, and solid contaminants of organic origin. Bioremedial technologies for soils, sludge, and dredged materials are applicable in situ or ex situ depending on whether the waste matrix involved is in its natural setting or is removed and transported into a reactor.

Potential advantages of in situ bioremediation methods include minimal site disruption, simultaneous treatment of contaminated soil and groundwater, minimal exposure of public and site personnel, and relatively low costs compared to other decontamination practices. Disadvantages and limitations of in situ systems include:

- Longer treatment times compared to other remedial technologies due to slow groundwater movement, circulation, mixing, and other mass-transfer limitations
- Seasonal variations of microbial activity resulting from direct exposure to prevailing environmental factors (pH, temperature, alkalinity) and lack of control of these factors
- Problematic application of treatment additives (co-substrates, nutrients, surfactants, and oxygen) for soils of low (less than 10^{-4} cm/s) or inconsistent permeability

Biological Treatment of Hazardous Wastes, Edited by Gordon A. Lewandowski and Louis J. DeFilippi
ISBN 0-471-04861-5   ©1998 John Wiley & Sons, Inc.

- Costly hydraulic structures for plume containment and control (i.e., slurry walls, sheet piles, etc.) may be necessary to minimize potential contaminant mobilization.

Most in situ systems are still in the developmental stage and only limited experience of successfully engineered full-scale treatment systems is available.

Ex situ bioremediation technologies address most of the disadvantages and limitations mentioned above in a satisfactory manner; however, they suffer from significant costs associated with solids handling processes including excavation, screening and fractionation, mixing and homogenization, and final disposal.

Based on phase considerations of the contaminated material under treatment, ex situ bioremediation processes are classified as:

- Solid-phase systems including land treatment and soil piles (composting)
- Slurry-phase systems involving treatment of solid–liquid suspensions in bioreactors or impoundments.

The selection of the applicable process depends on bioremediation goals and objectives to be met, physical and chemical characteristics of the materials involved, site characteristics, handling of materials, equipment requirements, and economic factors. This chapter focuses on the use of slurry-phase treatment systems for bioremediation of contaminated soils and sludges.

SLURRY-PHASE TREATMENT FUNDAMENTALS

Slurry-phase treatment is in some aspects analogous to conventional suspended-growth biological systems (e.g., activated sludge). Contaminated solid materials (soils, dredged sediment, etc.), microorganisms, and water formulated into a consistent slurry are brought in contact within an engineered containment (bioreactor). Bioslurry reactors are designed to:

1. Alleviate microbial growth-limiting factors in soil environments including substrate, nutrient, and oxygen availability
2. Promote suitable and stable environmental conditions for bacterial growth including moisture, temperature, and pH
3. Enhance uniformity and thus reduce toxicity due to localized contaminant pockets
4. Minimize mass-transfer limitations and facilitate phenomena such as desorption of organics from the soil matrix

In essence, slurry-phase treatment is a triphasic system involving three major components: water, suspended particulate matter, and air. Water serves primarily as the suspending medium where supplementary nutrients, other trace elements needed for bacterial growth, as well as chemical conditioners (e.g., pH adjustment chemicals, desorption enhancers, etc.) and desorbed contaminants are dissolved. Suspended particulate matter consists of a biologically inert substratum bearing the contaminants (soil particles or aggregates) and biomass attached to the soil matrix or free in the suspending medium. Air, in the form of bubbles, provides the oxygen necessary for bacterial growth and supplements mixing.

The interaction of the components described above as well as numerous physical and biochemical phenomena involved, make slurry-phase treatment a complex and difficult system to formulate. Examples of possible mechanisms influencing the system include adsorption/desorption, dissolution/precipitation, ion exchange, complexation, oxygen transfer, volatilization, particle size reduction, and bio-degradation. It is noteworthy that some of these mechanisms may be occurring simultaneously, often with conflicting effects on principal process variables. The rate-limiting step approach is used to relieve process complexity to some extent. For example, for certain hydrophobic contaminants such as wood-treating chemicals (i.e., pentachlorophenol, creosote) that exhibit a high affinity for soil particles, desorption is the rate-limiting step while biodegradation proceeds at reasonable rates if the contaminants become bioavailable. Consequently, such systems must be designed in a way that optimizes the rate of desorption and maximizes contaminant bioavailability. Bioavailability of contaminants may be achieved by methods such as soil washing and chemical extraction ahead of the bioreactor by means of a pretreatment step. Also since the bulk contaminants are preferentially adsorbed on fine solids, size fractionation of the contaminated soil to exclude coarse material may be desired prior to bioavailability enhancement. It is therefore evident that process integration along with identification of the rate-limiting step can be used to resolve conflicting effects and mitigate process complexity.

Process Description

Unlike solid-phase treatment systems that have been used for several decades primarily by the oil industry, slurry-phase reactors are a relatively new development in the field of bioremediation. Generally, there are three types of slurry-phase bioreactors (Brox, 1989; Castaldi and Ford, 1992; Griffin et al., 1990; Hanify et al., 1993; Irvine et al., 1993a,b; Lewis, 1993; Tang and Fan, 1987, 1989; USEPA, 1993a,c, 1989):

1. Aerated lagoons
2. Low-shear airlift reactors
3. Fluidized-bed soil reactors

The first two types have been used in a variety of full-scale bioremediation applications, while the fluidized-bed soil reactors are still in the research and development stages (Satija and Fan, 1985; Tang and Fan, 1987, 1989). Although each reactor type has its own operational peculiarities, certain principles are common to all.

Aerated Lagoons The simplest form of a slurry-phase system is a lined impoundment very similar to aerated lagoons used for treatment of small community municipal wastewater. Figure 3.1 illustrates a conceptual schematic of a slurry-phase lagoon system. Nutrients, an active culture of microorganisms and aeration, are added to the reactor. Batch is the most commonly practiced mode of operation. The process may be used as a single-stage or a multistage configuration depending on pre- and post-treatment requirements of the contaminated soil or sediment.

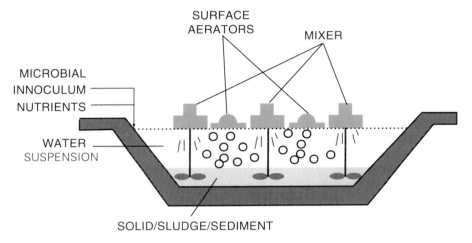

Figure 3.1 Schematic of a slurry-phase lagoon.

Turbine mixers and surface aerators provide mixing and aeration of the lagoon contents. The mixing power requirements of slurry-phase lagoons are considerably higher than those of municipal wastewater treatment lagoons since heavier particles must be kept in suspension in the former systems. Surface aerators alone provide limited suspension capabilities, and mixing must be supplemented with turbine mixers. Mixing deficiencies and reactor geometry may result in dead spaces that adversely affect treatment performance.

Slurry-phase lagoons are subject to operational instability due to variations in weather conditions such as precipitation and temperature. This results in poor definition of process control. Open slurry-phase lagoons are not appropriate for wastes containing regulated volatile components unless they are covered and a vapor control system is provided. Lagoon covers and vapor-phase treatment systems add significantly to overall remediation cost.

Low-Shear Airlift Reactors (LSARs) In situations where the waste contains volatile components, and tight process control and increased efficiencies are desired, low-shear airlift bioreactors can be adopted. LSARs are cylindrical tanks made of stainless steel or other relatively inert material. As closed reactors LSAR limit abiotic losses of contaminants and allow separate treatment of the off-gases by means of carbon adsorption or biofilters. Process control with respect to pH, temperature, nutrient addition, mixing, and oxygen transfer is readily adaptable to such systems.

Intimate contact of the microbial community with the targeted contaminants, and homogeneity between all phases involved, is accomplished by mixing. Mechanical agitation and aeration, or most commonly a combination of both along with proper baffle placement, make the hydrodynamic behavior of slurry-phase bioreactors more defined and superior to that of lagoons. Motor-driven shaft setups equipped with single or multiple impellers are the most common method of mechanical agitation. Ordinary shaft designs also include rake arms with blades for resuspension of coarse material that tends to settle on the bottom of the bioreactor. Diffuser arrangements are ordinarily spaced radially along the rake arm. Certain configurations provide internal top-to-bottom circulation of the reactor contents by means of airlift. External circulation systems may also be used. Figure 3.2 illustrates various features of a low-shear airlift bioreactor. A number of designs are currently commercially available under various trade names (USEPA, 1992a,b,c, 1993b). Additional benefits of this type of reactor include:

- Nonpermanent installation for various on-site applications by means of mobile trailers
- Easy integration into various pre- and post-treatment process schemes
- Versatility in all modes of operation including batch, sequential batch, and continuous.

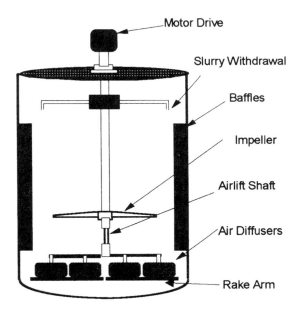

Figure 3.2 Basic features of a low-shear airlift slurry-phase bioreactor.

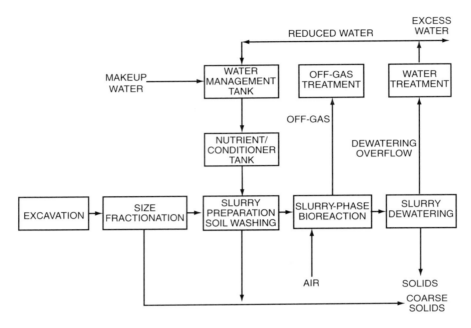

Figure 3.3 Conceptual flowchart of an integrated slurry-phase treatment system.

Pretreatment processes include size fractionation to exclude oversized solids, soil washing, milling to reduce soil particle size, and slurry preparation (USEPA, 1993b). Pre- and post-treatment generate additional waste streams that may require treatment. In such cases, a liquid treatment process train must be devised and integrated into the overall treatment scheme. Similarly, off-gases containing volatilized contaminants from the bioreactor must be treated prior to atmospheric discharge. Post-treatment includes slurry dewatering as well as treatment of the dewatering overflow liquid stream. An integrated process flowchart for slurry-phase bioremediation is illustrated in Figure 3.3.

Adsorption/Desorption

Adsorption/desorption is a mechanism of paramount importance to bioslurry remediation as it affects bioavailability and the success of the entire process (Aronstein et al., 1991; Christodoulatos et al., 1994; Talimcioglu et al., 1993). Solute adsorption is an interfacial phenomenon resulting in solute accumulation at a solid–liquid interface. The solute molecules attach to the surface where they are held by electrostatic forces or chemical binding, and a dynamic equilibirium is established between the liquid and solid phases. In slurry-phase systems adsorption is favored by the large interfaces formed by the tiny soil particles that resist solute release to the aqueous phase. The bulk of the contaminants is attached to the fine soil fraction, which is comprised of particles with an approximate diameter of 60 μm or less. Desorption of organic substances from the soil matrix is a function of soil and solution properties, and in many instances, it is strongly affected by irreversible phenomena.

The partitioning of the solute in the liquid and solid phases is often described by an equilibrium curve called adsorption isotherm. The shape of the equilibrium curve depends on the solid and solute properties. Both linear and nonlinear isotherms have been observed in environmental systems. For dilute systems a linear model relating the amount of solute sorbed to the concentration in the liquid phase at equilibrium takes the form:

$$q_e = K_d C_e \tag{1}$$

where q_e is the concentration of solute in the sorbed phase, C_e is the equilibrium concentration of the solute in the liquid phase, and K_d is the adsorption partition coefficient. The most widely used nonlinear adsorption model for solid–liquid systems is the Freundlich equation:

$$q_e = K_f C_e^{1/n} \tag{2}$$

where K_f and $1/n$ are characteristic constants. Although Freundlich's equation is empirical, the coefficient K_f is related to the adsorption capacity, and the exponent $1/n$ is characteristic of the energy distribution of the adsorption sites.

Another sorption parameter frequently used is the normalized partition coefficient with respect to the organic carbon content of the soil, K_{oc}, defined as:

$$K_{oc} = \frac{K_d}{f_{oc}} \tag{3}$$

where f_{oc} is the fraction of organic carbon in the soil. The above equation defines a constant that depends on the solute alone. An empirical equation that relates the organic carbon partition coefficient to the octanol/water partition coefficient for hydrophobic compounds is:

$$\log K_{oc} = a \log K_{ow} + b \tag{4}$$

where a and b are empirically determined coefficients (Karickhoff, 1984). This equation has been used to model the sorption behavior of both nonionizable and ionizable compounds. However, for ionizable compounds only an apparent partition coefficient can be obtained since the concentrations of the different forms of the solute cannot be directly measured by conventional chromatographic techniques. In addition, K_{oc} does not adequately describe the sorption equilibria, which are influenced by other parameters such as grain size distribution, pH, clay and silt content, ionic strength, total organic carbon (TOC) content, temperature, cation exchange capacity (CEC), presence of co-solvents and surfactants, anion exchange capacity (AEC), solubility, contaminant aging, and soil mineralogy. In addition, the rate and extent of desorption depend on the formation of solute complexes with ions present in solution, the concentration of solids in the slurry, and the thickness of the electrical double layer that is formed as a result of charge distribution between the phases.

Due to the complexity of the system, one can only hope to develop statistical correlations between the various parameters affecting the process and construct an empirical model. Christodoulatos and Mohiuddin (1995) proposed nonlinear multiple regression models for prediction of the apparent distribution ratio, K, of ionogenic substances that include soil and solution parameters. K is defined as the ratio of the amount of solute in the liquid phase to the amount adsorbed on the solid phase. For example, the apparent distribution ratio of pentachlorophenol (PCP), a weak organic acid, in various slurry environments and pH domains can be described by the following equation:

$$K = \frac{213.80(SS)^{0.09}\ (CEC)^{1.09}\ (f)^{-0.60}\ (f_{oc})^{0.47}}{1 + 10^{pH-4.75}} \tag{5}$$

where SS is the soil-to-solution ratio, f is the fraction of silt and clay of the soil, and CEC in meq/100 g. Equations of this form are very useful for the design and operation of soil remediation systems since they include parameters such as pH and solids concentration, which can be directly manipulated and controlled. Equation (5) quantifies the effects of the system variables on the distribution ratio and can be used to optimize the conditions favoring desorption and system performance in general. Thus pH adjustments during soil pretreatment mobilizes adsorbed contaminants and enhances their bioavailability in subsequent treatment stages.

Sorption processes are generally slow, and days or even weeks may be necessary for equilibrium to be attained. When sorption kinetics are limiting the process throughput, the desorption rates should be carefully considered and incorporated in the design of bioslurry systems.

Role of Surfactants in Desorption Enhancement

Role of Surfactants in Desorption Enhancement Bioremediation and other remediation technologies such as soil washing rely on desorption phenomena to transfer the contaminants from the solid to the aqueous phase (Aronstein et al., 1991; Edwards et al. 1994a,b). In many instances desorption processes are limited by the aqueous solubility of the contaminants. Hydrophobic substances have low affinity for water and tend to accumulate at interfaces. Solvents and surface-active agents (surfactants) promote desorption by modifying the free energy of the surface and by increasing the aqueous solubility of the compounds. Solubilization facilitates microbial oxidation and enhances the rate of degradation of many hydrophobic compounds of environmental relevance.

Surfactants are substances that, when present in small concentrations in an aqueous system, have the property of adsorbing on the surfaces or interfaces of the system and by so doing altering the interfacial tension between the two phases. Surfactants are strongly amphiphilic compounds, either natural or synthetic, with polar and nonpolar regions. A fundamental property of surfactants is the formation of colloidal size clusters called micelles. Micelles are formed when the bulk solution concentration reaches a threshold value above which excess surfactant monomers aggregate with their hydrophilic groups in the bulk solution and the hydrophobic moieties in the center of the micelle. The concentration at which micellization occurs is called the critical micelle concentration (CMC) or in the case of bilayer lamellae-forming surfactants the critical aggregate concentration (CAC).

Surfactants are classified according to the nature of their surface-active portion as anionic, cationic, zwitterionic (where both positive and negative charges are present in the surface-active portion), and nonionic. All four types have potential environmental applications depending on the nature of the contaminant and the properties of the contaminated matrix. For instance, ionic surfactants may be used for solubilization of heavy metals, whereas nonionic surfactants may be added to soil–water slurries to increase the rate of release of polycyclic aromatic hydrocarbons (PAHs).

The surface tension of aqueous solutions decreases upon addition of surfactant and attains a steady value at concentrations above CMC. If hydrophobic organic compounds (HOCs) are present in the system as a separate (undissolved) phase, the observed drop in surface tension is accompanied by a rise in the solubility of the HOCs in the surfactant solution. In general, the solubility of HOCs increases as the surfactant dose increases, with a sharp linear rise at surfactant doses above the CMC.

In the presence of soil, surfactant monomers are sorbed at the solid–liquid interfaces and a dynamic equilibrium is established between the two phases. Therefore, the amount of surfactant required to reach the critical micelle concentration, at which micellization and solubilization commence, is substantially higher in soil–water slurries than in aqueous solutions. Maximum adsorption of the surfactant occurs in the vicinity of the critical micelle concentration and can be ascribed to both monolayer and bilayer formation. Surfactant type and structure, presence of electrolytes, surface charge, and solution pH are among the factors determining the extent of surface adsorption.

In light of the above discussion it becomes apparent that design of bioslurry systems, which utilize additives to enhance mass transfer of the contaminants into the liquid, requires some knowledge of the partitioning of the hydrophobic substances among the various compartments of the system. Slurries containing surfactant above the CMC form three distinct compartments: the soil phase, the aqueous pseudophase, and the micellar pseudophase. The HOC, assuming the absence of a separate phase, exists dissolved in the aqueous pseudophase, solubilized in the micellar pseudophase, and sorbed on the soil surface. The distribution of the HOC in the three compartments is governed by the equilibria established between the phases. Edwards et al. (1994a,b), working with nonionic surfactants and individual (PAHs) in aqueous slurries, developed a single-solute model that relates the mass fraction of solute present in the bulk surfactant solution F:

$$F = \frac{A}{A + B} \tag{6}$$

where

$$A = 1 + \frac{K_m V_w (m_{\text{tot}} - Q_{\max} W_{\text{soil}} - v_{\text{aq}} \text{ CMC})}{v_{\text{aq}} (1 - K_m V_w C_{aq})} \tag{7}$$

$$B = \frac{K_{d.\text{cmc}} W_{\text{soil}}}{v_{\text{aq}}} \tag{8}$$

where K_m is the micellar pseudophase/aqueous pseudophase HOC partition coefficient, V_w is the molar volume of water, m_{tot} is total mass of surfactant in the system, $Q_{\max}$ is the maximum mass of surfactant sorbed per unit weight of soil, W_{soil} is weight of soil, v_{aq} is volume of aqueous solution, C_{aq} is the HOC concentration in the aqueous pseudophase, and $K_{d, \text{cmc}}$ is the soil-phase/aqueous-pseudophase HOC partition coefficient at surfactant concentration equal to CMC. The parameters K_m, $Q_{\max}$, CMC, K_d, $K_{d,\text{cmc}}$ and C_{aq}, can be measured or estimated from independent experiments. This deterministic model explains some of the underlying mechanisms influencing the sorption process, and it is applicable to one-component systems and nonionizable solutes. It does not account for pH effects and other factors such as grain size distribution, mineral adsorption, and solids concentration, which are found to greatly influence sorption of many hydrophobic substances.

Another approach to modeling sorption in the presence of surfactants would be application of the linear or Freundlich isotherms to determine the apparent adsorption constants. For instance, the sorption of perylene on a soil with 0.92% organic content in Makon 10 solutions of varying concentration can be described by the Freundlich model. Makon 10 is an alkylphenoxypolyoxyethelene ethanol with polar hydrophilic ethylene oxide moieties and nonpolar hydrophobic nonyl phenol

Table 3.1   Freundlich Constants for Sorption of Perylene on a Soil with 0.92% Organic Carbon in the Presence of Nonionic Surfactant Makon-10

Surfactant Concentration (%)	K_f $(\text{mL/g})^{1/n}$	$1/n$
0.2	4.31	0.60
0.5	3.14	0.77
0.75	2.51	0.90

From Korfiatis et al., 1994.

moieties. It is water soluble, it has a mole ratio of hydrophobic to hydrophilic groups of 10, an average molecular weight of 660, and forms micelles at concentrations above the CMC. Makon 10 has an estimated CMC of 6.06×10^{-5} mol/L and increases the aqueous solubility of perylene from 0.0004 mg/L to 1.42 mg/L. The estimated Freundlich constants are given in Table 3.1. In general, as the concentration of the surfactant increases, contaminant solubility increases, adsorption becomes less favorable, and the isotherms approach linear behavior. However, since surfactants themselves adsorb on the soil, they can effectively increase the organic carbon content, which in some instances can actually increase the soil adsorption capacity. Therefore, depending on the soil characteristics and type of contaminant, surfactants can increase the release rates of the contaminants or they can enhance adsorption. The action of a surfactant may be regulated by close control of its concentration in the system.

Biodegradation

Kinetics of Contaminant Biodegradation   The effectiveness of the bioslurry reactor can be estimated utilizing substrate disappearance data from bench- or pilot-scale treatability studies. Microbially mediated destruction of contaminants in slurry reactors is an extremely complex process governed by a large number of interacting variables. Biokinetic models range from simple first-order decay to elaborate computer codes. First-order decay models are derived from substrate concentration–time profiles and are subsequently used to predict the rate and extent of contaminant destruction. While their simplicity makes them attractive for quick estimation of the biodegradation rates, the disadvantage of first-order models is that they neglect the effect of biomass concentration on the degradation process. First-order decay assumes that the rate of substrate depletion is proportional to its concentration (S) as described by the equation:

$$\frac{dS}{dt} = -k_1 S \tag{9}$$

where k_1 is a first-order rate constant determined by least–squares regression of experimental data. Integration of Eq. (9) results in the following equation (where S_0 is the initial concentration):

$$S = S_0 \exp(-k_1 t) \tag{10}$$

This equation allows calculation of the time required for a specified treatment goal to be realized. Moreover, the half-life of a contaminant, the time it takes for the initial concentration (S_0) to drop to one-half its value, can be estimated by substitution of $S_0/2$ for S. The half-life is given by:

$$t_{1/2} = \frac{\ln(2)}{k_1} \tag{11}$$

Another simple analytical model widely used to describe the rate of microbial growth and substrate utilization is the Monod equations. The Monod model was developed to describe growth of pure microbial cultures in the presence of a single, noninhibitory, limiting substrate. However, it has been used extensively in the environmental field to model systems violating the single substrate and the pure culture assumptions. For instance, the design and control of activated-sludge wastewater treatment plants is based on Monod kinetics. In these systems the multitude of substrates is collectively represented by a single measurement, namely the biochemical oxygen demand (BOD) of the wastewater, and the diverse microbial ecosystem is treated as a pure culture. Being aware of the limitations of the Monod model, one can apply it to interpret data from slurry reactors. The equations of the rate of growth (R_g) and substrate utilization (R), assuming a negligible death rate, have the form:

$$R_g = \frac{dX}{dt} = \frac{\mu_{max} XS}{K_s + S} \tag{12}$$

$$R = \frac{dS}{dt} = -\frac{1}{Y} \frac{\mu_{max} \, XS}{K_s + S} \tag{13}$$

The substrate depletion rate and the microbial growth rate at any given time are functions of the cell mass (X) and the substrate concentration in the system (S). Application of the model requires knowledge of the constants μ_{max}, K_s, and Y (which is termed the "yield coefficient"). Estimation of the Monod constants is extremely difficult in slurry systems because there are no reliable techniques to accurately quantify the active biomass, X. However, elimination of time from the model and integration from the initial conditions (X_0, S_0) gives:

$$X = X_0 + Y (S_0 - S) \tag{14}$$

which substituted into Eq. (13) yields:

$$R = \frac{dS}{dt} = -\frac{kS[X_0 + Y (S_0 - S)]}{K_s + S} \tag{15}$$

where $k = \mu_{max}/Y$. Equation (15) expresses the substrate utilization rate as a function of S, and is applicable to systems in which S is the only growth-limiting substrate (all other nutrients, including oxygen, being in excess), and in which S does not inhibit cell growth. Simplified forms of the substrate depletion rate are:

- First order with respect to substrate concentration ($S << K_s$):

$$\frac{dS}{dt} = -kXS \tag{16}$$

- Zero order with respect to substrate concentration ($S >> K_s$):

$$\frac{dS}{dt} = -kX \tag{17}$$

It should be noted that the form of the rate expression is not important as long as it explains the variability of the kinetic data. However, these models should be applied within the data range and conditions under which they were developed.

Role of Surfactants in Biodegradation Enhancement
Surfactant-induced desorption and solubilization of hydrophobic compounds have received considerable attention the last decade, but the mechanisms by which biodegradation is stimulated in soil slurries are poorly understood. The ability of surfactants to increase the rate of biodegradation of a large number of HOCs is well documented. However, to effectively solubilize hydrophobic substances, surfactants must be added to the slurry at concentrations that exceed the critical micelle concentration. High surfactant concentrations, besides their high costs, cause foaming, impart toxicity to the system, inhibit growth, and subsequently result in hindering or cessation of the biodegradation of the contaminant. Therefore, it is critical to optimize the surfactant concentration.

Recent research on the interaction among suspended particles, surfactants, and bacteria has shown that low concentrations of biodegradable surface-active additives are capable of significant increases in the biodegradation rates of HOCs even when desorption is not appreciable (Hatzinger and Alexander, 1995). Aronstein et al. (1991) studied the microbial destruction of phenanthrene and biphenyl sorbed on organic and mineral soils in low concentrations of nonionic alcohol ethoxylate surfactants, Alfonic 810–60 and Novel II 1412–56. Both surfactants when added in quantities below their CMCs had a profound stimulatory effect on the biodegradation of phenantherene in both organic and mineral soils. On the other hand, mineralization of the pollutants was inhibited at high surfactant dosages. Since only a small fraction of the HOCs was solubilized at surfactant dosages below CMC, most of the HOC must have been degraded by microorganisms attached to the soil surfaces. These findings are corroborated by the work of White (1994), who studied the three-way interactions between surface-active agents, bacteria, and sediment. The events that follow addition of small amounts (below CMC) of biodegradable surfactant in a slurry are as follows: surfactant adsorption on soil and biodegradation by free or attached microorganisms, stimulation of bacterial attachment (the ratio of attached/free cells markedly increases), accelerated decomposition of surfactant by attached bacteria, consumption of surfactant, and detachment of bacteria. Nonbiodegradable surfactants do not stimulate attachment. Similarly, readily biodegradable substrates with unfavorable adsorption characteristics substantially increase the population density but do not induce microbial attachment.

The effect of nonionic surfactants (0.2% Adsee 799) on the biodegradation of anthracene, pyrene, and perylene in artificially contaminated soils of low and high organic carbon content is illustrated in Table 3.2 (Christodoulatos et al., 1995). One or

Table 3.2   Observed Percent Degradation of Three Model Compounds in Soil Slurries

	Percent Degraded after 14 Days of Incubation		
	Without Surfactant	With Surfactant	Control
Soil 1: TOC = 0.118%			
Anthracene	42.2	82.1	4.15
Pyrene	50.5	82.1	16.3
Perylene	83.0	89.9	8.21
Soil 2: TOC = 2.69%			
Anthracene	84.7	83.8	8.11
Pyrene	79.4	82.7	−0.10
Perylene	88.4	92.5	16.2

more of the following mechanisms are believed to affect biodegradation in freshly contaminated soils: (1) the surfactant is utilized as a carbon source and acts as a co-substrate that induces the production of beneficial enzymes that enhance the degradation of the target compounds; (2) the biomass in the system grows at a higher rate due to the utilization of the surfactant, thereby increasing the total biomass concentration resulting at higher hydrocarbon uptake rates; and (3) sorption of the surfactant on the soil reduces its effectiveness in solubilizing the PAHs. Therefore, an excess of surfactant is needed to satisfy the immediate sorption demand and leave a residual concentration that is above the critical concentration required for solubility enhancement. Depletion of the surfactant by any of these mechanisms may result in a reduction of the rate of biodegradation of the contaminant.

FACTORS AFFECTING SLURRY BIODEGRADATION

Bioslurry performance is affected by several factors pertinent to the various system components. These factors may be physical, chemical, and biological in nature. A convenient classification that facilitates their study is based on the various components involved and separates them into:

- System parameters
- Contaminant properties
- Soil properties

As noted previously, a bioslurry system consists of three principal components, namely suspending medium, suspended particulate matter, and air. The behavior of the system is subject to interactions among the principal components within the granted operational setup. Tables 3.3–3.6 provide summaries of the most important factors affecting slurry biodegradation.

Table 3.3   System Factors Affecting Slurry Degradation

System Factor	Related Phenomena	Comments
pH	• Contaminant solubility • Adsorption/desorption • Biodegradation	• Optimum pH range for microbial degradation 5.5–8.5
Moisture content	• Contaminant solubility • Microbial growth • Slurry rheology • Mixing • Oxygen transfer	• Also directly related to slurry solids content • Slurry solids content range of 5–40%
Temperature	• Rate of biodegradation • Rate of desorption • Oxygen transfer/solubility • Slurry rheology • Contaminant volatilization	• Temperature range in most applications 10–30°C • Optimum temperature for microbial degradation 20–30°C • Temperature control infrequently practiced in bioslurry systems
Oxygen	• Microbial growth • Supplementary mixing	• Aerobic metabolism is preferred • Anoxic/anaerobic systems require presence of other terminal electron acceptors (e.g., NO_3^-, Fe^{2+}, etc.)
Aging	• Adsorption/desorption • Bioavailability	• Results in lower release and biodegradation rates
Mixing	• Homogenization • Inhibition reduction • Microbial growth • Mass transfer	• Primarily mechanical mixing supplemented by air mixing and internal or external circulation
Nutrients	• Microbial growth and activity	• Nitrogen, phosphorous, and micronutrients are needed for optimal microbial growth • Recommended C/N/P ratio in the range of 100/10/1–100/1/0.5
Microbial population	• Biodegradation rate	• Naturally occurring organisms are satisfactory if suitable environmental conditions exist • Acclimation may be necessary • Genetically engineered organisms for degradation of target compounds may be added
Reactor operation	• Process efficiency	• Batch, SBR, and continuous operation modes have been used successfully in field applications
Residence time	• Process efficiency	• System dependent usually derived from bench- and pilot-scale studies

System Factors

pH   Three mechanisms that transpire in a bioslurry system are strongly affected by pH, namely contaminant solubilization, adsorption/desorption, and biodegradation. The optimum pH range for bacterial degradation is known to be around the neutral region of 6.5–7.5. Moderate to low degradation rates can be achieved in the range of 5.5–8.5. Outside this pH range biodegradation is possible, but limited. It should be noted that pH is an organism-specific parameter. For example, fungal degradation can proceed at satisfactory rates at acidic pHs as low as 3.5. Therefore, pH control may be required in order to maintain optimal microbial growth conditions.

The aqueous solubility of many neutral and polar molecules in water is strongly influenced by pH. For example, the solubility of PCP in water at pH 4.5 is about 4 mg/L while at pH 8.0 it increases to 14,960, mg/L (Arcand et al., 1995).

Desorption of neutral and polar organic molecules is strongly dependent on the organic content of the soil. However, for ionogenic substances the pH is a primary factor affecting adsorption. PCP, a major soil contaminant in wood-preserving sites, is a weak organic acid with $pK_a = 4.75$. At low pH the undissociated form of the acid predominates and PCP adsorbs strongly on the soil particles. In alkaline environments, where the pentachlorophenate ion is the predominant species, very little adsorption is observed. The dependence of PCP partitioning on pH is illustrated in Figure 3.4 (Talimcioglu et al., 1993).

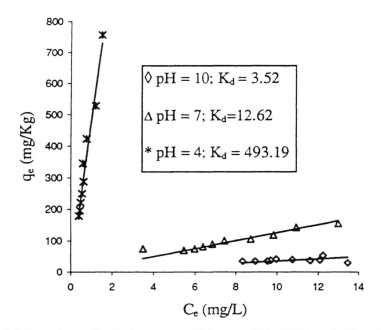

Figure 3.4 Dependence of soil adsorption coefficient on pH for PCP in soils ($f_{oc} = 2.96\%$).

Water–Solids Ratio   Soil water serves as the vehicle for various physicochemical interactions between soil solids and microbial mass. Thus water can be perceived as carrier and transport medium through which many nutrients, inorganic and organic ions, and molecules diffuse into the microbial cells, while metabolic end products are carried away from the cells and into the bulk solution.

For saturated and oversaturated soil systems such as those encountered in slurry treatment, the relative amounts of soil and water are expressed in terms of solids rather than moisture content. Accordingly in many studies conducted in bench, pilot plant, and full-scale systems, the solids content of slurry systems varies in the range of 5–50% solids by weight. Slurry systems with solids contents higher than 50% are subject to mixing and conveyance limitations (Borow, 1989; Irvine et al., 1993a,b; Jerger and Woodhull, 1994; Lauch et al., 1992; Lewis, 1993; Mahaffey and Sanford, 1991; Middleton et al., 1991; USEPA, 1993d; Woodhull and Jerger, 1994a,b; Yare et al., 1989). On the other hand, solids contents less than 5% may result in low solids loadings and high reactor volumetric requirements, thus imposing a heavy economic burden on the treatment system. It appears, therefore, that an appropriate range should be about 10–40% solids.

Temperature   Temperature is a parameter that has a pronounced effect on rates of principal physical, chemical, and biological phenomena occurring in slurry-phase treatment systems, particularly oxygen transfer, desorption, and biodegradation. In general, as temperature increases, the rate of these processes increase. This principle is also applicable within the three temperature ranges of microbial metabolism, namely psychrophilic, mesophilic, and thermophilic. In general, biodegradation rates are doubled for every 10°C increase, within the range of 5–30°C. Although temperature control is practiced infrequently, bioslurry reactors can be insulated to keep reactor temperature 5–10°C above ambient.

In addition, temperature affects a number of system/component properties including density, viscosity, vapor pressure, solubility, and surface tension. In turn these properties have a profound effect on secondary mechanisms such as mixing, conveyance, and volatilization.

Oxygen   In slurry-phase bioremediation systems, aerobic metabolism is preferred over anaerobic for a number of reasons:

- Reaction rates of aerobic processes are much faster than those of anaerobic systems
- Aerobic processes generally induce production of the most oxidized end products
- Anaerobic processes are often associated with production of odorous gases
- Anaerobic organisms may be more sensitive to heavy-metal toxicity

For saturated soil systems, the oxygen pool is limited to that of dissolved oxygen in the soil water. The dissolved oxygen solubility in soil water could be significantly less than that of fresh water as salt content and presence of organics in the soil affect it inversely. It is thus important that in oversaturated slurries external oxygen be provided to satisfy the oxygen demand of the aerobic organisms.

Aeration systems equipped with diffusers or membranes provide oxygen in slurry-phase reactors. Mechanical mixing provides homogeneous conditions and enhances oxygen transfer. It should be noted that oxygen transfer and solubility in slurry systems are more limited compared to clean water or dilute water suspensions. Consequently, the oxygen transfer efficiency must be determined under the actual field conditions or serious oxygen limitations may arise. The dissolved oxygen should be monitored routinely and should never be allowed to drop below 2.0 mg/L (USEPA, 1993c). Biological activity may be monitored using oxygen uptake rate. Dissolved oxygen uptake rates (DOUR) and specific oxygen uptake rates (SOUR) of 6.0 mg/L·h and 0.25 μmols/mg·min, respectively, have been reported in the literature (Mahaffey and Sanford, 1991; USEPA, 1993c).

Aging   A process that renders sorbed chemicals increasingly desorption resistant is chemical aging or weathering. Sorption of organics by soil is initially fast and reversible. The initial short period is followed by an extended period of slow and irreversible sorption. This slow sorption period has a profound effect on the process leading to a chemical fraction resistant to desorption. Aging refers to structurally intact molecules only and does not include reactions that change the chemical structure of the compound such as polymerization and covalent binding to humic substances.

The mechanisms responsible for chemical aging in soils and sediments remain largely unknown. Among the processes believed to influence chemical aging are partitioning and slow diffusion within components of the organic matter, diffusion and entrapment into soil micropores and aggregates with subsequent partitioning between the pore water and organic matter on the pore walls, and hydrogen bonding of the hydrophobic substance to humic materials.

Mineralization and extractability of aged phenathrene and 4-nitrophenol in soils were studied by Hatzinger and Alexander (1995). It was demonstrated that biodegradation is significantly lower in soils with aged compounds than in freshly amended soils. The aging effect is more pronounced in soils with high organic carbon content, and the biodegradability of chemicals decreases significantly with time of aging. Disruption of the soil structure by means of sonication, or comminution of the soil by prolonged vigorous agitation, aids the release of chemicals entrapped in stable microaggregates and increases bioavailability, thereby resulting in higher degradation rates.

Mixing   The design and operation of the mixing system is probably the single most critical factor for the success of slurry-phase treatment. Much of the experience on efficient soil mixing originates from the mineral ore processing industry. Typical features of such a system are shown in Figure 3.2. Axial impellers and rake arms provide the bulk of mixing energy in slurry bioreactors.

Rotational speeds for impellers and rake arms in the range of 20–30 rpm and 2–4 rpm, respectively, are used. Although mixing and agitation have been investigated extensively over the years, not much information exists for soil–water slurries. Vessel, impeller geometry/size, as well as number and placement of baffles are needed along with physical properties of the slurry for determination of the mixing requirements. Pertinent physical properties of the slurry include:

1. Solution and solid densities
2. Slurry solids content (wt %)
3. Solids size distribution

An empirical procedure for the design of slurry agitation systems has been reported by Walas (1988).

Nutrients   Depending on the amount required by microorganisms, which in turn is dictated by elemental cell composition, nutrients are classified as macronutrients (nitrogen, phosphorous) and micronutrients (trace metals and minerals). Nitrogen (N) and phosphorous (P) are considered limiting nutrients of major importance and must be externally provided. The soil normally provides all other minor nutrients and trace elements.

The amount of N and P that should be externally added in a soil bioremediation system is governed by:

1. The amount of N and P present in the contaminated soil
2. The amount of biodegradable carbon present

Alternatively, the amount of N and P can be determined from synthesis and cell growth oxidation/reduction reactions as suggested by McCarty (1972). Another estimation of the N and P requirements is based on empirical C/N/P ratios. Accordingly, a C/N/P weight ratio of 120/10/1 has been suggested by Sims et al., (1990) while a more typical range of 100/10/1–100/1/0.5 has been proposed by USEPA (1993d).

Microbial Population   Natural top soil formations comprise a diversified microbial ecosystem, including: aerobic, facultative, and anaerobic bacteria; fungi; actinomycetes; protozoa; earthworms and higher forms of life. It is estimated that the organism density is at least one million organisms per gram of agricultural soil (Brock et al., 1984). The indigenous organisms are capable of degrading a wide array of organics present in soil. In many instances, biodegradation proceeds due to a synergistic effect of the various microbial species present in the mixed culture. In addition, indigenous cultures are well adapted and acclimated to existing conditions and wastes present. Rarely, there is a need for addition of exogenous organisms. The degradation potential of indigenous organisms can be established by means of treatability studies (Clark, 1965; Ghiorse and Balkwill, 1985; Rogers et al., 1993; USEPA, 1989a), and can be enhanced by acclimation.

Residence Time   The average time a slurry spends in a reactor is called residence time, or hydraulic residence time (HRT). The soil remains in the reactor as long as it is necessary to achieve the specified treatment goals. Thus, the residence time is determined by the biodegradation rates of the contaminants. Residence times range from days to months depending on the nature of the contaminant and the characteristics of the soil. Soils contaminated with easily degraded compounds require shorter residence times while recalcitrant substances need longer residence times in order to reach the desired treatment objective.

Contaminant Properties

Transformation of most organic contaminants proceeds through a combination of biotic and abiotic mechanisms including biodegradation, volatilization, chemical degradation, photodegradation, and the like. The rate and extent of transformation depend on properties related to the chemical structure of the target compound. Compound solubility, volatility, biodegradability, and toxicity are of paramount importance and are discussed in the following sections (Table 3.4).

Table 3.4   Contaminant Characteristics Affecting Slurry Degradation

Contaminant Properties	Related Phenomena	Comments
Solubility	• Contaminant partitioning • Bioavailability	• Solubility of hydrophobic compounds can be enhanced with chemical extractants or surfactants
Volatility	• Volatilization (abiotic losses) • Vapor-phase treatment	• Vapor-phase treatment systems (e.g., carbon adsorption, biofilters) are necessary for volatile components
Biodegradability	• Rate of biodegradation • Biodegradation pathways • Type of metabolism (aerobic vs. anaerobic) • Acclimation	• Microcosm and bench-scale tests are necessary to determine reaction rate constants, half-lives, need for a co-substrate and intermediates
Toxicity	• Biodegradation • Acclimation	• Toxic substances include heavy metals, highly chlorinated organics, certain pesticides and herbicides, inorganic salts at high concentrations • Pretreatment may be needed for toxicity reduction

Solubility   Water solubility of organic compounds along with soil adsorption properties plays a decisive role on the relative partitioning of contaminants between the phases involved and as a result determines their bioavailability, as discussed in a previous section. Aqueous solubility and the octanol-water partition coefficient (K_{ow}), shown in Table 3.5 for selected organics, are used as measures of a contaminant's mobility and availability.

Volatility   Volatilization is one of the principal mechanisms controlling abiotic losses. The release of organic contaminants into the gaseous phase in a bioslurry system is controlled by two mechanisms, namely volatilization and gas stripping. Both mechanisms depend on the relative partitioning of contaminants between the liquid and gas phase. In general, contaminants with high vapor pressures tend to escape into the gaseous phase posing air pollution concerns particularly in open systems (e.g., lagoons). In closed reactor systems air pollution control devices may be necessary. A number of air pollution control technologies are available, including carbon adsorption, biofiltration, and incineration.

Biodegradability   A major complication often encountered in bioremediation systems is that various contaminants biodegrade at vastly different rates. In addition, pure compound degradation rates in pure cultures are of limited value, as in most soil systems a mixture of contaminants is handled by a mixed culture. Interactions

Table 3.5   Solubility, Volatility, and Degradation Data of Selected Organic Compounds

Compound	Solubility (mg/L)	Vap. Pressure (mmHg)	K_{ow}	Half-life ($t_{1/2}$, days)
Benzene	1787	76	135	0.1–1.0
Toluene	515	22	540	6.4
o-Xylene	213	10	1,320	93.9
p-Xylene	185	10	1,410	8.5–14.7
m-Xylene	146	10	1,585	8.5–14.7
Ethylbenzene	110	7	1,410	6.1
Trichloroethene	1100	60	260	
Pentachlorophenol	14	40	132,000	5.5–8.5
Naphthalene	31	1	2,000	8–30
Acenaphthylene	3.93		5,500	105
Acenaphthene	3.42		8,300	10–60
Fluorene	1.98		15,100	40–55
Phenanthrene	1.29		28,800	23–69
Fluoranthene	0.26	0.01	79,400	40
Pyrene	0.135		75,900	10–100
Anthracene	0.066	1	28,200	9–53
Benzo(a)pyrene	0.0038		933,000	160–860
Chrysene	0.002		407,000	5.5–116

From Metcalf and Eddy, 1991; Sims, 1990; Verschueren, K., 1983.

between contaminants in mixed culture systems may trigger phenomena such as co-metabolism, toxicity, synergism, and antagonism with very profound effects on the overall biodegradation rates. Pure compound kinetic constants and half-lives should be handled cautiously and kinetic studies should be conducted for determination of realistic biodegradation constants. However, some general conclusions can be drawn on the biodegradability of various organic compounds based on field experience. The following generalized sequence of decreasing biodegradability has been reported (Pollard et al., 1994):

> *n*-alkanes > branched chain alkanes > branched alkenes >
> low-molecular-weight *n*-alkyl aromatics > monoaromatics >
> cyclic alkanes, > polynuclear aromatics >>> asphaltenes

Petroleum hydrocarbons are considered fairly degradable while heavy petroleum fractions including asphalt, tars, and waxes are harder to degrade. Short-chain aliphatics, alcohols, ketones, as well as aromatic compounds such as creosote, phenols, and volatile aromatics including benzene, toluene, ethylbenzene, and xylene (BTEX) are considered reasonably degradable (Dibble and Bartha, 1979; Hrudey and Polland, 1993; Lackey et al., 1993; Mueller et al., 1991a,b; Nishino et al., 1994; Stroo, 1989). Light halogenated aromatics including chlorinated benzenes, PCP, some PAHs, and energetic materials such as trinitrotoluene (TNT) have been treated successfully (Montemagno and Irvine, 1990). Compounds that appear to be less prone to biodegradation are high-molecular-weight, highly substituted, branched, or cyclic compounds, highly chlorinated pesticides (e.g., DDT), high-molecular-weight PAHs, dioxins, and polychlorinated biphenyls (PCB) (Nowell and Panto, 1994). The compound with the slowest degradation rate determines the required residence time, and therefore the treatment cost.

Toxicity   Toxicity is referred to as the condition that hinders some or all of the basic cell functions. It must be noted in this regard that the Monod kinetic model assumes the absence of significant toxicity; thus it is not applicable for substrates that are known or are suspected of exhibiting inhibition. Substrate utilization and organism growth rate for inhibitory systems are frequently described by the following equation:

$$R = \frac{dS}{dt} = -\frac{1}{Y}\left(\frac{\hat{\mu}XS}{K_s + S + \dfrac{S^2}{K_i}}\right) \tag{18}$$

$$R_g = \frac{dX}{dt} = \frac{\hat{\mu}XS}{K_s + S + \dfrac{S^2}{K_i}} \tag{19}$$

where K_i is the inhibition constant.

Bioassays used to measure and quantify toxicity effects of a substance on test species under specified conditions are laborious, time consuming, and sometimes inconclusive (Sims et al., 1990). Microtox™ is a standardized, instrumental toxicity test based on the response of marine luminescence bacteria to toxicity. Toxicity results are presented as EC_{50} values defined as the contaminant concentration causing 50% luminescence reduction. Substances with low EC_{50} values, exhibit high toxicity. To overcome the conceptual problem of low EC_{50} associated with high toxicity, a soil toxicity unit (TU) is defined as:

$$TU = \frac{400}{EC_{50}} \tag{20}$$

In soil bioremediation systems where biological transformation of complex organic molecules proceeds through production of a number of intermediates, detoxification is not assured by monitoring only the concentration of the parent compound. Degradation pathways must be known and assessment of the toxicity of intermediates is sometimes required. However, soil bioremediation systems are known to induce a net toxicity reduction (Dassapa and Loehr, 1991; Matthews and Hastings, 1987).

Soil Properties

The properties of the soil play a very important role in bioslurry systems as all removal mechanisms are affected by them. Initial soil characterization is thus as necessary as contaminant characterization. Table 3.6 presents a summary of the most important soil parameters and the various mechanisms affected.

Of all soil characteristics, particle size is by far the most important. The rate and efficiency of most, if not all, physical, chemical, and biological actions taking place in a bioslurry system including pre- and post-treatment operations are strongly dependent on soil/slurry particle size distribution. Examples include fractionation, mixing and homogenization, size reduction, sedimentation, dewatering and filtration, adsorption/desorption, and chemical and biological reactivity. The effect and implications of particle size distribution on various treatment processes may be conflicting (Levine, 1985; Koutsospyros, 1990). For example, a particle size decrease favors mixing and homogenization, adsorption/desorption, and chemical and biological reactivity but impairs the performance of settling and dewatering facilities.

Size fractionation is a process used to eliminate large, heavy particles that require high amounts of energy to keep in suspension. Contaminants are preferentially adsorbed on fine rather than coarse particles. For example, in a soil contaminated with PAHs it was demonstrated that although the fine fraction represented only 37% of the soil dry mass, it contained 94% of the contamination (LaGrega et al., 1994). Thus, along with a reduction in mixing requirements, exclusion of the coarse material renders the added benefits of contaminant concentration and waste volume

Table 3.6 Soil Characteristics Affecting Slurry Degradation

Soil Properties	Affected Phenomena	Comments
Particle size	• Mixing, homogenization • Size fractionation (pre-treatment) • Size reduction • Desorption • Dewatering (post-treatment)	• Bulk of contaminants is absorbed by soil particles less than 60 μm • Fractionation process remove coarse soil particles • Size reduction may be needed if the soil contains coarse solids or aggregates • Fine fractions are hard to destabilize, coagulants may be needed to assist dewatering
Soil composition (sand—silt—clay content)	• Mixing • Rheological properties of the slurry • Slurry density • Adsorption/desorption	• Clayey and silty soils are easier to disperse than sandy soils • Clayey particles less than 20 μm are hard to destabilize • Presence of clayey solids may cause aggregate formation thus affecting particle size distribution
Cation/anion exchange capacity	• Adsorption/desorption • Complexation	
Organic carbon content	• Adsorption/desorption	• Desorption of neutral and polar organic molecules depends strongly on the organic carbon content of the soil

reduction. Particle size management by way of fractionation or size reduction by mechanical means has been shown to be beneficial in biological systems such as anaerobic sludge digestion (Koutsospyros, 1992) and bioslurry systems (Bachmann and Zehnder, 1989). In this latter study, the rate of degradation of various xenobiotics in soil systems was shown to increase several times as the soil particle size decreased from 60 to 30 μm.

Gravitational separation processes (e.g., sedimentation) are also known to be particle size dependent. Small size particles possess slow settling velocities thus increasing the residence time requirements of separation facilities. The design and efficiency of biomass/soil–liquid separation facilities (e.g., settling tanks, clarifiers) is very important for lagoon and continuous operational schemes resembling activated sludge. The dewatering properties (specific resistance to filtration, capillary suction time) of sludge and slurries are also affected adversely by the presence of fine particulate matter. Pre-treatment practices, such as size fractionation and/or size reduction, may result in significant enhancement of the fines fraction. The performance of gravitational separation and dewatering units may be improved by using coagulants and filtration aids, respectively.

DESIGN CONSIDERATIONS

The design and scale-up of bioslurry systems is mostly based on experience, engineering judgment, and trial and error. Pertinent reactor parameters are estimated from a sequence of laboratory-scale and pilot-scale studies (Korfiatis and Christodoulatos, 1993; LaGrega et al., 1994; Loehr, 1993; Rogers et al., 1993). The laboratory-scale studies are intended to derive data on desorption and biodegradation rates. The pilot-scale studies are performed in order to obtain ranges of operating parameters, such as solids loading, aeration rates, solids residence time, mixing efficiency, with the intent to optimize the performance of the system. Lab- and pilot-scale experiments are conducted with artificially contaminated soil or with real soil from the contaminated site under consideration. Working with real contaminated soils is preferable since the system's behavior is closer to that expected in the field during implementation of the technology. The protocol shown in Figure 3.5 outlines the steps involved in the design process of bioslurry systems.

Most bioslurry reactors in field applications are operated in one of three different modes, namely batch, sequencing batch (SB), and continuous. Selection of the operating mode is based on a number of factors among which the most important are:

- Treatment objectives
- Amount of waste to be treated
- Initial concentration of the target contaminants
- Reaction kinetics
- Integration into existing pre- and post-treatment schemes
- Economic considerations

Batch reactors are the most common in slurry-phase treatment. They operate at a constant slurry volume and are easier to control than semicontinuous or continuous stirred tank reactors (CSTRs), which require continuous premixing and hydraulic transport of the slurry. The residence time in the bioreactor depends on the nature of the soil and the contaminant, the degree of contamination, the rates of biodegradation, and the treatment goals. The reactor volume can be estimated from the half-life of the target compounds and the amount of soil to be treated. Continuous reactors with or without solids recycle can be sized by performing material balances of the key components around the system.

Another reactor type that has recently been used for slurry-phase remediation is the sequencing batch reactor (SBR) (Irvine et al., 1993 a,b). An SBR is a fill-and-draw system that operates in five basic process steps carried out in the following sequence: (1) fill, (2) react, (3) settle, (4) decant, and (5) idle. During the fill step aeration is on or off depending on the process (aerobic, anoxic, or anaerobic). Once the tank is filled, the react step commences and the SBR is operated as a batch reactor. In slurry-phase systems the settling step is omitted since resuspension of the solids may be difficult.

PHASE I: Preliminary Bioslurry Evaluation

1. Collection of representative soil samples
2. Physical, chemical, and microbial soil characterization
   - Soil properties (grain size distribution, pH, mineralogy, soluble salts, CEC, redox potential, organic carbon content)
   - Concentration of contaminants
   - Microbial enumeration, isolation, and identification of contaminant degraders
3. Identification of key components on which system performance will be based and selection of biological process (aerobic, anaerobic, anoxic)
4. Assessment of key component(s) solubility, bioavailability, and biodegradability (test-tube/shaker flask studies)
   - Adsorption/desorption studies with actual site samples to determine the rate of release and extent of desorption
   - Biodegradation studies to determine rates and extent of biodegradation in aqueous solutions and soil–water slurries. Selection of suitable microorganisms.
   - Toxicity effects (tolerance limits of microbial consortia to elevated contaminant concentrations)
   - Determination of the role of additives (surfactants, co-substrates) on desorption and biodegradation
   - Optimization of desorption and biodegradation rates (soil-to-solution ratio, pH and temperature effects, surfactant dosage, nutrient requirements, toxicity evaluation biodegradability of additives.
5. Identification and biodegradability of intermediates
6. Data evaluation and modeling
   - Adsorption isotherms, desorption kinetics
   - Contaminant degradation kinetics
7. Conceptual design of bench-scale studies

PHASE II: Bench-Scale Testing of the Bioslurry System

1. Development of experimental protocol for the treatability study
   - Construction of a bench-scale unit(s)
   - Assessment of pretreatment requirements
   - Design of data collection schedule for performance evaluation
2. Execution of experimental protocol
   - Determination of soil-handling capacity
   - Measurement of oxygen uptake rates and evaluation of oxygen transfer limitations
   - Evaluation of mixing techniques (mechanical, airlift, hydraulic, etc.)
   - Evaluation of operating modes (continuous or batch)
   - Data collection
3. Data analysis, modeling, and process performance evaluation
4. Parametric optimization of bench-scale units and selection of reactor type (batch, CSTR, SBR)
5. Preliminary design of pilot-scale units.

PHASE III: Pilot-Scale Testing

1. Construction of pilot-scale unit
2. Scheduling and execution of pilot runs
   - Soil handling and pretreatment (excavation, screening, milling, washing)
   - Preparation of extracting solution and growth media
   - Parameter monitoring and sampling schedules
   - Post-treatment requirements (disposal, soil dewatering)
3. Parameter estimation for scale-up and full implementation
4. Cost analysis

PHASE IV: Field Implementation

Figure 3.5 Design protocol for bioslurry remediation.

Table 3.7 Case Studies of Bioslurry Applications

Medium	Contaminant	Reactor Type	Destruction Efficiency	Cost ($/yd³)
Refinery sludge (5000 m³)[a]	PAHs, VOCs	Batch airlift, 10% solids	> 90% in 56 days	50.5
Soil and sludges[a]	PAHs, BETXs, PCBs, volatile chlorinated organics	10,000-L CSTR (three in series) 25% solids	> 80%, HRT=42 days	–
Sludge and soil (70-million gallons)[a]	PCBs, benzo(a)pyrene, benzene, vinyl chloride, arsenic	In situ (lagoon) 5–20% solids	Cleanup objectives reached in 300 days	234
Soils[b]	TNT	Suspended solids SBR 50% solids	67% in 42 days	30–150
Soils[c]	PCP	Batch, 5–40% solids	Below detection in 14 days	80–150
Soil[d] (14,140 tons)	Wood treating wastes	Batch, 20% solids	95% in 10 days	105–110
Soil[e]	Manufactured gas plant wastes	Batch, 10% solids	74% in 14 days	125–130
Sludge[f,g]	PAHs	Batch, 30% solids	89% in 9 weeks	–
Sludge[g]	PCBs	Batch	99.8% in 4 months	200–600
	PAHs,PCPs	Batch	50% in 90 days	–
Soil[h]	PAHs	CSTR 25–30% solids	98% in 12 days	110–145

[a] USEPA, 1993c.
[b] Montemagno and Irvine, 1990.
[c] Ross, 1990/91.
[d] Woodhull and Jerger, Summer 1994.
[e] USEPA, 1992c.
[f] USEPA, 1990.
[g] USEPA, 1989b.
[h] USEPA, 1992a.

During the decant period, most of the slurry is removed, with a small amount of treated slurry left in the tank to provide the seed for the next cycle. This may or may not be followed by an idle period, depending on reactor scheduling. Bioslurry applications on various media and wastes, and performance characteristics of several reactor configurations with associated costs, are listed in Table 3.7.

COST FACTORS

Factors affecting the total soil remediation cost are site and process specific. Site-specific factors include: excavation, soil characteristics, quantity of contaminated soil, extent of contamination, treatment objectives, and permitting requirements. Process-specific costs are related to desorption enhancers, nutrient requirements, oxygen supply, pH adjustments, mixing, and chemical analysis. Analytical costs depend on the nature of the target compounds and can be a substantial percent of the total cost. For long-duration projects a field analytical laboratory may be cost effective. Moreover, the costs of soil dewatering and disposal of liquid wastes and soil must be incorporated in the project costs. The capital equipment costs are associated with excavation, screening and milling equipment, the reactor and its accessories, thickening and dewatering facilities, off-gas treatment system, and monitoring and analysis equipment. In full-scale implementation, labor is a major expense and is typically 50% of the total remediation cost. Table 3.8 summarizes various capital and operating cost contributing tasks and activities.

Average costs for various bioslurry systems are given in Table 3.7. System parameters related to the remediation cost are the solids concentration in the slurry reactor, the slurry residence time, and the percent of the material (i.e., sand and gravel) removed in slurry preparation. Separation and washing of the coarse size fractions of the soil reduces the volume requiring treatment and results in substantial project savings. Project costs decrease with increasing solids concentration and percent material removed in slurry preparation, and increase proportionally as the slurry residence time increases (Ross, 1990/91; Ross et al., 1988; USEPA, 1993c; Woodhull and Jerger, 1994a,b).

In addition to capital and operating costs a contingency cost is added to account for unforeseen adversity that may arise over the life of the project. The contingency cost is proportional to the remediation uncertainty. The total bioslurry remediation

Table 3.8 Summary of Cost-Contributing Tasks and Activities

Preliminary	Pre-treatment	Treatment	Post-treatment
Site assessment	Site preparation	Bioslurry reactor	Thickening/dewatering
Characterization	Excavation and transport	and accessories	Sidestream treatment
Bench-scale testing	Screening/fractionation	Off-gas treatment	Final disposal
Pilot-scale testing	Size reduction	Monitoring and	Site closure
Design	Slurry preparation	analysis	
Permitting	Soil washing	Compliance	

cost can be obtained by adding the three component costs, namely capital, operating, and contingency costs. The overall cost depends on the type of waste matrix. Generally, in-place sludge and dredged material treatment in lagoons is less expensive than treatment of soil, as the former matrices require limited or no pretreatment operations (site preparation, excavation and transport, screening, and size reduction). Typically, bioslurry remediation costs range from $100 to $200 per cubic yard. However, when difficult to degrade compounds such as PCBs are treated, the cost can be as high as $600 per cubic yard. Bioslurry treatment remediation costs are higher than those of solid-phase biological treatment, typically varying in the range of $50–$80 and $60–$100 per cubic yard for landfarming and composting, respectively. Slurry-phase treatment costs (per ton of waste) compare favorably with other on-site, ex situ treatment technologies such as incineration ($400–$1000), chemical oxidation ($200–$500), stabilization/solidification ($120–$520), and solvent extraction ($100–$500).

REFERENCES

Arcand, Y., J. Hawari and S. R. Guiot (1995). Solubility of Pentachlorophenol in Aqueous Solutions: The pH Effect, *Wat. Res.*, **29**(1):131–136.

Aronstein, D. A., Y. M. Calvillo, and M. Alexander (1991). Effect of Surfactant at Low Concentrations on the Desorption and Biodegradation of Sorbed Aromatic Compounds in Soil, *Environ. Sci. Tech.*, **25**(10):1728–1731.

Bachmann, A. and A. J. B. Zehnder (1989). Overview of Selected Bioremediation Technologies, *Hazardous Waste and Hazardous Materials Conference*, New Orleans, LA, April 12–14.

Borow, H. S. (1989). Biological Cleanup of Extensive Pesticide Contamination in Soil and Groundwater Proceedings of HSMRI's 2nd National Conference on Biotreatment, Washington DC, Nov. 27–29, pp. 51–56.

Brock, T. D., D. W. Smith, and M. T. Madigan (1984). *Biology of Microorganisms*, 4th ed., Prentice-Hall, Englewood Cliffs, NJ, pp. 250–265.

Brox, G. H. (1989). A New Solid/Liquid Contact Bioslurry Reactor Making Bioremediation More Cost Competitive, *Proceedings of the 10th National Conference-Superfund '89*, HMCRI, Washington DC, Nov. 27–29.

Castaldi, F. J. and D. L. Ford (1992). Slurry Bioremediation of Petrochemical Waste Sludges, *Wat. Sci. Tech.*, **25**(3):207–213.

Christodoulatos, C. and M. Mohiuddin (1995). Generalized Models for Prediction of Pentachlorophenol Adsorption by Natural Soils, *Wat. Environ. Res.* **68**(3):370–378.

Christodoulatos, C., M. Talimcioglu, G. P. Korfiatis, and M. Mohiuddin (1994). Adsorption of Pentachlorophenol by Natural Soils, *J. Environ. Sci. Hth. Part A—Environ. Sci. Eng.*, **A29**(5):883–898.

Christodoulatos, C., N. Pal, and R. Vemuri (1995). Biodegradation of PAHs in Soil-Water Slurries with Enhanced Bioavailability, *GEOENVIRONMENT 2000: Characterization, Containment, Remediation and Performance, in Environment Geotechnics*, An ASCE Specialty Conference sponsored by the Geotechnical and Environmental Engineering Divisions, New Orleans, Louisiana, February 14–16.

Clark F. (1965). Agar Plate Method for Total Microbial Count, in *Methods of Soil Analysis*, American Society of Agronomy, Madison, WI, Vol. 2, p. 1460.

Dasappa, S. M. and R. C. Loehr (1991). Toxicity Reduction in Contaminated Soil Bioremediation Processes, *Wat. Res.*, **25**(9):1121–1130.

Dibble, J. T. and R. Bartha (1979). Effect of Environmental Parameters on the Biodegradation of Oil Sludge, *Appl. Environ. Microbiol.*, **37**:729.

Edwards, D. A., Z. Liu, and R. G. Luthy (1994a). Experimental Modeling for Surfactant Micelles, HOCs, and Soil, *J. Environ. Eng.*, **120**(1):23–41.

Edwards, D. A., Z. Liu, and R. G. Luthy (1994b). Surfactant Solubilization of Organic Compounds in Soil/Aqueous Systems, *J. Environ. Eng.*, **120**(1):5–22.

Ghiorse, W. C. and D. L. Balkwill (1985). Microbial Characterization of Subsurface Environments, in *Ground Water Quality*, C. H. Ward and P. L. McCarty Eds., Wiley, New York.

Griffin, E. A., G., Bronx, and M. Brown (1990). Bioreactor Development with Respect to Process Constraints Imposed by Bio-Oxidation and Waste Remediation, *Appl. Biochem. Biotechn.*, **24/25**:627–635.

Hanify, D. E., S. P. Duncan, R. C. Emmett, G. H. Brox, and L. T. O'Connor (1993). Bioslurry Reactor for Treatment of Slurries containing Minerals, Soils and Sludges, US Pat. 005–227–136A, July 13.

Hatzinger, P. B. and M. Alexander (1995). Effect of Aging of Chemicals in Soil on Their Biodegradability and Extractability, *Environ. Sci. Techn.*, **29**(2):537–545.

Hrudey, S. E. and S. J. Pollard (1993). The Challenge of Contaminated Sites: Remediation Approaches in North America, *Environ. Rev.*, **1**(1):55–72.

Irvine, R. L., J. P. Earley, G. J. Kehrberger, and B. T. Delaney (1993a). Bioremediation of Soils Contaminated with Bis-(2-etylhexyl) Phthalate (BEHP) in a Soil Slurry-Sequencing Batch Reactor, *Environ. Prog.*, **12**(1):39–44.

Irvine, R. L., P. S. Yocum, J. P. Early, and R. Ghozick (1993b). Periodic Processes for In Situ and On-Site Bioremediation of Leachates and Soils, *Wat. Sci. Tech.*, **27**(7):97–104.

Jerger, D. E. and P. M. Woodhull (1994). Full-Scale Bioslurry Reactor Treatment of Wood Preserving Wastes at a Superfund Site, extended abstract presented at *the I&EC Special Symposium American Chemical Society*, Atlanta, GA, September, pp 1370–1373.

Karickhoff, S. W. (1984). Organic Pollutant Sorption in Aquatic Systems, *J. Hydraulic Eng.*, **110**(6):707.

Korfiatis, G. P. and C. Christodoulatos (1993). Treatability Studies as a Remedial Option Screening Tool for Contaminated Soil, *Remediation*, **3**(4):395–412.

Korfiatis, G. P., D. Dermatas, and C. Christodoulatos (1994). Selection and Use of Surfactants for NAPL Flushing, extended abstract presented at *the I&EC Special Symposium American Chemical Society*, Atlanta, GA, September, pp 109–112.

Koutsospyros A. D. (1990). Effect of Particle Size on Anaerobic Sludge Digestion, Ph.D. Thesis, Polytechnic University, Brooklyn, New York.

Koutsospyros A. D. (1992). Ultrasonic Pretreatment of Primary Sludge Solids to Enhance the Efficiency of Anaerobic Digestion, Proceedings of International Conference on *Restoration and Protection of the Environment in the 90s*, G. P. Korfiatis, D. Dermatas, and A. C. Demetracopoulos Eds., Thessaloniki, Greece, August 19–21, pp. 173–182.

Lackey, L. W., T. J. Phelps, P. R. Bienkowski, and D. C. White (1993). Biodegradation of Chlorinated Aliphatic Hydrocarbon Mixtures in a Single-Pass Packed-Bed Reactor, *Appl. Biochem. Biotech.*, **39/40**:701–713.

LaGrega, M. D., P. L. Buckingham, and J. C. Evans (1994). *Hazardous Waste Management*, McGraw-Hill, New York, pp. 603–615.

Lauch, P. R., J. G. Hermann, W. R. Mahaffey, A. B. Jones, M. Dosani, and J. Hessling (1992). Removal of Creosote from Soil by Bioslurry Reactors, *Environ. Prog.*, **11**(4):265–271.

Levine, A. D. 1985. Characterization of the Size Distribution of Contaminants in Wastewater: Treatment and Reuse Implications, *J. Wat. Pollut. Cont. Fed.*, **57**:805.

Lewis, F. R. (1993). SITE Demonstration of Slurry-Phase Biodegradation of PAH Contaminated Soil, *Air Waste*, **43**:503–508.

Loehr, R. C. (1993). Bioremediation of Soils, in *Geotechnical Practice for Waste Disposal*, D. E. Daniel, Ed., Chapman & Hall, London, pp. 520–549.

Mahaffey, W. R. and R. A. Sanford (1991). Bioremediation of PCP-Contaminated Soil: Bench to Full-Scale Implementation, *Remediation*, (3):305–323.

Metcalf & Eddy, C. (1991). *Wastewater Engineering Treatment Disposal Reuse*, 3rd ed., McGraw-Hill, New York. (1987).

Matthews, J. E. and L. Hastings, (1987). Evaluation of Toxicity Test Procedure for Screening Treatability Potential of Waste in Soil, *Toxic. Assess.*, **2**:265–281.

McCarty, P. L. (1972). Stoichiometry of Biological Reactions, Presented at the International Conference, *Toward a Unified Concept of Biological Waste Treatment Design*, Atlanta, GA.

Middleton, A. C., D. V., Nakles and D. G. Linz (1991). The Influence of Soil Composition on Bioremediation of PAH-contaminated Soils, *Remediation*, **1**(4):391–406.

Montemagno, C. D. and R. L. Irvine (1990). Feasibility of Biodegradating TNT-Contaminated Soils in a Slurry Reactor, Final Report prepared for U.S. Army Toxic and Hazardous Materials Agency, Aberdeen Proving Ground, Maryland, June.

Mueller, J. G., S. E. Lantz, B. O. Blattmann, and P. J. Chapman (1991a). Bench-Scale Evaluation of Alternative Biological Treatment Processes for the Remediation of Pentachlorophenol- and Creosote-Contaminated Materials: Solid-Phase Bioremediation, *Environ. Sci. Tech.*, **25**(6):1045–1055.

Mueller, J. G., S. E. Lantz, R. J. Colvin, D. P. Middaugh, P. H. Pritcard (1991b). Bench-Scale Evaluation of Alternative Biological Treatment Processes for the Remediation of Pentachlorophenol- and Creosote-Contaminated Materials: Slurry-Phase Bioremediation, *Environ. Sci. Tech.*, **25**(6):1055–1061.

Nishino, S. F., J. C. Spain and C. A. Pettigrew (1994). Biodegradation of Chlorobenzene by Indigenous Bacteria, *Environ. Toxico. Chem.*, **13**(6):871–877

Nowell, C. A. and A. T. Panto (1994). Field Application of Anaerobic Soil Slurry Reactor Systems for Degradation of Nitoaromatic Compounds and Other Herbicides and Pesticides, *Proceedings of the 18th Annual Army Environmental Technology Symposium*, Williamsburg, VA, June 28–30, pp. 236–244.

Pollard, S. J. T., S. E. Hrudley, and P. M. Fedorak (1994). Bioremediation of Petroleum and Creosote-Contaminated Soils: A Review of Constraints, *Waste Manage. Res.*, **12**:173–194.

Rogers, J. A., D. J. Tedaldi, M. C. Kavanaugh (1993). A Screening Protocol for Bioremediation of Contaminated Soil, *Environ. Prog.*, **12**(2):146–156.

Ross, D. (1990/91). Slurry-Phase Bioremediation: Case-Studies and Cost Comparisons, Remediation, **1**(1):61–74.

Ross, D., T. P. Maziarz, and A. W. Bourquin (1988). Bioremediation of Hazardous Waste Sites in the USA: Case Histories, *Proceedings of the 9th National Conference, Superfund '88*, HSMRI, Washington DC, pp. 395–397, November.

Satija, S. and L-S, Fan (1985). Characteristics of Slugging Regime and Transition to Turbulent Regime for Fluidized Beds of Large Coarse Particles, *AIChE J.*, **31**(9):1554–1562.

Sims, J. L., R. C. Sims, and J. E. Matthews (1990). Approach to Bioremediation of Contaminated Soil, *Hazard, Wastes Hazard. Mat.*, **7**(2):117–149.

Stroo, H. F. (1989). Bioremediation of Hydrocarbon-Contaminated Soil Solids Using Liquid/ Solids Contact Reactors, *Proceedings of the 10th National Conference-Superfund '89*, HMCRI, Washington, DC, Nov. 27.

Talimcioglu, M. N., C. Christodoulatos, G. Korfiatis, M. Mohiuddin, and S. Mazzoni (1993). Adsorption Effects on the Transport of Pentachlorophenol in Soils, Joint CSCE-ASCE National Conference on Environmental Engineering Proceedings, Montreal, pp. 1257–1264.

Tang, W-T and L.-S, Fan (1989). Hydrodynamics of a Three-Phase Fluidized Bed Containing Low-Density Particles, *AIChE J.* **35**(3):355–364.

Tang, W-T and L.-S, Fan (1987). Steady Sate Phenol Degradation in a Draft-Tube, Gas-Liquid-Solid Fluidized-Bed Bioreactor, *AIChE J.*, **33**(2):239–249.

USEPA (1989a). Guide for Conducting Treatability Studies under CERCLA, EPA/540/2–89/058, Office of Solid Waste and Emergency Response, Washington, DC.

USEPA (1989b). Innovative Technology: Slurry Phase Biodegradation, Directive 9200.5–252FS. Solid Waste Emergency Response, Cincinnati.

USEPA (1990). Slurry Biodegradation, *Eng. Bull.*, EPA/540/2–90/016, Office of Emergency and Remedial Response, Washington, DC, Sept.

USEPA (1992a). Bio Trol Soil Washing System for Treatment of a Wood Preserving Site, Superfund Innovative Technology Evaluation, Technology Demonstration Summary.

USEPA (1992b). Biological Treatment of Wood Preserving SITE Groundwater by Biotrol, Inc. Superfund Innovative Technology Evaluation, Technology Demonstration Summary, EPA/540/S5–91/001, Jan.

USEPA (1992c). Slurry Biodegradation, Superfund Innovative Technology Evaluation, Demonstration Bulletin, EPA/540/M5–91/009, February.

USEPA (1993a). Bioremediation of Hazardous Waste Sites: Practical Approaches to Implementation, Seminars, EPA/600/K–93/002, Office of Research and Development, Washington, DC, April.

USEPA (1993b). Pilot-Scale Demonstration of a Slurry-Phase Biological Reactor for Creosote-Contaminated Soil, Superfund Innovative Technology Evaluation, Technology Demonstration Summary, EPA/540/S5–91/009, September.

USEPA (1993c). Pilot-Scale Demonstration of a Slurry-Phase Biological Reactor for Creosote-Contaminated Soil, Applications Analysis Report, EPA/540/A5–91/009, January.

USEPA (1993d). Selecting Remediation Techniques for Contaminated Sediment, Office of Water, EPA–823–B93–001, June.

Verschueren, K. (1983). *Handbook of Environmental Data of Organic Chemicals*, 2nd edn., Van Nostrand-Reinhold,

Walas, S. M. (1988). *Chemical Process Equipment—Selection and Design*, Butterworths, Stoneham, MA, pp. 287–300.

White, G. F. (1994). Multiple Interactions in Riverine Biofilms—Surfactant Adsorption, Bacterial Attachment and Biodegradation, *International IAWQ Research Seminar on: Biological Degradation of Organic Chemical Pollutants in Biofilm Systems*, Copenhagen, Denmark, May.

Woodhull, P. M. and D. E. Jerger (1994a). Bioremediation Using a Commercial Slurry-Phase Biological Treatment System: Site-Specific Applications and Costs, *Remediation*, **4**(3):353–362.

Woodhull, P. M. and D. E. Jerger (1994b). Slurry-Phase Biological Treatment for Remediation of Contaminated Material from Former MGP Sites, extended abstract presented at the *I&EC Special Symposium American Chemical Society*, Atlanta, GA, pp. 1034–1037, September.

Yare, B. S., W. J. Adams, and E. G. Valines (1989). Pilot-Scale Bioremediation of Chlorinated Solvent and PNA-Containing Soils, Proceedings of HMCRI's 2nd National Conference on Biotreatment, Washington, DC, Nov. 27–29.

CHAPTER 4

Membrane Biofilm Reactors

Peter A. Wilderer, Frank R. Kolb, and Matthias Kniebusch

Wassergütewirtschaft, Technische Universität München, D-85748 Garching, Germany (P.A.W., F.R.K.), Technische Universität Hamburg, Hamburg, Germany (M.K.)

Organic hazardous constituents of groundwaters and of industrial process waste-waters are, in most cases, biodegradable under aerobic conditions, although cell yields are mostly limited either because of poor nutritive quality of the substrates or because of a low supply of substrate per unit time. The latter is the case when dilute groundwaters are to be treated.

To accomplish biodegradation under those conditions, biofilm reactors appear to be a good choice because they retain the biomass in the reactor. As soon as biofilm mass has been accumulated, the reactor volume can be favorably exploited, even at high hydraulic loading conditions (as opposed to suspended-growth reactors in which the biomass can wash out at high hydraulic loadings). An important precondition of long-term process economy is, however, that the microorganisms living in the biofilm are simultaneously and adequately supplied—throughout the reactor—with both substrate and oxygen. Mass transfer to the biofilm organisms is controlled by the hydrodynamic conditions in the reactor. Therefore, major attention has to be placed on hydrodynamic aspects in reactor development and design.

The task of operating aerobic biofilm reactors in an economic way gets especially difficult when the pollutants to be degraded are volatile. Bubble aeration of the biofilm reactor would cause stripping losses, unless the gas-transfer rate into the liquid is balanced with the actual gas consumption rate at each location in the reactor.

Alternatively, membrane biofilm reactors (MBR) can be applied. In MBRs, oxygen is transferred across a membrane directly to the biofilm microorganisms. The biological oxygen demand can be adjusted to the actual needs by adjustment of the oxygen partial pressure at the gas side of the membrane. Transfer of volatile organic substrates to the gas phase is limited by transfer resistance of the biofilm due to the metabolic activity of the biofilm microorganisms.

Biological Treatment of Hazardous Wastes, Edited by Gordon A. Lewandowski and Louis J. DeFilippi
ISBN 0-471-04861-5   ©1998 John Wiley & Sons, Inc.

In the following discussion, the MBR concepts are described and preliminary results of experimental studies are presented.

CONCEPTS

In an MBR, gas-permeable membranes serve as substratum for biofilm growth. The membranes separate a series of gas compartments containing oxygen from liquid compartments containing the pollutants to be degraded. Oxygen permeates through the membrane and supports the microorganisms living in the biofilm. In turn, CO_2 exhausted by the bacteria permeates partially into the liquid compartment, partially into the gas compartment from where it is then vented off the reactor.

Figure 4.1 provides a cross section of an MBR system element, consisting of the membrane (which may be either nonporous or porous), the biofilm, the liquid, and the gas compartments.

As membrane material, either flat membranes can be used (Fig. 4.2) or tubing (Fig. 4.3), through which either the polluted water or the oxygen containing gas is driven.

The MBR concept was introduced by Wilderer and Märkl in 1989 and was further elaborated by Kniebusch et al. (1989). Advanced applications are described in a patent of Irvine et al. (1992). The applicability of the MBR in combination with granular activated carbon was investigated by Jaar (1991) and Chozick (1992).

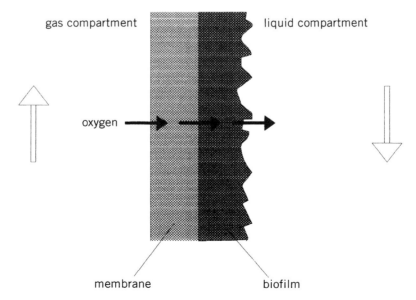

Figure 4.1 Cross section of a membrane biofilm system element.

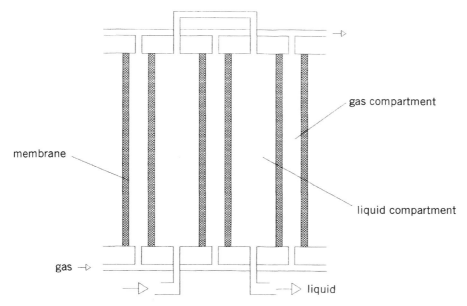

Figure 4.2 Schematic representation of a flat membrane biofilm reactor element.

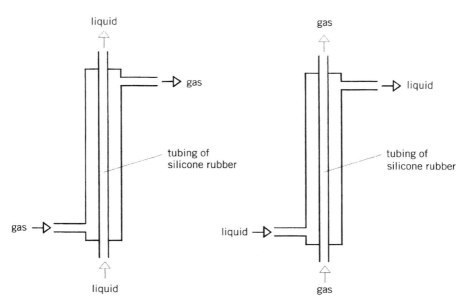

Figure 4.3 Schematic representation of a tubular reactor, with either the wastewater to be treated flowing through the tubing (left) or the oxygen containing gas flowing through the tubing (right).

MEMBRANE BIOFILM SYSTEMS

Membrane Materials

Various kinds of membrane materials are applicable for MBR construction. Among these are nonporous materials, such as silicone rubber. The gas dissolves in these membrane materials. Solution, diffusion, and desorption processes are responsible for mass transfer.

Alternatively, porous materials can be used. Gas transfer occurs in this case at the gas–liquid interface, which develops inside of the pores. Candidates for porous membrane materials are fabrics made of hydrophobic fibers, such as Teflon, and flat membranes, such as polyetherimide. The latter material can be manufactured in a way that the pore size changes systematically across the membrane (Strahtmann, 1986).

In Figure 4.4, a cross section of a membrane of polyetherimide is shown, which provides large caverns open to the liquid compartment. These caverns are large enough for microorganisms to invade and to establish a biofilm inside of the membrane.

In Table 4.1, the oxygen permeability of selected membrane materials is listed. Silicone rubber tubing has been used under laboratory conditions by various authors

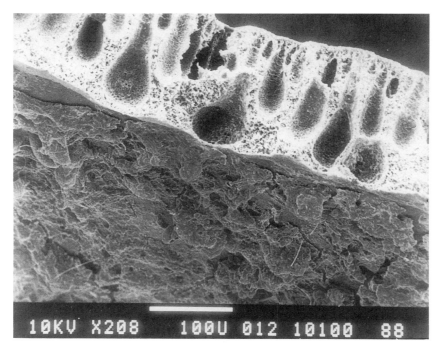

Figure 4.4 Cross section of a porous membrane with fingerlike caverns for bacterial colonization (scanning electron micrograph).

Table 4.1 Permeability of Various Membrane
Materials at 25°C

Membrane Material	Permeability [g(mm)/m^2 (h) bar]
Teflon (PTFE)	0.016
Natural rubber	0.092
PVC	0.093
Silicone rubber (General Electric)	1.929
Silastic 500–1 (Dow Corning)	2.226

(e.g., Debus and Wanner, 1992; Chozick, 1992). The tubing was utilized both with and without glass fiber reinforcement. Reinforcement allows application of variable gas pressures (and therefore, variable oxygen transfer rates) across the tubing. Results of studies on the application of integral asymmetric membranes of polyetherimid were reported by Kniebusch (1989).

Mass Transfer Across the Membrane

The driving force for transport of gaseous substances across the membrane is the difference between concentrations and partial pressures at both sides of the membrane (Müller, 1986). Oxygen partial pressure in the gas compartment is controlled by the gas composition and the applied gas pressure. The oxygen sink is in the biofilm growing at the membrane, in which microorganisms utilize oxygen as an electron acceptor. The principal goal of oxygen transfer is satisfaction of the oxygen demand of those microorganisms.

The change in dissolved oxygen concentration over time in the bulk liquid may be described by the following equation [Eq. (1)]:

$$\frac{dC}{dt} = K_L a \, (C_S - C) - \text{OUR}$$

where a = specific area of the transfer surface
     C = concentration of dissolved oxygen
     C_S = saturation concentration
     K_L = overall oxygen transfer coefficient
     OUR = oxygen uptake rate by the biofilm microorganisms

The overall oxygen transfer coefficient depends on the permeablility of the membrane, the transfer coefficients for the various transfer layers, and the thickness of those layers. Within the biofilm—or in the case of porous membranes, within the membrane itself—the oxygen that is imported is consumed. The consumption rate at any location within the biofilm depends on the substrate specific biological

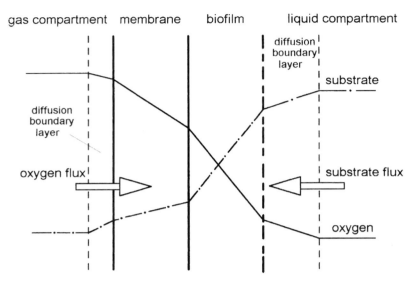

Figure 4.5 Concentration gradients across a membrane–biofilm system.

oxygen demand, the local substrate concentration, the local pH, and other factors. As a result, the steepness of the oxygen concentration gradient across the biofilm (Fig. 4.5) depends on:

- The steepness of the substrate concentration gradient
- The steepness of the pH gradient
- The density of the centers of microbial activity (cells) in the biofilm compartment

A mathematical description of the transfer process requires quantitative information on all of these factors, which has yet to be acquired through intensive research.

Biofilm Structure and Stability

Biofilms at the surface of gas-permeable membranes are formed when bacteria attach to the surface (substratum) and grow. Extracellular polymeric substances (EPS) play an important role in the attachment of cells to the substratum and among each other. The binding is controlled by van der Waals forces, hydrogen bridging, hydrophobic interactions, or electrical forces.

During primary colonization, the surface of the substratum and its properties have strong effects on the strength of adhesion. The surface gets conditioned by EPS adsorption, and the surface properties of the primary EPS layer become dominant (Characklis and Marshall, 1990). To achieve strongly attached biofilms, special attention has to be given to the initial process of surface coating. In special cases, coating with selected natural or artificial polymeric substances might be advisable.

Hydrodynamic shear has a significant effect on the strength of adhesion, as well as on the density of the biofilm. Under low shear, the bacteria tend to grow in fluffy structures. Erosion of cells from the biofilm and sloughing of parts of the biofilm occur as soon as shear even slightly increases. To maintain process stability, hydrodynamic shear should be set high from the very beginning of biofilm growth. This will result in high mass-transfer rates and also minimize future fluctuations in biofilm thickness.

When bacteria are grown in the caverns of porous membranes, they are protected against erosion and sloughing. Increases in the flow rate of the bulk liquid may shear off the biofilm on the membrane surface, but biological activity remains fairly constant because the bacterial cells remaining in the pores serve as a starter culture that is always in place for regeneration of the biofilm (Fig. 4.6).

When specific strains of bacteria are required to achieve degradation of hazardous substances, porous membranes should be inoculated with these specials

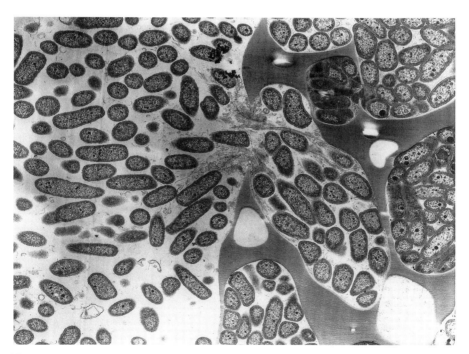

Figure 4.6 TEM of a porous membrane colonized by bacteria. Bacterial cells occupy most of the pores of the membrane and form a biofilm at the surface of the membrane. (Picture taken by Anne Camper, Montana State University, USA)

strains first, and the specialists should be allowed to take over the entire inner space of the membrane. Only then should the reactor be brought into service to achieve long-term process stability.

APPLICATIONS

Treatment of PAH-Containing Groundwater

Polycyclic aromatic hydrocarbons (PAH) are frequently found in groundwaters polluted by industrial sites. The biodegradability of these substances, as well as cell yield, are in general very low and become worse with increasing number of aromatic rings in the molecule. Because of the low cell yields, biofilm reactors, particularly MBRs, appear well-suited to treat PAH-contaminated waters.

The applicability of the MBR was studied in bench-scale units. The reactor system used for the experiments is shown in Figure 4.7. The MBR was equipped with a flat membrane of polyetherimid that was manufactured so that the pore size (caverns) at the interface between the membrane surface and water was between 3 and 8 μm. The surface-to-volume ratio (surface of the membrane vs. volume of the MBR and the reservoir) was $16\,m^{-1}$. The Reynolds number in the bulk liquid was set to 200.

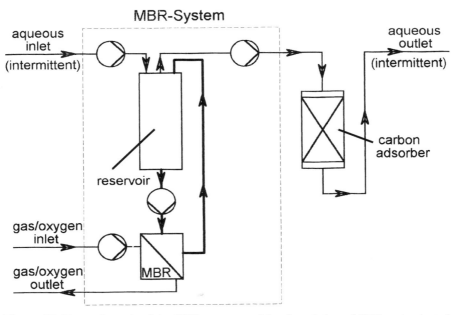

Figure 4.7 Flow schematic of the MBR system used for degradation of PAH-contaminated groundwater.

The water to be treated was collected in a holding tank and periodically introduced into the reactor system (sequencing batch operation). During each cycle, the water was recirculated through the MBR and the reservoir. At the end of the cycle, the reservoir was partially drained, and the effluent was passed through an activated carbon column for final polishing.

To investigate the potential of the reactor system, experiments were conducted with pyrene and *n*-hexadecane as model substances. A mineral medium was added to satisfy the nutrient and trace substrate requirements of the bacteria. Cycle time was 24 h. The MBR was inoculated with activated sludge from a municipal wastewater treatment plant. The concentration change in the reactor was monitored by gas chromatography.

A typical result of track studies is presented in Figures 4.8 and 4.9. The highest reaction rates were achieved at a temperature of 30°C. Both substrates, pyrene and *n*-hexadecane, were simultaneously taken up. At 30°C, the required cycle time (under the conditions of the experiment) was only 8 h.

Treatment of Process Wastewater Containing Volatile Organics

The process wastewaters in chemical and metalworking industries contain cyclic hydrocarbons and their halogenated derivates in highly variable, generally high, concentrations. Substrate inhibition might easily be encountered by the micro-organisms. To allow successful application of an MBR for the treatment of these

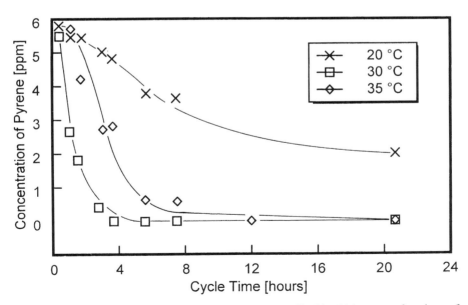

Figure 4.8 Concentration of pyrene in the MBR (liquid side) as a function of temperature.

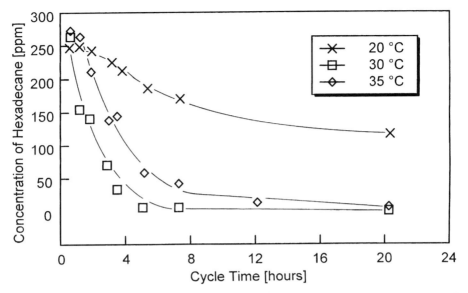

Figure 4.9 Concentration of hexadecane in the MBR (liquid side) as a function of temperature.

wastewaters, the substrate concentration has to be decreased below the toxicity limits.

One solution of this problem is to couple the MBR to an adsorber. This coupling can be realized in two ways:

1. External coupling of the adsorber to the MBR
2. Integration of the MBR inside the adsorber

In the second case, the adsorbent surface is additionally employed as substratum for the biomass. Under unfavorable conditions, the adsorbent may be overgrown with biomass, thus losing its adsorptive properties. With an externally coupled adsorber, this tendency can be suppressed by minimizing the dissolved oxygen in the liquid phase.

Activated carbon has proven to be an adsorbent of high effectiveness for various cyclic hydrocarbons. In order to maintain long-term capacity of the adsorber, an in situ regeneration of the activated carbon has to be achieved. For various reasons, sequencing batch operation appears to be the most favorable process for this purpose. During the fill phase, the concentration of the substrate is reduced by adsorption, potentially below the toxicity level. During the recirculation phase, the adsorber serves as a source of substrate for the biofilm microorganisms, and the adsorbent is gradually regenerated. Either powdered or granular activated carbon may be employed, but granular is preferred. The process is thus called "granular activated-carbon sequencing batch biofilm reactor" (GAC-SBBR) (Irvine et al., 1992).

A flow schematic of that system is presented in Figure 4.10. Two versions of the GAC-SBBR are currently being tested: a system with the MBR and the GAC adsorber separated (reactor A) and a system with the MBR integrated into the GAC adsorber (reactor B). Silicone tubing is used as a membrane and as substratum for biofilm growth. The GAC surface amounts to $900 \, m^2/g$, and the packing has a weight of $1400 \, g$. The membrane surface area exposed to the liquid is $0.44 \, mm^2$.

The fill volume of the reactors currently used by the authors for experimental purposes is 10 l. Biodegradation of 2-chlorophenol (2-CP) and benzene is being investigated. The concentration of these substances in the bulk liquid circulating in the reactor is continuously measured by means of on-line gas chromatographs.

Figures 4.11 and 4.12 contain the result of a track study. In Figure 4.11, the concentration of 2-CP, benzene, dissolved O_2, and CO_2 in reactor A are shown. Figure 4.12 shows the results for reactor B.

The influent concentration of 2-CP and benzene was 875 and 653 mg/L, respectively. During the fill phase, 2-CP was reduced to 614 mg/L, and benzene to

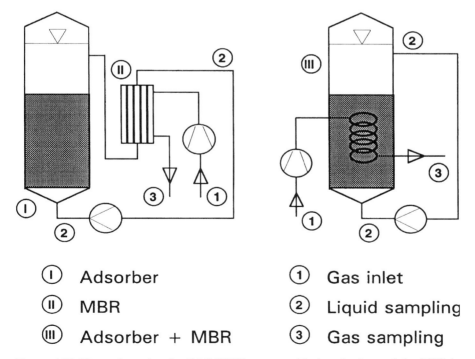

Figure 4.10 Flow schematic of a GAC-SBBR system with the adsorber and the MBR in series (reactor A, left–hand side) and MBR integrated in the GAC packing (reactor B, right-hand side)

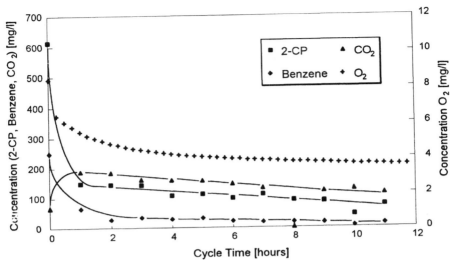

Figure 4.11 Concentration of 2-CP, benzene, O_2, and CO_2 in reactor A (MBR and GAC separated).

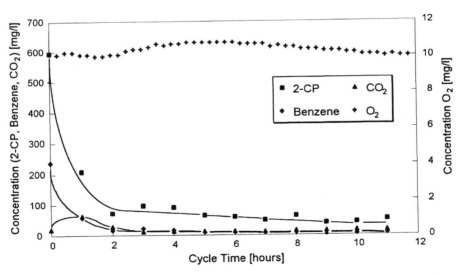

Figure 4.12 Concentration of 2-CP, benzene, O_2, and CO_2 in reactor B (MBR integrated into the GAC adsorber).

249 mg/L, by adsorption on the activated carbon. The subsequent reduction of the pollutant concentration is on on the one hand caused by further adsorption and on the other by biological degradation.

The concentration of dissolved CO_2 is increased by biological activity. On an average, the CO_2 concentration in reactor A increased by a factor of 10 compared to reactor B. (In the aeration system of both reactors, the CO_2 concentration in the gas phase is about the same, i.e., less than 20 mg/L.) The improved biological activity in reactor A can only be explained as the reult of the consecutive arrangement of MBR and GA adsorber. It has to be assumed that oxygen limitation affected the performance of reactor B. The reduction of the pollutant concentration in reactor B is mainly achieved by means of adsorption.

Figure 4.13 illustrates the de-inhibition effect of the adsorber within the system presented. For 3 days running, the influent concentration was changed from 2234, to 1103, to 243 mg/L of 2-CP. In both systems (A and B), the substrate concentration was effectively reduced by adsorption, and after a recirculation time of 3 h, the concentration was independent of the initial concentration.

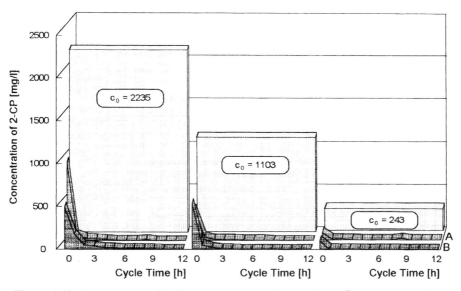

Figure 4.13 Concentration of 2-CP in reactor A and B at higher influent concentrations.

Figure 4.14 shows the influent and effluent chemical oxygen demand (COD) concentrations over the course of the entire experimental program. Despite rather large deviations in the influent concentration, the effluent COD concentrations were generally lower than 20 mg/L. After a running time of 80 days, the discharge concentrations were less than 10 mg/L.

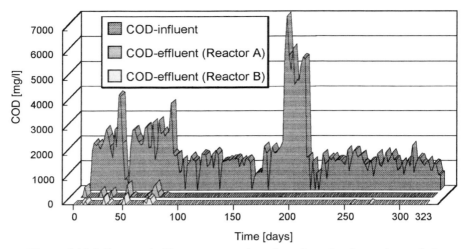

Figure 4.14 Influent and effluent concentrations throughout the observation period.

From these results it can be stated that the MBR-GAC coupling provides a great potential for the treatment of hazardous substances. Goals can be effectively met such as:

- Biological degradation of xenobiotics
- In situ regeneration of the adsorbent, in order to increase the in-service time of the adsorber
- Damping of impact loads

FINAL REMARKS

Full-scale applicability of the MBR concept depends on one major problem yet to be solved. The efficacy of the MBR is mainly influenced by the surface-to-volume ratio. It is necessary to provide a maximum of membrane surface area per reactor volume or per volumetric loading of the reactor, to limit the costs in relation to the benefits. A variety of technical solutions are currently being discussed and investigated. Only after completion of scale-up attempts can economic assessments be made, and further recommendations for practical applications given.

REFERENCES

Characklis, W. G. and K. C. Marshall (1990). *Biofilms*, Wiley, New York.

Chozick, R. (1992). Operation and Performance of Membrane-Aerated Sequencing Batch Biofilm Reactors for the Treatment of Wastewater Containing Volatile Organic Compounds. Dissertation, University of Notre Dame, Indiana.

Debus, O. and O. Wanner (1992). Degradation of Xylene by a Biofilm Growing on a Gas-Permeable Membrane. *Wat. Sci. Tech.*, **26**: 607–616.

Irvine R. L., L. H. Ketchum, P. A. Wilderer, and C. D. Montemagno (1992). Granular Activated Carbon—Sequencing Batch Biofilm Reactor (GAC-SBBR). U. S. Pat. 5, 126, 050.

Jaar, M. A. (1991). Biologische Regeneration schadstoffbeladener Aktivkohle am Beispiel der Modellsubstanzen 3-Chlorbenzoesaure und Thioglykolsäure. Dissertation, TU Hamburg-Harburg, Germany.

Kniebusch M. M., P. A. Wilderer, and R. D. Behling (1989). Immobilization of Cells at Gas Permeable Membranes, in *Physiology of Immobilized Cells*, J. A. M. de Bont et al., Eds., Elsevier Sc. Publ., Amsterdam, pp. 149–160.

Müller, N. (1986). Berechnungsverfahren und Anwendungsbeispiele zum Sauerstoffeintrag in Wasser und Abwasser über porenfreie Membrane. Dissertation, TU Hamburg-Harburg, Germany.

Strahtmann, H. (1986). Preparation of Microporous Membranes by Phase Inversion Processes, in *Membrane and Membrane Processes*, E. Drioli et al., Eds., Plenum, New York.

Wilderer, P. A. and H. Märkl (1989). Innovative Reactor Design to Treat Hazardous Wastes, in *Biotechnology Applications in Hazardous Waste Treatment*, G. Lewandowski et al., Eds., United Engineering Trustees, Inc., NY.

CHAPTER 5

Biofiltration of VOC Vapors

Basil C. Baltzis

Department of Chemical Engineering, Chemistry, and Environmental Science, New Jersey Institute of Technology, Newark, NJ 07102

INTRODUCTION

There are a number of existing technologies for controlling the emission of volatile organic compounds (VOCs): incineration, catalytic oxidation, activated-carbon adsorption, and the like. These are well-established technologies, but they present a number of problems, among which are high cost and little public acceptance. These shortcomings, combined with stricter environmental regulations and an increased public awareness, have motivated research efforts to develop new methods to deal with VOCs. Biofiltration, ultraviolet oxidation, use of membranes, and use of corona plasma for VOC destruction are among the new methods being considered in recent years.

Biofiltration refers to biological oxidation of VOC vapors by microorganisms immobilized in biofilms on a solid support material. These solids are placed in open or closed structures known as biofilters. The ultimate objective is to optimize biofilter design to achieve maximum VOC removal at a minimal cost. A schematic of the biofiltration principle is shown in Figure 5.1.

Biofiltration is not really a new idea. It has been used for years, especially in Europe, for the removal of odoriferous compounds from airstreams (Pomeroy, 1982; Bohn and Bohn, 1988; Leson and Winer, 1991; Bohn, 1992). Relatively new, however, is the idea of using biofiltration for treatment of VOCs in general and hazardous air pollutants (HAPs) in particular. This idea originated in Europe and was first made known to the English-speaking world in 1983 by the pioneering work of Simon Ottengraf (Ottengraf and van den Oever, 1983). It is worth mentioning that in a 1991 review paper by Leson and Winer almost all the cited literature is either

Biological Treatment of Hazardous Wastes, Edited by Gordon A. Lewandowski and Louis J. DeFilippi
ISBN 0-471-04861-5 ©1998 John Wiley & Sons, Inc.

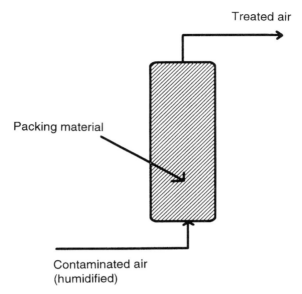

Figure 5.1 Simple schematic of the biofiltration principle. A VOC-laden airstream passes through a bed packed with solids on which microbial biofilms are attached. The organisms degrade the VOCs and the airstream exits the biofilter pollutant-free.

in German or Dutch, while the few papers in English were produced by Ottengraf and his associates (Ottengraf, 1986, 1987; Ottengraf et al., 1986; van Lith et al., 1990).

Since 1990, understandably coinciding with the amendments to the Clean Air Act, a lot of research and development work on biofiltration has been performed in the United States. The literature on this topic, especially in conference proceedings, has increased dramatically. The interest is so intense that in 1995 alone, three sessions of the Air and Waste Management Association Annual Meeting were devoted to biofiltration, while a special international conference on biofiltration was held in Los Angeles (Hodge and Reynolds, 1995).

The interest in biofiltration stems from the fact that it is a non-energy-intensive technology that leads to VOC destruction without the potential for concomitant production of pollutants more hazardous than the original VOCs. The latter, of course, requires proper selection of microorganisms and process conditions. In general, biofiltration is viewed as an environmentally friendly technology that has the potential of providing cost-effective solutions to the VOC control problem. The low-energy requirements of biofiltration probably explain why this technology has been more popular in Europe than in the United States. As mentioned by Reynolds and Hodge (1995) fuel costs in the United States are 20% of those in Western Europe, which somewhat reduces the incentives to employ biofiltration in the United States.

Despite the widespread interest in biofiltration, and its existing commercial applications, this technology is not yet well understood. Even the term *biofiltration* is inappropriate since *filtration* implies a physical rather than a biochemical process. There are a number of factors, discussed in the following section, that significantly affect the design and operation of biofilters and yet have not been properly investigated. With relatively few exceptions, most of the existing studies use a "black-box" approach (such as that of Fig. 5.1) and try to scale-up and design industrial units based on, at best, semi-empirical approaches.

Biofiltration, although simple in principle, is really a very complex process. It involves physical and biochemical steps, fluid flow and diffusion, properties of microbial communities and materials (solid support), and the like. A fundamental process understanding is needed that can then be translated to appropriate mathematical models for rational and optimal design of industrial units. Progress in this direction has been made in recent years.

It should be mentioned that the terms *biofiltration* and *biofilters* are used in this chapter with Figure 5.1 in mind. Perhaps more appropriate would have been the use of the qualification "classical" or "conventional" in front of both terms, as they are also used for alternate reactor configurations more specifically known as *biotrickling filters* and *bioscrubbers*. The latter involve a continuous liquid phase flowing downward through the packing. Biotrickling filters are more recent developments and are only briefly discussed here. Classical (or simply) biofilters involve the direct contact of airstreams (gas phase) with biofilms and their solid supports.

FACTORS AFFECTING BIOFILTRATION

A complete discussion of the various factors affecting biofiltration is not possible in a single chapter. However, an attempt is made to give better insight into the most important factors for design and long-term performance of biofilters.

The key process in biofiltration is oxidation of pollutants by microorganisms. The microorganisms act as catalysts for the process, and in many instances biomass is also a product of the reaction. Hence, as with any type of biodegradation, biofiltration is a catalytic, or autocatalytic, process. Consequently, biofiltration is affected by reaction kinetics and type of catalyst (biomass).

Biological activity is strongly affected by properties of the reaction environment. The presence of water, for example, is necessary for biological activity to occur, and thus the moisture (water) content of a biofilter is a very important factor. Another factor affecting biological activity, and consequently biofiltration, is the pH of the environment. Microorganisms cannot utilize or degrade VOCs unless oxygen and other nutrients are also present in their environment. It becomes clear then that biofiltration is affected by chemicals other than the target compounds (i.e., nutrients and oxygen). Although these chemicals determine in part the kinetics of the process, their influence on biofiltration is discussed here separately because of their importance.

As mentioned earlier, biofiltration employs microorganisms immobilized on solid particles. This solid support affects the process in a number of different ways. For example, microorganisms may or may not form biofilms on a particular solid. Furthermore, the solids may adsorb VOCs thus introducing a new component (physical adsorption) to the overall biofiltration process. These and other factors related to the properties of the solids are discussed in more detail later in this chapter.

Microorganisms and Kinetics

With the exception of very few studies on biofiltration that utilized a pure culture (Zilli et al., 1993; Mirpuri et al., 1995), biofiltration is based on the use of microbial consortia (mixed cultures). Although mixed cultures invariably involve interactions among species of various populations, a consortium is viewed as a single functional population, and the kinetics are studied and described under the assumption that the consortium is stable. Stability here implies that the biomass has a constant composition with respect to the presence of various species in it. With this in mind, biofiltration requires a consortium of microorganisms (referred to as just *microorganisms* or *biomass*, in the remaining part of this chapter) capable of biodegrading, or more specifically completely mineralizing the target pollutants. Complete mineralization requires that a VOC is transformed to carbon dioxide, water, other inorganic compounds (e.g., hydrochloric acid in the case of chlorinated VOCs), and biomass. The mere disappearance of a compound does not imply its complete mineralization. For example, Oh et al. (1994) have reported results of BTX (benzene, toluene, xylene) vapor removal by both a pure culture and a consortium. They found that with both cultures benzene and toluene could be completely mineralized. On the other hand, xylene was only transformed to an organic intermediate. When the same consortium was used in a biofilter, BTX removal occurred initially and then virtually ceased (Oh, 1993). This was probably due to accumulation of a stable organic product of xylene degradation, which was polymeric in nature, and inhibited further transport of BTX to the supported biofilm. However, complete mineralization of all BTX constituents is feasible with properly selected microbial consortia. As with any biodegradation process, proper selection of the biomass is of paramount importance. It needs to be also mentioned here that if a given biomass is found to be capable of completely mineralizing a VOC under certain conditions, it does not mean that this capability is ensured under other operating conditions. For example, ethanol vapors have been reported to be successfully removed in a large biofilter only under certain operating conditions (Leson et al., 1993). Under other conditions, ethanol was transformed to acetic acid, probably due to a lack of sufficient oxygen within the biofilms. This incomplete ethanol mineralization led to biofilter failure due to the pH drop caused by the acetic acid.

Once microorganisms capable of completely mineralizing the target VOCs have been isolated/developed, one needs to know the rate at which they are capable of destroying these VOCs. These reaction rates (kinetics) depend on the concentration of the VOC, oxygen, other nutrients, pH, and temperature.

To discuss the kinetics of the biofiltration process one needs to consider two cases. The first involves situations where the airstream carries a single VOC, while the second case refers to mixtures of VOC vapors. In either case, the kinetics are generally determined in experiments with suspended cultures rather than with biofilms (which are the form that biomass takes in a biofilter). Kinetic experimentation with biofilms is difficult to perform and is a technique in its infancy. The commonly made assumption, discussed by Karel et al. (1985), is that the inherent biodegradation kinetics are the same regardless of whether the microbial cells are suspended in a nutrient medium or immobilized on a solid support as is the case with biofilters.

For the case of single VOCs, Ottengraf and van den Oever (1983) have described the kinetics with either zero- or first-order expressions for high and low VOC concentrations, respectively. These expressions can be viewed as approximations of the well-known model of Monod (1942). Due to its simplicity, the approach of Ottengraf and van den Oever has been used in a number of studies that attempt to describe biofiltration in terms of mathematical models. However, in a biofilter the VOC concentration changes as the airstream moves from the entrance to the exit of the unit. This implies that using the approach of Ottengraf and van den Oever, a different kinetic expression needs to be used in different segments of the biofilter. In addition, the kinetics cannot abruptly switch from zero- to first-order. There must be a VOC concentration range (thus, a segment in the biofilter unit) where the reaction order is between zero and one as implied by the actual Monod expression. Furthermore, a high VOC concentration may imply low reaction rates due to inhibition as described by the Andrews (1968) model. Shareefdeen et al. (1993) were the first to use an Andrews-type expression for describing biofiltration of methanol vapor. The same kinetic expression has been used to describe biofiltration of toluene (Shareefdeen and Baltzis, 1994a,b; Mirpuri et al., 1995), butanol, and ethanol (Baltzis and Androutsopoulou, 1994). An explicit Monod-type kinetic expression has also been used to describe biofiltration of benzene (Shareefdeen and Baltzis, 1994b), methyl ethyl ketone (MEK), and methyl isobutyl ketone (MIBK) vapors (Deshusses et al., 1995a). The use of kinetic expressions involving more than one kinetic constant, that is, expressions implying a dependence other than first or zero order with respect to the VOC concentration, complicates significantly the description of biofiltration. However, this complication appears to be necessary as sensitivity studies have shown that at least two kinetic constants are needed to accurately describe the process (Shareefdeen, 1994; Baltzis and Shareefdeen, 1993; Androutsopoulou, 1994; Mirpuri et al., 1995).

The kinetics of biofiltration of mixed VOCs is an even more complicated issue. Ottengraf and van den Oever (1983), who studied the removal of mixtures of toluene, ethyl acetate, butanol, and butyl acetate (all constituents of emissions from laqueries), have assumed that the kinetics of biodegradation of each compound remain unaltered by the presence of other VOCs. However, recent biodegradation studies with mixtures have shown that the kinetics of a compound in the presence and absence of other pollutants are different. In fact, structurally similar compounds such as benzene and toluene have been found to be involved in a competitive cross inhibition during their biodegradation (Chang et al., 1993; Oh et al., 1994). Kinetic expressions accounting for cross inhibition have been used in describing biofiltration of benzene–toluene mixtures (Shareefdeen, 1994; Baltzis and Share-efdeen, 1994), as well as mixtures of MEK and MIBK (Deshusses et al., 1995a,b). Since in most cases, biofilters are expected to be used for removing mixtures of VOCs, it becomes clear from the foregoing discussion that determination of appropriate kinetic expressions is an important and difficult task.

Reaction rates (kinetics) essentially determine the size of the processing unit. Hence, optimal sizing and design of a biofilter unit are strongly related to the degree of knowledge of the kinetics of the process. Complex kinetics may have interesting and counterintuitive implications for optimal biofilter design. Such an example, taken from Androutsopoulou (1994) is shown in Figure 5.2. This figure is based on computational studies and shows the volume of biofilter required to achieve 96% removal of ethanol vapor from airstreams as a function of the ethanol concentration in the untreated air. Furthermore, this diagram has been prepared under the assumption that the rate of ethanol mass supply (load) to the biofilter is constant. As can be seen from the graph, it is predicted that the biofilter volume reaches a minimum at a particular value of inlet ethanol concentration. The implication is that if the ethanol concentration in the polluted airstream is higher than that at which the

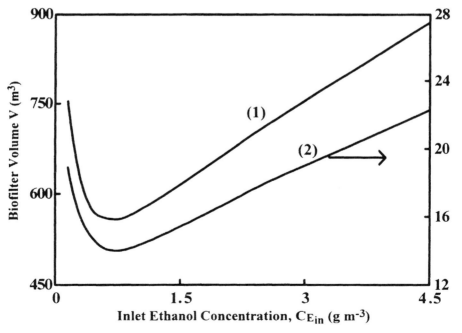

Figure 5.2 Model-predicted biofilter size for achieving 96% removal of ethanol vapor as a function of inlet ethanol concentration. Curves represent constant-load isoclines, (1) 21.1 kg/h, (2) 0.56 kg/h. From Androutsopoulou (1994).

minimum biofilter volume occurs, the airstream can be diluted with clean air so that treatment is achieved with a smaller biofilter, thus reducing capital expenditure for the facility. This is counter to intuition and is a direct consequence of the inhibitory kinetics of ethanol biodegradation (Baltzis and Androutsopoulou, 1994).

Oxygen Availability

Since biofiltration is an aerobic process, the presence of oxygen in the reaction environment is essential. In particular, oxygen must be present in the biofilm formed on the surface of the solid support particles. This is an important point that seems to have created a good deal of confusion. In most existing studies, the availability of oxygen is neglected as a factor explicitly affecting biofiltration, probably based on the argument that oxygen is present in abundance in the airstreams that carry low amounts of VOCs. However, this does not necessarily mean that oxygen is present in excess in the environment where the reaction actually takes place (i.e., the biofilm). Whether oxygen is present in excess in the biofilm or not depends on the concentration of VOCs in the air and their partition coefficient between air and water (biofilm). Computational studies have shown that in most cases oxygen in the biolayer is present in excess relative to the VOC only if the VOC is a hydrophobic (or highly volatile) substance such as benzene or toluene, as shown in Figure 5.3*a* (Shareefdeen and Baltzis, 1994b). For hydrophilic compounds such as methanol (Shareefdeen et al., 1993), ethanol, and butanol (Baltzis and Androutsopoulou, 1994), oxygen is predicted to be depleted much faster than the VOC in the biolayer as shown in Figure 5.3*b*.

As mentioned earlier, a pilot-scale biofilter experienced serious problems in removing ethanol vapors, probably due to lack of sufficient oxygen (Leson et al., 1993). Baltzis and Sareefdeen (1993) have shown through the use of computer simulations that in cases of hydrophilic solvents removal rates can increase if the contaminated airstream is enriched with oxygen. Addition of oxygen is an operating cost that, however, may be offset by the reduction in capital cost resulting from a smaller biofilter unit. The importance of oxygen in biofiltration has been also discussed at a qualitative level by Bohn (1993).

One can conclude that since oxygen availability significantly affects biofiltration, it has to be explicitly accounted for in models attempting to describe the process. Although this is done in many of the recent studies (e.g., Shareefdeen et al., 1993; Shareefdeen and Baltzis, 1994a; Baltzis and Androutsopoulou, 1994; Mirpuri et al., 1995), most published studies continue to neglect oxygen effects.

Moisture and Temperature

As a biological process, biofiltration cannot occur in the absence of water. To begin with, the packing material must have the capacity to retain water within its pores. Furthermore, the contaminated airstream must be humidified before it enters the biofilter, since otherwise it dries the packing material. Thus, a humidification tower is an integral part of a biofiltration facility.

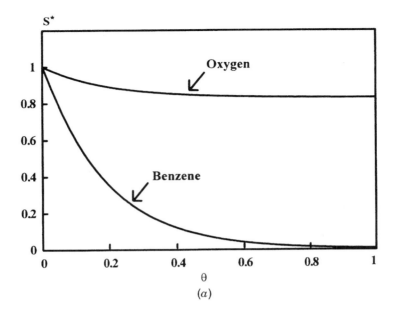

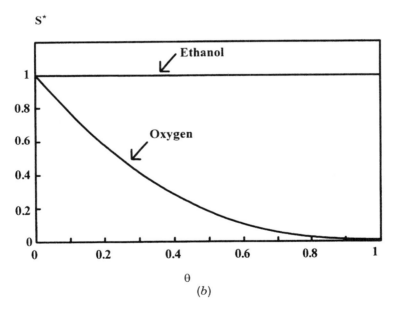

Figure 5.3 Characteristic model-predicted VOC and oxygen concentration profiles in the biofilm at a location along the biofilter bed. (*a*) hydrophobic VOCs, from Shareefdeen and Baltzis (1994b); (*b*) hydrophilic VOCs, from Baltzis and Androutsopoulou (1994). S^* is concentration of VOC and oxygen normalized with the corresponding concentration in the air; θ is position in the biofilm normalized with actual effective biofilm thickness.

In many instances, humidification of the airstream is not enough to maintain the required moisture level. Biodegradation is an exothermic process, and a rising temperature can also dry out the packing. Hence, periodic addition of water to the biofilter bed through sprinkling at the top of the unit appears to be necessary. Water addition has to be carefully designed since it can lead to creation of lumps of packing material with a consequent decrease in contact surface between air and biofilm (van Lith, 1989), or creation of anaerobic zones due to difficulties in oxygen transfer in flooded areas (Ottengraf et al., 1984).

These factors are particularly important at the bottom of the biofilter, where the entering vapor may be imperfectly humidified and where the VOC concentrations (and hence exothermic heat effects) are highest. To address this problem, via water addition, either the contaminated airstream should be fed from the top of the bed (as shown in Fig. 5.4) or the biofilter bed should be divided in segments (as shown in Fig. 5.5). A segmented bed allows sprinkling of water in the area where water is actually needed.

Significant moisture loss also results in uneven changes in the column void fraction, which leads to channeling of the gas stream, and a further reduction in volumetric efficiency (Ottengraf et al., 1984).

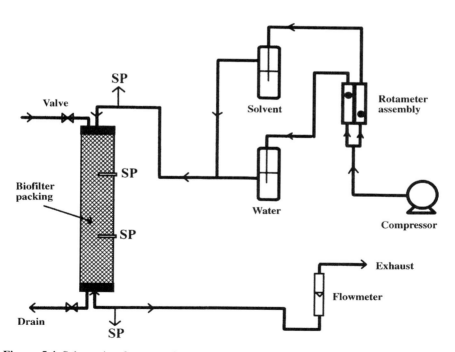

Figure 5.4 Schematic of an experimental biofilter unit. The main part of the airstream is humidified before it is mixed with a slip stream carrying the solvent vapor. The airstream is fed to the biofilter from the top for better moisture control. The biofilter bed is unsegmented. SP stands for sampling port.

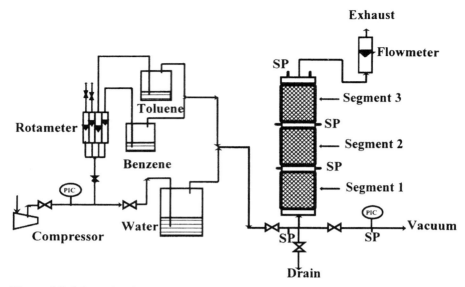

Figure 5.5 Schematic of a biofilter unit for removal of benzene–toluene mixtures (from Baltzis and Shareefdeen, 1994). The filter-bed consists of three equal segments. This design allows for better (local) moisture control and reduces the chance for compaction of the packing material.

Temperature effects can also reduce the biological activity if the biofilter is not operated at the optimal temperature of the microbial consortium (van Lith et al., 1990).

Temperature and moisture effects on biofiltration have been discussed by Leson et al. (1993) and in a more quantitative fashion by Pinnette et al. (1995). Often, the amount of water needed in a biofilter bed is estimated to be enough to fill 50% of the pore space of the packing material (Shareefdeen et al., 1993; Zilli et al., 1993).

pH

Biological activity is a strong function of the pH of the environment. In most cases, biological activity is almost nonexistent if the pH is outside of the range from 6 to 8. Hence, biofiltration of compounds such as chlorinated hydrocarbons, which lead to the formation of acidic products, is problematic.

Biofiltration of chlorinated VOCs has been attempted in biofilters containing calcium carbonate. Such units not only would have a limited life span due to exhaustion of $CaCO_3$ but also present other problems. For example, calcium chloride, which is formed from the reaction of $CaCO_3$ with HCl, is an extremely hydroscopic compound that takes up moisture from the biofilter packing and clogs the void space of the bed.

It is safe to say that (classical) biofilters, which are discussed here, cannot be employed for treating VOCs that result in a significant alteration of the pH. For such compounds, a modified process (biotrickling filter) needs to be used, as discussed later.

Other Nutrients

The basis of biofiltration is that the target pollutants are used as carbon and energy sources by the microorganisms that are immobilized on the solid packing material. However, the presence of carbon/energy sources and oxygen (which serves as terminal electron acceptor) are not enough to ensure biological activity. Micro-organisms also need sufficient nitrogen, phosphorous, and various other elements (micronutrients) in order to function. These additional nutrients are usually not supplied, except at process start-up, to biofilters. The packing material (peat moss, compost) can serve as a source for these nutrients. However, it appears that over the long run, these nutrients are recycled by the biomass itself through lysis. The fact that biomass undergoes lysis can be inferred from the absence of any significant pressure drop buildup during biofilter operation. If biomass was not undergoing lysis, its continuous accumulation would result in clogging of the biofilter.

One could argue that addition of nutrients should increase VOC removal rates. However, this may not be the case. Even if nutrients were properly supplied and distributed, they would lead to excess biomass formation, and consequently to clogging, channeling, and pressure drop problems. Additional biomass is only of benefit if it increases coverage of the surface of the solid support particles. If the additional biomass only increases the biofilm thickness, there would be no improvement of the treatment rate, since either the VOC (carbon source) or oxygen would be depleted in the outer layers of the biofilm, and the inner layers would be ineffective (Shareefdeen et al., 1993; Shareefdeen and Baltzis, 1994a).

Although addition of supplemental nutrients is still under investigation, there is not enough evidence to unequivocally conclude whether it really affects the process or not.

Packing Material

The packing material must allow organisms to attach to it and form biofilms. Shareefdeen et al. (1993) and Oh (1993) examined the ability of materials such as peat, perlite, vermiculite, and polyurethane foam, alone and in mixtures, to serve as biofilter packing. Experiments with methanol vapor showed that a mixture of peat moss and perlite gave the best removal rates with an insignificant pressure drop. Actually, peat moss alone would be the best packing as it also contains nutrients, but perlite proved a good bulking agent.

The solids used must also have a good capability to hold water (see discussion of moisture effects above) and provide sufficient void volume for vapor flow.

Another way the solid packing affects biofiltration is through its ability to adsorb VOCs. The term *adsorption* is used loosely here since it does not only imply actual adsorption on the solid surface but also dissolution of VOCs in the water retained by the solids. Adsorption is a reversible process and leads to an equilibrium distribution of the VOCs between the air and the solids. Examples of such equilibrium curves are shown in Figure 5.6. Individual VOCs usually follow the Freundlich isotherm, as has been shown for the cases of ethanol and butanol (Wojdyla, 1996), toluene (Shareefdeen and Baltzis, 1994a; Wojdyla, 1996), and benzene (Wojdyla, 1996). In all aforementioned studies, the solid support consisted of a peat and perlite mixture (2:3 by volume). Hodge (1995) has investigated other packing materials and expressed the VOC distribution between air and solids in terms of a single parameter (partition coefficient). As can be seen from the curves of Figure 5.6, hydrophilic (polar) compounds adsorb more on the packing than hydrophobic VOCs. This is because, as mentioned earlier, adsorption also accounts for dissolution in the water held by the solids. When the air contains more than one VOC, adsorption equilibrium becomes more complicated since the various compounds are involved in a competitive interaction. This has been shown for benzene and toluene mixtures (Stuart et al., 1991; Wojdyla, 1996).

The adsorption/desorption phenomena particularly affect biofiltration during transient operation (Baltzis and Androutsopoulou, 1994; Shareefdeen and Baltzis, 1994a; Baltzis and Wojdyla, 1995a). An increase of the VOC concentration in the polluted airstream leads to temporarily high removal rates. However, this removal is not only due to biofiltration (destruction) of the VOCs. It is also the consequence of adsorption on the solid packing. Conversely, a concentration decrease of the VOCs in the air can lead to temporarily low removal rates due to desorption of VOCs from the packing material. In fact, desorption may be so intense that the concentration of pollutant at the exit of the biofilter may be higher than in the inlet stream. Examples of these phenomena, observed in experimental units (Baltzis and Androutsopoulou, 1994), are shown in Figure 5.7.

Clearly, different types of packing material have different adsorption characteristics. Adsorption can be desirable since it leads to temporary storage of excess VOCs, but desorption should occur slowly so that the VOCs are biodegraded before they reach the biofilter exit. Biofilters should be sized based on the expected fluctuations in VOC concentrations.

Other Factors

In addition to the parameters discussed there are additional factors affecting biofiltration. These include the void fraction of the biofilter bed, the density and thickness of the biofilms formed around the solids, the equilibrium distribution (Henry's constant) of VOCs and oxygen between the air and biofilm, the diffusivity of pollutants and oxygen in the biofilm, the biofilm–air interfacial area, the mass-transfer coefficient of VOCs onto (during adsorption) and from (during desorption) the solid packing, and the like.

The effect of the aforementioned parameters is not yet fully understood as many of them are extremely hard to measure. Only recently were efforts undertaken, for example, to measure mass-transfer coefficients (Hodge, 1995; Wojdyla, 1996).

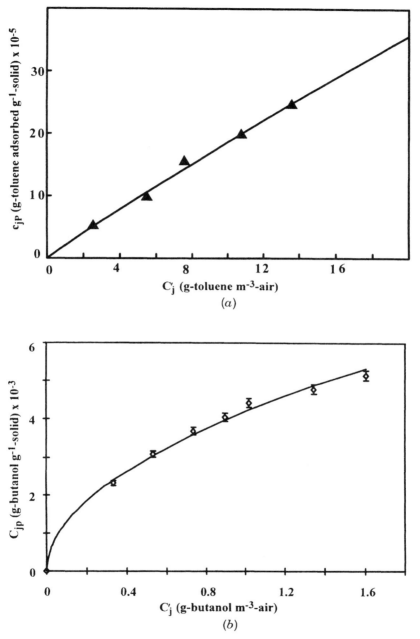

Figure 5.6 Equilibrium distributions of (*a*) hydrophobic and (*b*) hydrophilic VOC between the air and a biofilter packing material made of peat and perlite (2:3, v/v). Curves represent fitted Freundlich isotherms. (*a*) From Shareefdeen and Baltzis (1994a) and (*b*) from Wojdyla (1996).

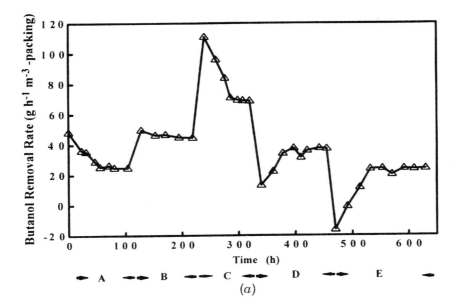

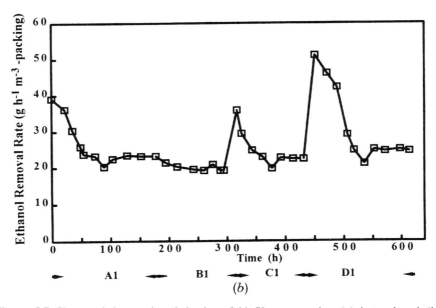

Figure 5.7 Characteristic transient behavior of biofilters removing (*a*) butanol and (*b*) ethanol vapors under constant air residence time (τ) and varying solvent concentrations in the airstream fed to the biofilter. For (*a*), $\tau = 2$ min and inlet butanol concentrations (g/m^3): $A = 1.59$, $B = 2.61$, $C = 5.09$, $D = 3.13$, $E = 0.98$. For (*b*), $\tau = 2.33$ min and inlet ethanol concentrations (g/m^3): $A1 = 1.52$, $B1 = 1.61$, $C1 = 2.34$, $D1 = 3.63$. From Baltzis and Androutsopoulou (1994) and Androustsopoulou (1994).

Probably the most important of all these additional factors (as predicted by numerical studies) is the air–biofilm interfacial area (Shareefdeen and Baltzis 1994b; Androutsopoulou, 1994; Mirpuri et al., 1995).

PROCESS MODELING

Based on the discussion in the preceding section regarding the various factors affecting biofiltration, it becomes clear that this is a very complicated process. Thus, attempting to model biofiltration is not an easy undertaking and requires a number of simplifying assumptions. First, a conceptual model is needed, an example of which (proposed by Shareefdeen and Baltzis, 1994a) is shown in Figure 5.8. This model is based on three levels of successive "magnification": from a section of the biofilter bed, to a solid support particle, to the biofilm formed on the solid surface (i.e., to the actual reaction environment). This conceptual model has been used in describing biofiltration of a single VOC under both transient and steady-state conditions. As discussed earlier, transient operation introduces a whole new process, namely adsorption of VOCs on the packing material. This model has proved successful in describing data from biofiltration of toluene (Shareefdeen and Baltzis, 1994a) and is generalized here for the case of mixed VOCs.

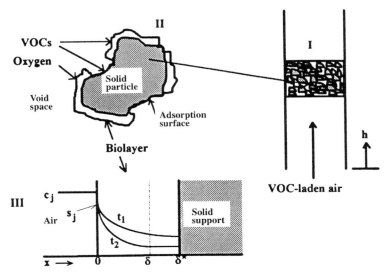

Figure 5.8 Schematic of the conceptual basis for modeling biofiltration, adapted from that of Shareefdeen and Baltzis (1994a). Successive magnifications are shown from the biofilter bed (I), to the packing particle (II), to the biofilm on the packing (III). The particle surface is partially covered with biofilm. VOCs are transferred to the biofilm and the solid packing while oxygen is transferred to the biofilm only.

The assumptions made in deriving the model equations are as follows:

1. The biofilms are formed on the exterior surface of the packing material. No biomass is present in the pores of the particles, and thus reaction does not take place in the pores.

2. The surface of the solid particles is not uniformly covered with biofilm. The biofilm is formed in patches on the surface of the solids, the uncovered parts of which are in direct contact with the airstream.

3. Adsorption of VOCs on the solid particles occurs only through the direct contact of the airstream with the (bare of biofilm) solid. Adsorption on the biofilm does not occur.

4. Adsorption of oxygen can be neglected.

5. The patches of biofilm may be made of the same or different types of biomass. The biomass may be different only in cases of mixed VOC treatment.

6. The thickness of the biofilm is small relative to the diameter of the solid particles, and thus, planar geometry can be used.

7. The surface (i.e., the extent) of the biofilm patch is much larger than its depth. Consequently, VOC and oxygen transfer into the biofilm through the side surfaces of the patch can be neglected, and diffusion/reaction in the biofilm can be described by using only the direction that is perpendicular to the main surface of the patch.

8. Adsorption of VOCs on the packing material is a reversible process and at equilibrium it follows adsorption isotherms.

9. VOCs and oxygen at the biofilm–air interface (i.e., at the main biofilm surface) are distributed between the two phases as dictated by Henry's law. Equilibrium conditions prevail at all times. Henry's law constants are the same as if the biofilm consisted of water only.

10. VOCs and/or oxygen are depleted in a fraction of the actual biofilm thickness called *effective biofilm thickness*.

11. If the effective biofilm thickness is determined via depletion of oxygen, anaerobic degradation of VOCs does not occur in the remaining part of the biofilm.

12. There is no transport of VOCs from the biofilm to the solid particles, or vice versa, through the biofilm–solid interface.

13. Diffusivities of VOCs and oxygen in the biofilm are equal to those in water, corrected by a factor depending on the biofilm density (using a correlation developed by Fan et al., 1990).

14. The biofilm density is constant throughout the biofilter at all times.

15. The extent of coverage of the packing material with biofilm is constant both temporally and spacially. Hence, the specific biofilm surface (biofilm surface area per unit volume of filter bed) is a constant parameter for the system.

16. There is no net accumulation of biomass in the filter bed. In other words, the biofilter is always at a quasi steady state with respect to biomass and thus, no biomass balance is needed for describing the biofiltration process.
17. The specific (i.e., per unit amount of biomass) reaction rates depend only on the availability of VOCs and oxygen, and their functional forms can be determined from suspended-culture experiments.
18. The airstream passes through the biofilter bed in plug flow.

Using the foregoing assumptions, and considering an airstream carrying n VOCs, the biofiltration process can be described by the following mass balances, which involve the biofilm, the air (gas phase), and the packing material (solid phase).

Mass Balances in the Biofilm

$$\frac{\partial s_j}{\partial t} = f(X_V) \, D_{jw} \frac{\partial^2 s_j}{\partial x^2} - \frac{X_V}{Y_j} \, \mu_j(s_1, \ldots, s_j, \ldots s_n, s_O), \qquad j = 1, \ldots, n \qquad (1)$$

$$\frac{\partial s_O}{\partial t} = f(X_V \, D_{Ow} \frac{\partial^2 s_O}{\partial x^2} - \sum_{j=1}^{n} \frac{X_V}{Y_{Oj}} \, \mu_j(s_1, \ldots, s_j, \ldots, s_n, s_O) \qquad (2)$$

with initial and boundary conditions,

$$t = 0, h = 0, x = 0: \qquad s_j = \frac{c_{ji,0}}{m_j}; \qquad j = 1, \ldots, n, \quad s_O = \frac{c_{Oi,0}}{m_O} \qquad (3)$$

$$t = 0, 0 < h \le H, x = 0: \qquad s_j = \frac{c_{j,0}(h)}{m_j}; \quad j = 1, \ldots, n, \quad s_O = \frac{c_{O,0}(h)}{m_O} \qquad (4)$$

$$t = 0, 0 < x \le \delta, 0 \le h \le H: \quad s_j = s_{j,0}(x); \qquad j = 1, \ldots, n, \quad s_O = s_{O,0}(x) \qquad (5)$$

$$t > 0, h = 0, x = 0: \qquad s_j = \frac{c_{ji}}{m_j}; \qquad j = 1, \ldots, n, \quad s_O = \frac{c_{Oi}}{m_O} \qquad (6)$$

$$0 < t < \tau, h > u_g t, x = 0: \qquad s_j = \frac{c_{j,0}(h)}{m_j}; \quad j = 1, \ldots, n, \quad s_O = \frac{c_{O,0}(h)}{m_O} \qquad (7)$$

$$0 < t < \tau, 0 < h < u_g t, x = 0: \; s_j = \frac{c_j(h)}{m_j}; \qquad j = 1, \ldots, n, \quad s_O = \frac{c_O(h)}{m_O} \qquad (8)$$

$$t > \tau, 0 < h \le H, x = 0: \qquad s_j = \frac{c_j(h)}{m_j}; \qquad j = 1, \ldots, n, \quad s_O = \frac{c_O(h)}{m_O} \qquad (9)$$

$$t \ge 0, 0 \le h \le H, x = \delta: \quad \frac{\partial s_j}{\partial x} = \frac{\partial s_O}{\partial x} = 0; \qquad j = 1, \ldots, n \qquad (10)$$

Mass Balances in the Gas Phase

$$v\frac{\partial c_j}{\partial t} = -u_g\frac{\partial c_j}{\partial h} + D_{jw}f(X_V)\varepsilon_j\alpha A_S\left(\frac{\partial s_j}{\partial x}\right)_{x=0} - k_{aj}(1-\alpha)A_S(c_j-c_j'), \quad j = 1,\ldots,n$$

$$\tag{11}$$

$$v\frac{\partial c_O}{\partial t} = -u_g\frac{\partial c_O}{\partial h} + D_{Ow}f(X_V)\alpha A_S\left(\frac{\partial s_O}{\partial x}\right)_{x=0}$$

$$\tag{12}$$

with initial and boundary conditions,

$$t = 0, h = 0: \qquad c_j = c_{ji,0}; \qquad j = 1,\ldots,n, \quad c_O = c_{Oi,0} \tag{13}$$

$$t = 0, 0 < h \le H: \qquad c_j = c_{j,0}(h); \qquad j = 1,\ldots,n, \quad c_O = c_{O,0}(h) \tag{14}$$

$$t > 0, h = 0: \qquad c_j = c_{ji}; \qquad j = 1,\ldots,n, \quad c_O = c_{Oi} \tag{15}$$

Mass Balances in the Solid Phase (Packing)

$$(1 - v)\rho_p\frac{\partial c_{jp}}{\partial t} = k_{aj}(1 - \alpha)A_S(c_j - c_j'), \qquad j = 1,\ldots,n \tag{16}$$

with initial conditions,

$$t = 0, 0 \le h \le H: \quad c_{jp} = c_{jp,0}(h); \quad j = 1,\ldots,n \tag{17}$$

Expressions $\mu_j(s_1,\ldots,s_j,\ldots,s_n,s_O)$ appearing in Eqs. (1) and (2) reflect assumption 17. The dependence of the specific reaction rates on the availability of the carbon sources (VOCs) can be separated from the dependence on the oxygen availability through the notion of interactive models (Bader, 1982). Hence,

$$\mu'_j(s_1,\ldots,s_j,\ldots,s_n)f(s_O) = \mu_j(s_1,\ldots,s_j,\ldots,s_n,s_O) \tag{18}$$

Usually, biodegradation rates have a Monod-type dependence on oxygen and thus,

$$f(s_O) = \frac{s_O}{K_{Oj} + s_O} \tag{19}$$

As mentioned earlier, most of the existing models do not account for oxygen availability, essentially implying that $f(s_O) = 1$. In such cases Eqs. (2) and (12), and their corresponding initial and boundary conditions, are not needed. Despite the resulting simplifications, oxygen has to be explicitly accounted for as explained in the preceding section.

The functional dependence of the specific biodegradation rates (μ'_j) on the concentrations of VOCs present ($s_1,\ldots,s_j,\ldots,s_n$) can take several different forms. For any VOC j that does not interact kinetically with other VOCs one can write

$$\mu'_j(s_j) = \frac{\mu^*_j s_j}{K_j + s_j + K'_{Ij} s_j^2} \tag{20}$$

while for a VOC j engaged in a competitive kinetic interaction with VOC q one can write

$$\mu'_j(s_j, s_q) = \frac{\mu^*_j s_j}{K_j + s_j + K'_{Ij} s_j^2 + K_{jq} s_q} \tag{21}$$

Examples of VOCs following expressions (20) and (21) during their biodegradation were given in the earlier discussion of microorganisms and kinetics. Expression (20) represents the Andrews (1968) model when $K'_{Ij} \neq 0$, and the Monod (1942) model when $K'_{Ij} = 0$.

Concentration c'_j appearing in Eqs. (11) and (16) is related to c_{jp} through the adsorption isotherm. If a single VOC is involved, and if the Freundlich isotherm is followed, one has

$$c_{jp} = k_{dj}(c'_j)^r \tag{22}$$

Factor ε_j, which appears in Eq. (11) represents the fraction of the biofilm–air interfacial area belonging to patches of biomass capable of biodegrading VOC j. If the biomass present in the biofilter is capable of biodegrading all VOCs present in the airstream, then $\varepsilon_j = 1$ for any j. This is the assumption made in all studies except a very recent one. Baltzis and Wojdyla (1995b) were able to successfully model data from steady-state biofiltration of ethanol and butanol mixtures only when they attributed a different value to ε_j for each compound. This approach implies biomass differentiation in the biofilter bed and needs further investigation.

The transients of biofiltration last over long times (even days) especially for hydrophilic VOCs, that is, VOCs that are not highly volatile. For this reason, as has been argued by Shareefdeen and Baltzis (1994a), there is no loss of accuracy if conditions (7) and (8) are omitted and conditions (9) are used for any $t > 0$ rather than for $t > \tau$ only. This is because the residence time τ ($\tau = V/F$) is very small when compared to the extent of the transients.

For biofiltration under steady-state conditions the model is significantly simplified as Eq. (16) and all initial conditions [(3)–(5), (7)–(9), (13), (14), and (17)] are not needed, while the last term in the right-hand side of Eq. (11) and the left-hand side of Eqs. (1), (2), (11) and (12) become equal to zero.

For a polluted airstream of given composition (i.e., c_{ji} values known) and volumetric flow rate (F), and for a given biofilter bed volume (V), solution of the model equations leads to determination of the VOC concentrations in the airstream at the exit of the unit (c_{je}). Knowledge of c_{je} allows for determination of the removal rate, which is the quantity used in the literature to describe biofilter performance. The removal rate is defined by the following equation:

$$R_j = \frac{F}{V}(c_{ji} - c_{je}) = \frac{c_{ji} - c_{je}}{\tau}, \quad j = 1, \ldots, n \tag{23}$$

Conversely, if one knows the values of c_{ji} and F and wants to achieve certain values (e.g., as specified by existing environmental regulations) for c_{je}, the model equations can be solved through trial and error in order to determine the size (V) of the required biofilter.

Table 5.1 Experimental and Model-Predicted Steady-State Removal Rates for Single Hydrophilic VOC Vapors

τ^a (min)	$V_p^{\,b}$ ($m^3 \times 10^{-6}$)	$c_{ji}^{\,c}$ (g/m^3)	$R_{exp}^{\,d}$ ($g/(h\ m^3$ packing))	$R_{pred}^{\,e}$	Error^f (%)
Methanol^g					
2.30	706	6.45	100.8	90.7	−10.0
2.85	706	6.44	94.1	89.1	−5.3
3.00	932	2.67	53.3	53.3	0.0
3.00	932	8.72	101.6	93.7	−7.8
3.34	932	6.11	92.8	85.8	−7.5
Ethanol^h					
0.47	188	0.31	40.0	35.7	−10.8
1.86	932	4.15	25.8	22.5	−12.8
2.33	932	2.34	22.5	25.3	+12.4
2.33	932	3.63	24.5	22.4	−8.5
3.72	932	4.34	19.0	20.9	+9.1
Butanol^h					
0.82	984	0.58	24.2	26.1	+7.9
0.89	984	0.78	21.6	25.0	+15.7
0.89	784	0.35	23.6	22.9	−0.03
1.63	984	1.06	25.4	24.6	−3.1
1.96	984	0.80	24.5	24.2	−1.2

^{a} Residence time of airstream in biofilter.
^{b} Volume of biofilter packing material.
^{c} Inlet VOC concentration.
^{d} Experimentally obtained removal rate.
^{e} Model-predicted value for removal rate.
^{f} Percent error in prediction of removal rate.
^{g} Data from Shareefdeen et al. (1993),
^{h} Data from Baltzis and Androutsopoulou (1994).

The steady-state version of the model equations has been used to describe experimental data obtained from laboratory units (Fig. 5.4 and 5.5). For the case of single VOCs, methanol (Shareefdeen et al., 1993), ethanol and butanol (Baltzis and Androutsopoulou, 1994), and benzene and toluene (Shareefdeen and Baltzis, 1994a, b) have been used as model compounds. A comparison between experimentally obtained and model-predicted values is shown in Tables 5.1 and 5.2, and schematically in Figure 5.9a. Biofiltration data for binary VOC mixtures under steady-state operation have been also modeled for benzene and toluene (Baltzis and Shareefdeen, 1994) and ethanol and butanol (Baltzis and Wojdyla, 1995b). Data and comparisons with model predictions for mixtures are shown in Tables 5.3 and 5.4, and as profiles along the biofilter bed in Figure 5.9b. The transient model has been solved only for single VOCs and validated with data from toluene biofiltration (Shareefdeen and Baltzis, 1994a). An example of transient profiles is shown in Figure 5.10. In almost all cases, the agreement between data and model predictions is very good, especially when one considers the complexity of the process and the number of simplifying assumptions involved.

Values for the model parameters used in describing the data shown in Tables 5.1–5.4 and Figures 5.9 and 5.10 can be found in the corresponding references. The numerical methodologies for solving the model equations are described in detail by Shareefdeen et al. (1993), Shareefdeen and Baltzis (1994a), and Shareefdeen (1994).

Table 5.2   Experimental and Model-Predicted Steady-State Removal Rates for Single, Hydrophobic VOC Vapors

τ^a (min)	$V_p{}^b$ ($m^3 \times 10^{-6}$)	$c_{ji}{}^c$ (g/m^3)	$R_{exp}{}^d$ ($g/(h\ m^3$ packing$))$	$R_{pred}{}^e$	Error^f (%)
		Benzene[g]			
2.7	4000	0.13	0.9	0.8	−11.1
4.1	4000	0.07	0.5	0.4	−20.0
4.1	4000	0.28	1.3	1.6	+23.1
4.5	4000	0.43	2.7	2.5	−7.4
4.7	4000	0.56	4.5	3.1	−31.1
		Toluene[h]			
2.7	5150	0.62	9.4	7.4	−21.3
4.2	5150	0.92	10.4	9.0	−13.5
6.3	5150	2.81	24.8	21.5	−13.3
7.7	5150	1.65	12.2	11.3	−7.4
8.6	5150	0.68	4.8	4.4	−8.3

[a—f] As in Table 5.1.
[g] Data from Shareefdeen and Baltzis (1994b).
[h] Data from Shareefdeen and Baltzis (1994a)

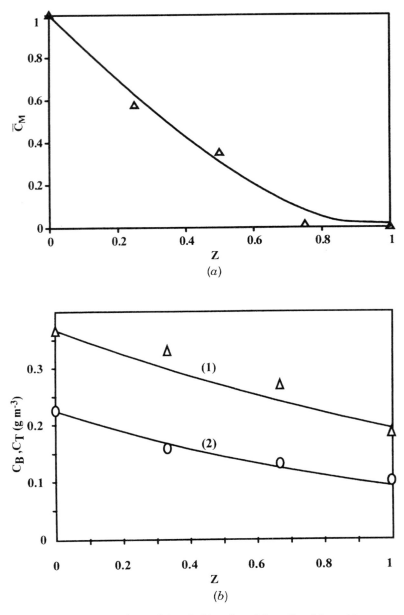

Figure 5.9 Steady-state experimental (symbols) and model-predicted (curves) concentration profiles along biofilter beds. The x axis shows the position on the bed normalized with the total height; $z = 0$: entrance, $z = 1$: exit position. Figure (a) is for methanol removal and the concentration is normalized with that at the entrance ($2.67\,g/m^3$); from Shareefdeen et al. (1993). Figure (b) is for removal of a mixture of benzene (curve 1) and toluene (curve 2); from Baltzis and Shareefdeen (1994).

Table 5.3   Experimental and Model-Predicted Steady-State Removal Rates for Benzene—Toluene (Hydrophobic VOCs) Mixtures[a]

τ^b (min)	VOC^c	$c_{ji}{}^d$ (g/m^3)	$R_{\text{exp}}{}^e$ (g/(h m^3 packing))	$R_{\text{pred}}{}^f$	Error^g (%)
1.3	B	0.21	1.9	2.1	+10.5
	T	0.40	6.3	5.5	−12.7
1.4	B	0.17	1.5	1.7	+13.3
	T	0.38	6.1	5.3	−13.1
1.5	B	0.19	1.8	2.0	+11.1
	T	0.27	3.4	3.7	+8.8
3.1	B	0.37	3.5	3.3	−5.7
	T	0.23	2.4	2.6	+8.3

[a] Data from Baltzis and Shareefdeen (1994). The volume of biofilter packing was 15.3 × 10^{-3} m^3.
[b] As in Table 5.1.
[c] Compounds: B = benzene, T = toluene.
[d—g] As in Table 5.1.

Table 5.4   Experimental and Model-Predicted Steady-State Removal Rates for Ethanol–Butanol (Hydrophilic VOCs) Mixtures[a]

τ^b (min)	VOC^c	$c_{ji}{}^d$ (g/m^3)	$R_{\text{exp}}{}^e$ (g/(h m^3 packing))	$R_{\text{pred}}{}^f$	Error^g (%)
0.86	E	2.75	3.1	2.8	−9.7
	B	1.07	4.5	6.6	+46.7
2.20	E	0.63	8.1	6.5	−19.8
	B	0.54	6.7	6.6	−1.5
2.28	E	0.92	7.3	6.0	−17.8
	B	0.26	4.4	5.6	+27.3
4.06	E	2.25	5.2	3.5	−32.7
	B	1.21	5.3	6.1	+15.1

[a] Data from Baltzis and Wojdyla (1995b). The volume of biofilter packing was 1.05 × 10^{-3} m^3.
[b] As in Table 5.1.
[c] Compounds: E = ethanol, B = butanol.
[d—g] As in Table 5.1.

Other general models for steady-state and transient-state biofiltration have been proposed by Deshusses et al. (1995a, b) and Hodge and Devinny (1995). Except for the fact that oxygen is not accounted for, these models have a lot of similarities with the one presented here.

The complexities inherent in the model discussed above are necessary in order to optimize the design and operation of biofilters and to diagnose problems in the field. In fact, additional complexities may be needed, if experience determines that the extensive list of assumptions on which the model is based are not justified in a particular application.

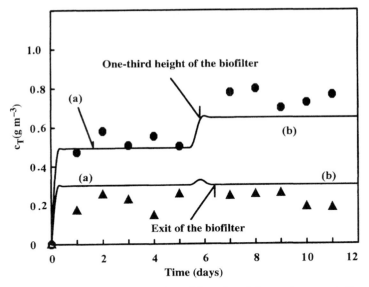

Figure 5.10 Experimental (symbols) and model-predicted (curves) concentration profiles under transient biofilter operation for removal of toluene. Profiles are shown at two locations along the biofilter bed. Parts (a) of profiles are for operation under residence time of 2.8 min and inlet toluene concentration of 0.63 g/m^3; conditions for parts (b) of profiles are, correspondingly, 4.3 min and 0.92 g/m^3. From Shareefdeen and Baltzis (1994a).

RESEARCH NEEDS AND TRENDS

Biofiltration has proven to be a feasible process in a number of studies that investigated a variety of compounds (alcohols, aldehydes, ketones, and simple and complex mixtures of gasoline constituents). However, optimizing the size and operating conditions of commercial biofilters still leaves much room for fundamental research.

Based on the existing conceptual models for the process, biodegradation kinetics and air–biofilm interfacial area appear to be the most important factors affecting the process, as has been earlier discussed. Determination of kinetic constants is based on the assumption that suspended and immobilized (biofilm-forming) cells have the same intrinsic kinetics. This is a crucial assumption that needs to be tested. With the increasing availability of microscale measurement methods (e.g., electrodes for oxygen and/or VOC concentration monitoring in biofilms), efforts have already started for testing this assumption via monitoring concentrations in the actual reaction environment, that is, the biofilm. An example is the work of Mirpuri et al. (1995) who actually measured the oxygen profiles in prototype biofilm reactors.

The actual surface area of the air–biofilm interface has never been measured. Estimated total particle–air interfacial areas (Wojdyla, 1996) indicate that coverage of particles with biofilm as estimated by a number of investigators (Shareefdeen et al., 1993; Shareefdeen and Baltzis, 1995a; Baltzis and Androutsopoulou, 1994; Zilli et al., 1993) is in the range of 3–8% of the total particle–air interfacial area. This seems to be in accord with reports from soil microbiology studies suggesting a 2–5% coverage of soil surfaces with biofilm clusters. Efforts need to be undertaken to increase the coverage of the packing materials with biomass, since this factor significantly affects biofilter performance, as discussed in an earlier section (e.g., Shareefdeen, 1994; Androutsopoulou, 1994).

All existing conceptual models of biofiltration assume complete biofilm uniformity. This is probably incorrect. As mentioned in the modeling section, a recent study (Baltzis and Wojdyla, 1995b) introduced some aspects of biofilm structure in which different organisms or consortia occupy different surfaces on the solid support, in order to explain data from biofiltration of ethanol–butanol mixtures. Even this modification did not account for possible heterogeneities in the direction of biofilm depth. An improved understanding and modeling of such structures is needed and will hopefully come from the advances in understanding biofilms in general, which is currently underway.

The coupling of microbiology and engineering is very essential to the success of such efforts. The objective would be to possibly select conditions leading to the formation of optimal biofilm structures to achieve the maximum removal rate. Biomass differentiation and biofilm structure may result even in cases where one starts with a homogeneous microbial population. For example, the notion of "injured" cells (Lebby et al., 1995) was used in a study involving toluene vapors (Mirpuri et al., 1995).

Biofilters are most likely operating under transient conditions at all times due to the variation in emission levels. Despite some efforts to understand and model the transient behavior of biofilters, further work is needed in this area. Adsorption of VOCs on the packing material needs to be characterized in more cases, while the fate of the adsorbed material needs to be further understood. The approach of Shareefdeen and Baltzis (1994a) assumes that adsorption is reversible, but the adsorbed material is not directly available to the biofilm. Similar is the approach of Deshusses et al. (1995a, b), while that of Hodge and Devinny (1995) assumes that adsorbed VOCs may be directly transferred to the biofilm. Work with transient biofilter operation is especially needed with airstreams carrying mixed VOCs.

Moisture content of the packing material is acknowledged as a major factor in biofiltration but has never really been studied in detail or modeled. Some efforts that account for water tension and the like are currently under way. Moisture affects not only biodegradation per se but also adsorption/desorption characteristics and the flow patterns of air through the biofilter bed.

Other issues that need investigation include channeling in the bed, compaction of the solid packing, temperature effects, air distribution, degree of axial mixing, delivery of supplemental nutrients, and the like.

General discussions of the status of biofiltration, and research needs, can also be found elsewhere in the literature (Devinny, 1995; Leson and Dharmavaram, 1995).

APPLICATIONS POTENTIAL

There are already a number of classical biofilters that operate very successfully in Europe and (although fewer in number) in the United States. These applications usually involve simple mixtures of easily biodegradable VOCs present at relatively low concentrations in airstreams. There are also biofilter applications for removing inorganic compounds (e.g., H_2S), a subject not covered here. The size of existing biofiltration units is usually large, and thus space availability and cost are important factors. Because of the required volume, staged and modular units have been introduced in the market by various vendors.

New biofilter applications for large industrial units in the United States will depend on the regulatory environment and the ability of biofilters to operate at a cost that is competitive with other technologies. Additional potential applications of classical biofilters include bakeries (ethanol emissions), spray painting facilities, and printing operations. These examples involve single or simple mixtures of VOCs that are easily degradable. However, these are not continuous processes and the issue of intermittent biofilter operation needs to be fully resolved. Intermittent operation implies that the microorganisms may be sitting in the column for an extended period of time without being exposed to VOCs or in fact any carbon source. Under those conditions, there are questions of organism survival and ability to readapt to the renewed presence of VOCs.

Another area of potential applications is that of tablet-coating operations in the pharmaceutical industry. This process utilizes simple volatile solvents (e.g., methanol) that can be easily treated in a biofilter. Nonetheless, such applications are not necessarily straightforward, as the off-gases may contain fine particles (dust) that can clog the biofilter and/or poison the microorganisms.

In situ treatment of contaminated soil also presents opportunities for biofilter applications. Soil vapor extraction requires VOC treatment prior to discharge. In cases where the total time for treating a particular site is not a critical issue, low vapor extraction flows can be used, which make biofilter application more attractive. In fact, potentially the largest application for classical biofilters in the United States may be clean up of gasoline-contaminated sites.

BIOTRICKLING FILTERS

Because of the original perception that biofiltration is a simple process and that biofilters could easily treat all types of VOC emissions, most of the research and development efforts in the past employed classical biofilters. As has been discussed here, biofiltration of hazardous air pollutants is anything but simple. The complexity of the issues that have not yet been resolved (microbiology, moisture content, biofilm coverage of packing, etc.) but also the intrinsic limitations of biofilters (e.g., pH-control for cases of chlorinated solvent emissions) have recently led to the initiation of efforts for modifying the process.

The most important modification that has already led to applications (e.g., Dharmavaram et al., 1995) is that of biotrickling filters. These units employ a well-specified, nonporous, inorganic packing material (e.g., polymeric materials and ceramics) in lieu of the complex and ill-characterized types of packing used in classical biofilters. In addition, a liquid stream is passed downward through the bed. This stream consists primarily of water, but various nutrients for the microorganisms and buffers for pH control are also present. Instrumentation and operating costs for biotrickling filters are higher than those for classical biofilters. However, VOC removal rates can substantially surpass those achieved with classical biofilters. Increased rates imply smaller units and thus, lower capital expenditure.

To date, studies with biotrickling filters have focused on chlorinated solvent emissions (e.g., Diks and Ottengraf, 1989, 1991) and on high loads of relatively easily degradable compounds such as toluene (Sorial et al., 1994, 1995).

Biotrickling filters appear to lead to biofilm–liquid interfacial areas much higher than the biofilm–air interfacial areas achieved in classical biofilters. This is definitely a big plus, as has been discussed in earlier sections, although one should keep in mind that in biotrickling filters there is an additional mass-transfer resistance. VOCs must be transferred from the air to the liquid (water) before they reach the biofilms. Furthermore, biotrickling filters can lead to excess biomass formation and require occasional removal of biomaterial to avoid clogging and severe pressure drop. Classical biofilters once installed require minimal supervision, while biotrickling filters require frequent and specialized (engineering) attendance. Thus, applications of biotrickling filters are expected to be primarily within industrial settings and not at small businesses (bakeries, dry cleaners, etc.). Preliminary results with biotrickling filters are very encouraging. However, these systems need further study and development, which appears to be the dominant research focus of the next few years, as classical biofilters gradually take second place.

The interest in biological treatment of VOC emissions is high, and this may lead in the future to drastic process modifications based on novel ideas. Such a totally new approach is that of Phipps and Ridgway (1995) who developed an experimental system for suspending biomass in a foam. This design does away with biofilms and increases the VOC–biomass contact area dramatically. However, this approach requires formation of an emulsion and the problems of its stability need considerable further work before pilot testing is undertaken.

CONCLUSIONS

Biofiltration is a relatively well-established technology for odor control purposes. For the case of VOCs, and in particular hazardous air pollutants, biofiltration is feasible, but its widespread applicability (especially in U.S. markets) will require further optimization and cost-reduction.

In this chapter, a detailed discussion of classical biofilters and an approach to process modeling were presented along with issues that need further study and development. Recent results have contributed significantly toward a better process understanding, while the earlier black-box approach is becoming less acceptable.

Despite its complexity and potential limitations, biofiltration applications are expected to increase as more imaginative, cost-effective methodologies are sought for control of volatile organic emissions.

NOTATION

A_S	surface area of biofilter packing material available for biofilm formation and VOC adsorption per unit volume of biofilter bed (m^{-1})
c_j	concentration of VOC j in the air passing through the biofilter; function of h (g/m^3)
c_j'	value of c_j under equilibrium conditions between the air and the biofilter packing material (g/m^3)
$c_{j,0}$	value of c_j at $t = 0$; function of h (g/m^3)
c_{je}	value of c_j at the biofilter bed exit, i.e., at $h = H$ (g/m^3)
c_{ji}	value of c_j at the biofilter bed entrance, i.e., at $h = 0$ (g/m^3)
$c_{ji,0}$	value of c_{ji} at $t = 0$ (g/m^3)
c_{jp}	concentration of VOC j in the biofilter packing material; function of h (g of VOC j adsorbed/g packing)
$c_{jp,0}$	value of c_{jp} at $t = 0$; function of h (g of VOC j adsorbed/g packing)
c_O	oxygen concentration in the air passing through the biofilter; function of h (g/m^3)
$c_{O,0}$	value of c_O at $t = 0$; function of h (g/m^3)
c_{Oi}	value of c_O at the biofilter bed entrance, i.e., at $h = 0$ (g/m^3)
$c_{Oi,0}$	value of c_{Oi} at $t = 0$ (g/m^3)
D_{jw}	diffusion coefficient of VOC j in water (m^2/h)
D_{Ow}	diffusion coefficient of oxygen in water (m^2/h)
F	volumetric flow rate of airstream supplied to biofilter (m^3/h)
$f(s_O)$	functional dependence, given by relation (19), of μ_j on oxygen (dimensionless)
$f(X_V)$	ratio of diffusivities in the biofilm and water, for VOCs and oxygen (dimensionless)
H	height of the biofilter bed (m)
h	position in the biofilter bed (m); $h = 0$: entrance, $h = H$: exit
K_j	kinetic constant in functions μ_j and μ'_j (g/m^3)
K'_{lj}	kinetic constant in functions μ_j and μ'_j (g/m^3)
K_{jq}	kinetic constant in functions μ_j and μ'_j expressing competitive inhibition of the removal of VOC j by the presence of VOC q (dimensionless)
K_{Oj}	kinetic constant in functions μ_j and $f(s_O)$ related to VOC j and oxygen (g/m^3)
k_{aj}	mass-transfer coefficient, between air and biofilter packing material, for VOC j (m/h)
k_{dj}	constant in expression (22); Freundlich isotherm parameter for VOC j (g VOC j/g packing)/(g/m^3)^V

m_j distribution coefficient (Henry's constant) between air and water for VOC j (dimensionless)

m_O distribution coefficient (Henry's constant) between air and water for oxygen (dimensionless)

R_j removal rate for VOC j (g/(h m^3-packing))

r constant in expression (22); Freundlich exponent for VOC j (dimensionless)

S cross-sectional area of the biofilter bed (m^2)

s_j concentration of VOC j in the biofilm; function of x and h (g/m^3)

s_q concentration of VOC q in the biofilm; function of x and h (g/m^3)

$s_{j,0}$ value of s_j at $t = 0$; function of x and h (g/m^3)

s_O oxygen concentration in the biofilm; function of x and h (g/m^3)

$s_{O,0}$ value of s_O at $t = 0$; function of x and h (g/m^3)

t time (h)

u_g superficial air velocity (F/S) in the biofilter (m/h)

V volume of the biofilter bed (m^3)

X_V biofilm density (g dry cells/m^3)

x position in the biofilm (m); $x = 0$: air–biofilm interface

Y_j biomass yield coefficient on VOC j (g biomass/g VOC j)

Y_{Oj} biomass yield coefficient on oxygen when VOC j is the carbon source (g biomass/g oxygen)

Greek Symbols

α fraction of A_S covered with biofilm (dimensionless)

δ effective biolayer thickness; value of x where oxygen and/or VOCs are depleted (m)

ε_j fraction of specific biofilm–packing interface (αA_S) covered with biomass capable of degrading VOC j (dimensionless)

μ_j kinetic expression (biomass specific growth rate) for removal of VOC j; function of s_j, $j = 1, \ldots, n$, s_q, s_O (h^{-1})

μ'_j functional dependence of μ_j on the availability of VOCs only, i.e., μ_j under excess oxygen conditions (h^{-1})

μ^*_j kinetic constant in functions μ_j and μ'_j (h^{-1})

ρ_p real density of biofilter packing (g/m^3-solids; not g/m^3 bed)

τ residence (or space) time of airstream in biofilter bed based on an empty unit, i.e., $\tau = V/F$ (h)

υ void fraction of biofilter bed (dimensionless)

REFERENCES

Andrews, J. F. (1968). A Mathematical Model for the Continuous Culture of Microorganisms Utilizing Inhibitory Substrates, *Biotech. Bioeng.*, **10**:707–723.

Androutsopoulou, H. (1994). A Study of the Biofiltration Process Under Shock-Loading Conditions, M. S. Thesis, New Jersey Institute of Technology, Newark, NJ.

Bader, F. G. (1982). Kinetics of Double-Substrate Limited Growth, in *Microbial Population Dynamics*, M. J. Bazin, (Ed.), CRC Press, Boca Raton, FL, pp.1–32.

Baltzis, B. C. and H. Androutsopoulou (1994). A Study on the Response of Biofilters to Shock-Loading, Paper no. 94-RP-115B.02, in *Proceedings of the 87th Annual A&WMA Meeting*, Cincinnati, OH.

Baltzis, B. C. and Z. Shareefdeen (1993). Modeling and Preliminary Design Criteria for Packed-Bed Biofilters, Paper no. 93-TP-52A.03, in *Proceedings of the 86h Annual A&WMA Meeting*, Denver, CO.

Baltzis, B. C. and Z. Shareefdeen (1994). Biofiltration of VOC Mixtures: Modeling and Pilot Scale Experimental Verification, Paper no. 94-TA-260.10P, in *Proceedings of the 87th Annual A&WMA Meeting*, Cincinnati, OH.

Baltzis, B. C. and S. M. Wojdyla (1995a). Characteristics of Biofiltration of Hydrophilic VOC Mixtures, in *Proceedings of the 4th Conference on Environmental Science and Technology* (Vol. B), Th. Lekkas, Ed., University of the Aegean, Lesvos, Greece, pp. 322–331.

Baltzis, B. C. and S. M. Wojdyla (1995b). Towards a Better Understanding of Biofiltration of VOC Mixtures, in *Proceedings of 1995 Conference on Biofiltration*, D. S. Hodge and F. E. Reynolds, Jr., Eds., Reynolds Group, Tustin, CA, pp. 131–138.

Bohn, H. (1992). Consider Biofiltration for Decontaminated Gases, *Chem. Eng. Prog.*, 4, April:34–40.

Bohn, H. (1993). Biofiltration: Design Principles and Pitfalls, Paper no. 93-TP-52A.01, in *Proceedings of the 86th Annual A&WMA Meeting*, Denver, CO.

Bohn, H. and R. Bohn (1988). Soil Beds Weed out Air Pollutants, *Chem. Eng.*, **95**(4):73–76.

Chang, M. K., T. C. Voice, and C. S. Criddle (1993). Kinetics of Competitive Inhibition and Cometabolism in the Biodegradation of Benzene, Toluene, and P-xylene by Two *Pseudomonas* Isolates, *Biotech. Bioeng.*, **41**:1057–1065.

Deshusses, M. A., G. Hamer, and I. J. Dunn (1995a). Behavior of Biofilters for Waste Air Biotreatment. 1. Dynamic Model Development, *Environ. Sci. Tech.*, **29**:1048–1058.

Deshusses, M. A., G. Hamer, and I. J. Dunn (1995b). Behavior of Biofilters for Waste Air Biotreatment. 2. Experimental Evaluation of a Dymamic Model, *Environ. Sci. Tech.*, **29**:1059–1068.

Devinny, J. S. (1995). Topics for Research in Biofiltration, in *Proceedings of 1995 Conference on Biofiltration*, D. S. Hodge and F. E. Reynolds, Jr., Eds., Reynolds Group, Tustin, CA. pp. vi–xiii.

Dharmavaram, S., U. C. Duursma, G. Rietbroek, and E. Waalewijn (1995). Use of a Biotrickling Filter (BFT) for Control of *N,N*-Dimethylacetamide Emissions, Paper no. 95-TA9B.01, in *Proceedings of the 88th Annual A&WMA Meeting*, San Antonio, TX.

Diks, R. M. M. and S. P. P. Ottengraf (1989). Process Technological View on the Elimination of Chlorinated Hydrocarbons from Waste Gases, in *Man and His Ecosystem*, L. J. Brasser and W. C. Mulder, Eds., Elsevier, Amsterdam, pp. 405–410.

Diks, R. M. M. and S. P. P. Ottengraf (1991). Verification Studies of a Simplified Model for Removal of Dichloromethane from Waste Gases Using a Biological Trickling Filter (Part. 1), *Bioprocess Eng.*, **6**:93–99.

Fan, L. S., R. Leyva-Ramos, K. D. Wisecarver, and B. J. Zehner (1990). Diffusion of Phenol through a Biofilm Grown on Activated Carbon Particles in a Draft-tube Three-phase Fluidized-bed Bioreactor, *Biotech. Bioeng.*, **35**:279–286.

Hodge, D. S. (1995). Determination of Mathematical Model Constants Using Specially Designed Mini-column Biofilters, in *Proceedings of 1995 Conference on Biofiltration*, D. S. Hodge and F. E. Reynolds, Jr., Eds., Reynolds Group, Tustin, CA, pp. 53–69.

Hodge, D. S. and J. Devinny (1995). Modeling Removal of Air Contaminants by Biofiltration, *ASCE J. Environ. Eng.*, **121**(1):21–44.

Hodge, D. S. and F. E. Reynolds, Jr., Eds. (1995). *Proceedings of 1995 Conference on Biofiltration*, Reynolds Group, Tustin, CA.

Karel, S. F., S. B. Libicki, and C. R. Robertson (1985). The Immobilization of Whole Cells: Engineering Principles. *Chem. Eng. Sci.*, **40**:1321–1354.

Lebby, M. B., D. W. Phipps, and H. F. Ridgway (1995). Catabolite-Mediated Mutations in Alternate Toluene Degradation Pathways in *Pseudomonas putida, J Bacteriol.*, **177**:4713–4720.

Leson, G. and S. Dharmavaram (1995). A Status Overview of Biological Air Pollution Control, Paper no. 95-MP9A.01, in *Proceedings of the 88th Annual A&WMA Meeting*, San Antonio, TX.

Leson, G. and A. M. Winer (1991). Biofiltration: An Innovative Air Pollution Control Technology for VOC Emissions, *J. Air Waste Manage. Assoc.*, **41**:1045–1054.

Leson, G., D. S. Hodge, F. Tabatabai, and A. M. Winer (1993). Biofilter Demonstration Projects for the Control of Ethanol Emissions, Paper no. 93-WP-52C.04, in *Proceedings of the 86th Annual A&WMA Meeting*, Denver, CO.

Mirpuri, R., W. Sharp, W. Jones, S. Villaverde, Z. Lewandowski, and A. Cunningham (1995). A Predictive Model for Toluene Degradation in a Flat Plate Vapor Phase Bioreactor, in *Proceedings of 1995 Conference on Biofiltration*, D. S. Hodge and F. E. Reynolds, Jr., Eds., Reynolds Group, Tustin, CA, pp. 71–84.

Monod, J. (1942). *Recherches sur la Croissance des Cultures Bacteriennes*, Herman et Cie., Paris, France.

Oh, Y. S. (1993). Biofiltration of Solvent Vapors from Air, Ph.D. Thesis, Rutgers University, New Brunswick, NJ.

Oh, Y. S., Z. Shareefdeen, B. C. Baltzis, and R. Bartha (1994). Interactions between Benzene Toluene, and *p*-xylene (BTX) During Their Biodegradation, *Biotech. Bioeng.* **44**:533–538.

Ottengraf, S. P. P. (1986). Exhaust Gas Purification, in *Biotechnology*, Vol. 8, W. Shonborn, Ed., VCH Verlagsgesellschaft, Weinheim, Germany, pp. 425–452.

Ottengraf, S. P. P. (1987). Biological Systems for Waste Gas Elimination, *Trends Biotech.*, **5**:132–136.

Ottengraf, S. P. P., J. J. P. Meesters, A. H. C. van den Oever, and H. R. Rozema (1986). Biological Elimination of Volatile Xenobiotic Compounds in Biofilters, *Bioprocess Eng.*, **1**:61–69.

Ottengraf, S. P. P., and A. H. C. van den Oever (1983). Kinetics of Organic Compound Removal from Waste Gases with a Biological Filter, *Biotech. Bioeng.*, **25**:3089–3102.

Ottengraf, S. P. P., A. H. C. van den Oever, and F. J. C. M. Kempennars (1984). Waste Gas Purification in a Biological Filter Bed, in *Innovations in Biotechnology*, E. H. Houwink and R. R. van dan Meer, Eds., Elsevier Science, Amsterdam, pp. 157–167.

Phipps, D. W. and H. F. Ridgway (1995). Using Biologically Activated Foam for the Degradation of Vapor Phase Volatile Organic Contaminants, in *Proceedings of 1995 Conference on Biofiltration*, D. S. Hodge and F. E. Reynolds, Jr., Eds., Reynolds Group, Tustin, CA, pp. 225–236.

Pinnete, J. R., C. A. Dwinal, M. D. Giggey, and G. E. Hendry (1995). Design Implications of the Biofilter Heat and Moisture Balance, in *Proceedings of the 1995 Conference on Biofiltration*, D. S. Hodge and F. E. Reynolds, Jr., Eds., Reynolds Group, Tustin, CA, pp. 85–98.

Pomeroy, D. (1982). Biological Treatment of Odorous Air, *J. Water Pollut. Contr. Fed.*, **54**:1541–1545.

Reynolds, Jr., F. E. and D. S. Hodge (1995). Forward, in: *Proceedings of 1995 Conference on Biofiltration*, D. S. Hodge and F. E. Reynolds, Jr., Eds., Reynolds Group, Tustin, CA, pp. iv–v.

Shareefdeen, Z. (1994). Engineering Analysis of a Packed-Bed Biofilter for Removal of Volatile Organic Compound (VOC) Emissions, Ph.D. Thesis, New Jersey Institute of Technology, Newark, NJ.

Shareefdeen, Z. and B. C. Baltzis (1994a). Biofiltration of Toluene Vapor under Steady-State and Transient Conditions: Theory and Experimental Results, *Chem. Eng. Sci.*, **49**:4347–4360.

Shareefdeen, Z. and B. C. Baltzis (1994b). Biological Removal of Hydrophobic Solvent Vapors from Airstreams, in *Advances in Bioprocess Engineering*, E. Galindo and O. T. Ramirez, Eds., Kluwer Academic, Dordrecht, The Netherlands, pp. 397–404.

Shareefdeen, Z., B. C. Baltzis, Y. S. Oh, and R. Bartha (1993). Biofiltration of Methanol Vapor, *Biotech. Bioeng.*, **41**:512–524.

Sorial, G. A., F. L. Smith, M. T. Suidan, P. Biswas, and R. C. Brenner (1994). Evaluation of Performance of Trickle-bed Biofilters—Impact of Periodic Removal of Accumulated Biomass, Paper no. 94-RA115A.05, in *Proceedings of the 87th Annual A&WMA Meeting*, Cincinnati, OH.

Sorial, G. A., F. L. Smith, A. Pandit, M. T. Suidan, P. Biswas, and R. C. Brenner (1995). Performance of Trickle Bed Biofilters under High Toluene Loading, Paper no. 95-TA9B.04, in *Proceedings of the 88th Annual A&WMA Meeting*, San Antonio, TX.

Stuart, B. J., G. F. Bowlen, and D. S. Kosson (1991). Competitive Sorption of Benzene, Toluene and Xylenes onto Soil, *Environ. Prog.*, **10**(2):104–109.

van Lith, C. (1989). Biofiltration: An Essential Technique in Air Pollution Control, in *Man and His Ecosystem*, L. J. Brasser and W. C. Mulder, Eds., Elsevier, pp. 393–399.

van Lith, C., S. L. David, and R. Marsh (1990). Design Criteria for Biofilters, *Trans. IChemE.*, **68**(Part B):127–132.

Wojdyla, S. M. (1996). Determination of Physical Parameters Affecting Biofiltration of VOCs, M. S. Thesis, New Jersey Institute of Technology, Newark, NJ.

Zilli, M., A. Converti, A. Lodi, M. del Borghi, and G. Ferraiolo (1993). Phenol Removal from Waste Gases with a Biological Filter by *Pseudomonas putida*, *Biotech. Bioeng.* **41**:693–699.

CHAPTER 6

Impact of Biokinetics and Population Dynamics on Engineering Analysis of Biodegradation of Hazardous Wastes

Basil C. Baltzis and Gordon A. Lewandowski

Department of Chemical Engineering, Chemistry, and Environmental Science, New Jersey Institute of Technology, Newark, New Jersey 07102

INTRODUCTION

Engineering Models

Industry relies on engineering models to estimate cost and feasibility before committing large amounts of capital (i.e., for design purposes) and to conserve as much of that capital as possible if something goes wrong in the field (i.e., for diagnostic purposes). The required accuracy of any engineering model is determined by the amount of money at risk. Furthermore, optimization of a process requires a detailed model. Otherwise, a potentially promising technology may be operated far from its optimum and thereby appear to be unattractive from an economic point of view.

In general, processes cannot be optimized solely on an experimental basis, particularly processes as complicated as those involved in biodegradation of hazardous wastes. There are too many variables to develop a traditional factorial experimental design. Field trials are expensive, and the permitting process is long and involved. Furthermore, existing predictive techniques that would otherwise be used to give at least an overall scope to the experimental plan are often unreliable. There is substantial need for better engineering tools (i.e., mathematical models) in the field of biodegradation in order to successfully deal with problems in an economical and technologically sound manner.

Biological Treatment of Hazardous Wastes, Edited by Gordon A. Lewandowski and Louis J. DeFilippi
ISBN 0-471-04861-5   ©1998 John Wiley & Sons, Inc.

Biodegradation as a Catalytic Process

In any complex process, the engineers developing a model must first have a physical picture of that process and identify the most important variables. The mathematics then follow from this physical picture.

Many organic compounds in dilute aqueous solution have thermodynamically favorable oxidation reactions at room temperature, but in sterile media the reaction rates are negligible. However, when microorganisms are present, oxidation rates accelerate considerably due to the catalytic action of enzymes, which are large protein molecules produced by the microorganisms.

Another way to accelerate such reactions is by elevating the temperature, for example, during incineration, but this is much less cost effective for aqueous waste treatment because of the large volumes of water that must also be heated up. A fundamental advantage of a catalytic process, such as biodegradation, is the much lower energy requirement needed to achieve the same objective.

Chemically, the catalytic oxidation of a dilute organic compound using microorganisms is much more complex than direct oxidation at an elevated temperature. Biodegradation of a single compound can involve dozens of individual reaction steps, each of which is mediated by a different enzyme. In order to simplify the mathematical treatment of such a large number of chemical reactions, the microorganisms (biomass) are used as an enzyme surrogate for the purpose of engineering analysis. That is, instead of specifying the concentrations of dozens of enzymes, along with their reaction pathways, we deal only with the concentration and growth of the biomass.

Use of Mixed Microbial Populations

In addition to low energy cost, a principal advantage of catalytic processes in general is their greater selectivity. That is, a particular product can be made from a specific raw material with fewer by-products.

However, selectivity is in general a disadvantage of catalysis as applied to waste treatment. Most hazardous wastes are mixtures of compounds, and the desired end product is usually CO_2 and (for chlorinated compounds) inorganic chlorides (i.e., "mineralization").

Mixed wastes require mixed microbial populations that can respond to varying concentrations of different organic compounds. In addition, the demands of mineralization may require a mixed population even for treatment of a single compound. Thus, biological treatment of waste generally involves the initial selection and growth of a mixture of microorganisms. As will be shown below, this complication leads to considerable difficulties in developing appropriate engineering tools for scale-up of biotreatment processes.

It should be noted that although mixed microbial populations often contain hundreds of different species, many of those species may not participate in a given degradation process. For example, in a study involving phenol and the mixed liquor from a municipal treatment plant (Varuntanya, 1986), 11 species dominated agar plate cultures after acclimation to 100 ppm phenol. However, when isolated in pure cultures, only three of those species could degrade phenol directly. Furthermore, the overall rate of phenol degradation by the original mixed culture was approximated by the rate of the three combined primary degraders. Thus, it may be possible to break down the usual "black-box" approach to mathematical analysis of mixed populations by substituting a more detailed analysis of the two or three primary degraders in the culture. This approach will be described in more detail below.

Specific Growth Rate

There are many models describing the kinetics of biodegradation. However, this chapter will focus on two models in particular, which are the most commonly used: the Monod model and the Andrews model. These models are applicable for a pure culture or a stable functional population, under growth conditions, with only one substrate (e.g., the pollutant) limiting the rate of reaction.

The Monod (1942) model assumes that the pollutant does not inhibit growth of the microorganisms at any concentration (see Fig. 6.1a):

$$\mu = \frac{\mu_m\, s}{K + s}$$

where
μ = specific growth rate of the microorganisms, typically expressed in units of inverse hours, which is the rate of growth of the biomass divided by the biomass concentration

s = substrate (in this case, pollutant) concentration, typically expressed in units of mg/L

μ_m, K = kinetic rate constants, which must be determined experimentally

The Andrews (1968) model considers those cases (which are most common with hazardous wastes) for which an increasing concentration of pollutant eventually inhibits the specific growth rate of the microorganisms (see Fig. 6.1b):

$$\mu = \frac{\mu^*\, s}{K + s + s^2/K_i}$$

where μ^*, K, and K_i are kinetic rate constants, which again must be determined experimentally. Note that when K_i is much larger than s^2, the Andrews model reduces to the Monod model. These two models have very different consequences for the analysis of biodegradation processes, as will be described further.

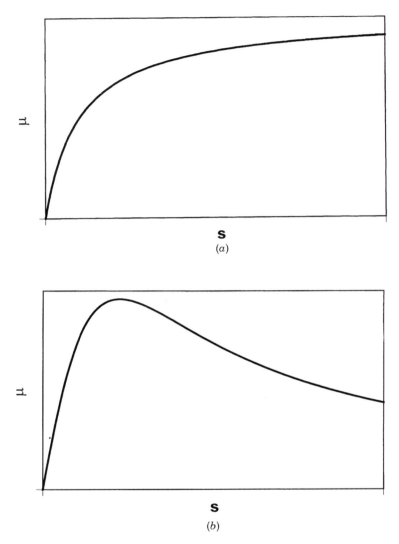

Figure 6.1 Schematic representations of (a) Monod model (noninhibitory kinetics): $\mu = \mu_m s/K + s$ and (b) Andrews model (inhibitory kinetics): $\mu = \mu s/(K + s + s^2/K_i)$

Problems with Current Methodologies for Kinetic Modeling of Biodegradation

There are two types of problems associated with current methodologies for kinetic modeling of biodegradation. The first is caused by an oversimplification of the kinetic expression for a given microbial population. The second is caused

Table 6.1   Variation in Andrews Kinetic Parameters for Phenol Degraders (as reported in the literature)

	μ^* (h^{-1})	K (mg/L)	K_i (mg/L)
Range	0.1–1.5	1–250	10–1200
Average	0.6	24	300

by the composition variation of that population with time. Each of these will be elaborated on in subsequent sections of this chapter.

Table 6.1 summarizes the range of kinetic parameters described in the literature (for a summary, see Dikshitulu, 1993) for a single pollutant (phenol) using a single kinetic model (Andrews). The parameters vary by up to three orders of magnitude.

Since the rate of a reaction directly affects the size of the reaction vessel, or the time it takes to complete the reaction, this table indicates that cost estimates for treatment can vary by at least an order of magnitude. As a result, kinetic parameters reported in the literature are generally of minimal value to an engineer seeking to scale a biotreatment process to commercial application.

The principal reason for this large variability in results is that each investigator is working with a different set of microorganisms, which produces a different set of catalysts.

Even if two investigators began with the same initial population of microorganisms, they could still observe very different rate constants if they differ in the methods by which they later use those populations. The relative composition of mixed populations can (and will) change with time.

This problem also extends to the same investigator working at different treatment scales. The usual engineering methodology involves successively testing a process at a laboratory and pilot scale prior to commercial development. But unless the initial ("seed") populations, and their use with time, are identical, results at different scales will be different, with the attendant frustration and inefficiency all too common with biological treatment processes.

What follows is a discussion of a few different systems and a suggested mathematical context for engineering analysis.

SINGLE POPULATION–SINGLE POLLUTANT

This section concerns cases where a single population is involved, and the waste contains a single pollutant exerting rate limitation on the process. The approach presented is applicable to cases employing either a pure culture or a mixture of microorganisms in which the relative species composition is constant with time.

A. Determination of Kinetics

The following equations describe consumption of the pollutant and production of biomass under batch conditions:

$$\frac{ds}{dt} = -\frac{1}{Y}\,\mu b \tag{1}$$

$$\frac{db}{dt} = \mu b - \mu_c b = \mu_{net} b \tag{2}$$

where s and b = pollutant and biomass concentrations (typically expressed in units of mg/L), respectively

Y = true yield coefficient (i.e., mg biomass produced/mg pollutant consumed)

μ = specific growth rate of biomass (as described previously)

μ_c = specific rate of biomass consumption for maintenance purposes (also known as rate of endogenous decay of biomass), which is generally treated as a constant (i.e., μ_c is independent of s)

μ_{net} = $\mu - \mu_c$

Equation (2) is written under the assumption that maintenance requirements are satisfied by self-oxidation of biomass as proposed by Herbert (1958).

Measurements of biomass and pollutant concentrations lead to determination of the apparent yield coefficient defined as

$$Y_{app} = \frac{db}{-ds} \tag{3}$$

If from various runs with different initial pollutant and biomass concentrations the value of Y_{app} is essentially constant, the implication is that maintenance can be neglected ($\mu_c = 0$) and the apparent and true yield coefficients are identical.

Using Eq. (2) and making the usual assumption that the specific growth rate is constant when organisms are growing in the exponential phase, one determines μ_{net} as the slope of the semilog plot of ln b vs. t. This slope is taken as the value of μ_{net} at the initial value of the pollutant concentration used in the particular experiment. However, it may be better to attribute μ_{net} to a pollutant concentration (s) equal to the average of s values in the linear region of the ln b vs. t plot (Dikshitulu et al., 1993). From Eqs. (1)–(3) it follows that

$$\frac{1}{Y_{app}} = \frac{1}{Y} + \frac{\mu_c}{Y}\frac{1}{\mu_{net}} \tag{4}$$

If maintenance can be neglected, $\mu = \mu_{net}$ and $Y_{app} = Y$. Otherwise, values for Y and μ_c can be obtained from Eq. (4) by linear regression of Y_{app} vs. μ_{net} data.

Once values of μ have been obtained at various pollutant concentrations (s), the data can be fitted to either a Monod or Andrews model in order to obtain the kinetic rate constants.

B. Biodegradation in a Single Continuous-Flow Reactor

Figure 6.2 shows a schematic of the case where a single well-stirred vessel is used with continuous flow of the waste stream. Two possibilities are shown. Figure 6.2a implies that there is no recycle to the reactor whereas Figure 6.2b shows a combined reactor–separator system. Separation allows for biomass (solids) settling and recycle back to the reactor.

Assuming that the recycle stream involves biomass only (e.g., the stream exiting the bottom of the settler is dewatered), and that the amount of biomass recycled to

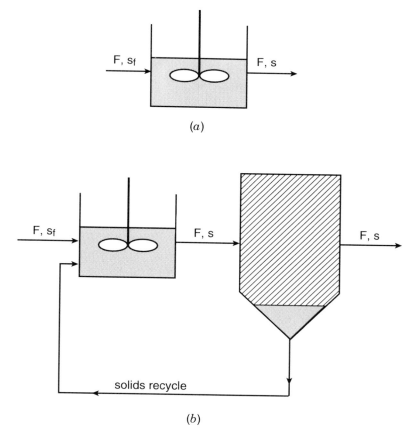

Figure 6.2 Schematic of continuous-flow biodegradation units operating (a) without and (b) with biomass recycle.

the reactor is a constant fraction, ϕ, of the biomass exiting the reactor, biodegradation can be described by the following equations:

$$\frac{db}{dt} = \frac{F}{V}(\phi - 1)b + \mu b - \mu_c b \tag{5}$$

$$\frac{ds}{dt} = \frac{F}{V}(s_f - s)b - \frac{1}{Y}\mu b \tag{6}$$

where s, b = concentrations of pollutant and biomass, respectively, in the reactor
        (and also the reactor outlet)
   F = volumetric flow rate of the waste stream
   V = volume of the reactor contents
   s_f = concentration of pollutant in the waste stream fed into the reactor

If there is no recycle of biomass (i.e., $\phi = 0$), Eqs. (5) and (6) describe the system of Figure 6.2a.

Equations (5) and (6) under steady-state conditions (i.e., when the derivatives are equal to zero) have two qualitatively different solutions. The first is a nonzero, finite value of the biomass concentration ($b \neq 0$, called "survival") in the reactor, resulting in biodegradation at steady state. The second solution is no biomass in the reactor ($b = 0$, called "washout"), resulting in no biodegradation at steady state.

In some cases, the state of survival exhibits multiplicity, that is, there are two or more survival states under the same operating conditions. There may also be survival steady-state solutions of Eqs. (5) and (6) that can not be physically achieved. This happens when a steady state is unstable. Unstable states can be transformed to stable ones (and thus, can be physically realized) if proper process control is applied.

From the design viewpoint, the operating conditions (F and s_f for the case considered here) have to be selected so as to prevent culture washout. Furthermore, if multiple survival states exist, the objective of process optimization is normally to achieve the lowest exit concentration of the pollutant.

For the system described by Eqs. (5) and (6) and when $\mu_c = 0$ and $\phi = 0$, two different cases are considered. In the first, Monod-type kinetics are examined, whereas in the second, biodegradation is assumed to follow Andrews kinetics. The details of the analysis are not presented, while its results are shown qualitatively in the diagrams of Figure 6.3. These diagrams are known as operating diagrams and show what state the system will reach under different F and s_f values, or equivalently, under different α and u_f values. Quantities α and u_f are equal to $F/(V\mu_m)$ [or $F/(V\mu^*)$] and s_f/K, respectively, and they are dimensionless versions of F and s_f.

The analysis shows that with Monod kinetics, the operating parameter plane $\alpha - u_f$ is divided in two regions as shown in Figure 6.3a. In region I culture washout

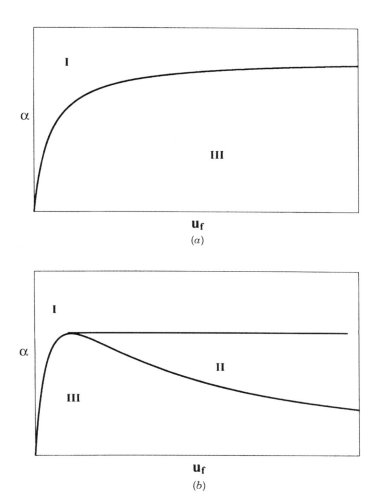

Figure 6.3 Operating diagram for the biodegradation of wastes containing a single pollutant by a single population in a continuous-flow mixed reactor without recycle of biomass. Biodegradation rates follow: (a) Monod kinetics and (b) Andrews kinetics.

occurs, whereas in region III culture survival in a single stable state occurs. The curve separating the two regions is described by the following equation:

$$F_c = \frac{V\mu_m s_f}{K + s_f} \tag{7}$$

where F_c stands for a critical value of the volumetric flow rate F.

Equation (7) is a design criterion. If the concentration of the pollutant is s_f and the volume of the treatment vessel is V, Eq. (7) dictates the maximum volumetric flow rate that can be used. Equivalently, Eq. (7) can be solved in terms of V, which is the minimum working volume of the reactor to be used when F and s_f are specified.

The picture changes drastically when biodegradation follows Andrews kinetics. As shown in Figure 6.3b, there are now three regions in the $\alpha-u_f$ plane. Regions I and III are the same as those in Figure 6.3a. In region II, two possible outcomes are physically attainable: survival of the biomass (in which case successful operation can proceed) or washout of the biomass (in which case treatment terminates). Which of the two outcomes is reached depends on how the process is started up. Furthermore, region II implies that even if culture survival is initially established, temporary operational upsets may later cause washout of the organisms. Thus, successful operation in region II requires careful consideration of process control.

The existence of region II is the fundamental difference between Monod and Andrews kinetics. The difference is not just quantitative, it is also qualitative. If biodegradation really follows Andrews kinetics, and incomplete initial kinetic studies have led to the erroneous conclusion that Monod kinetics are valid, serious problems can arise. Region II is not just a mathematical possibility. It is in fact realized in many actual treatment processes. In a subsequent section experimental data will be shown that under identical operating, but different startup, conditions the culture can reach either a washout or a survival state.

The curve separating regions I and III and II and III in Figure 6.3b is described by the following equation:

$$F_c = \frac{V\mu^* s_f}{K + s_f + s_f^2/K_i} \tag{8}$$

whereas the line separating regions I and II is described by

$$F_{c1} = \frac{V\mu^*}{1 + 2\sqrt{\dfrac{K}{K_i}}} \tag{9}$$

The existence of the washout region (I) in the operating diagrams is clearly undesirable. One would like to expand the region of culture survival as much as possible. This can be achieved through biomass recycling. Going back to Eqs. (5) and (6), and using nonzero values for ϕ, it can be shown (Saghafi, 1988) that the diagrams of Figure 6.3 do not change qualitatively. However, the boundaries between regions are shifted to higher α (or F) values. An example for the case of Monod kinetics is shown in Figure 6.4. Survival of the biomass occurs in the regions under the curves, as was also the case of Figure 6.3a. Recycle of 25% of the biomass (curve 2) increases the value of F_c dictated by Eq. (7) by 33%, while 50% recycle of the biomass (curve 3) increases F_c by 100%. Theoretically, if biomass is totally (100%) recycled, washout never occurs under Monod

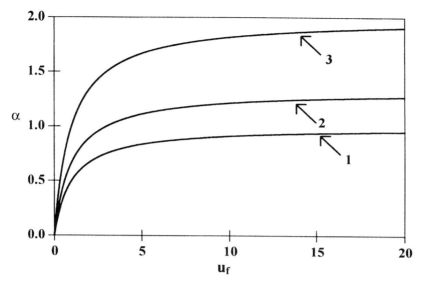

Figure 6.4 Operating diagram for the biodegradation of a single pollutant by a single biomass population in a continuous-flow mixed reactor under Monod kinetics. Operation under no biomass recycle (curve 1), 25 and 50% recycle of biomass (curves 2 and 3, respectively). The process requires selection of conditions defining points under the curves.

kinetics. Total recycle is not feasible, but, even if it were, one should keep in mind that very high biomass concentrations in the reactor lead to mass-transfer problems for both the pollutant and oxygen (if the process is aerobic). Mass-transfer limitations may result in lower overall reaction rates, and thus larger operating units.

For the case of Andrews kinetics, although region I can be essentially eliminated through biomass recycle, region II still remains at reasonable flow rates. However, the larger the biomass concentration, the higher are the fluctuations that can be tolerated in region II without leading to culture washout (Saghafi, 1988).

If maintenance requirements are important ($\mu_c \neq 0$), culture washout occurs at flow rates lower than those dictated by Eqs. (7)–(9).

C. Biodegradation in a Cascade of Two Vessels

A schematic of the process when two bioreactors in series are used is shown in Figure 6.5. For simplicity, it is assumed that there is no recycle from the second to the first vessel and no recycle of biomass to the reactor system. However, one can show that when these assumptions are relaxed, the qualitative behavior of the system remains nearly the same (Saghafi, 1988).

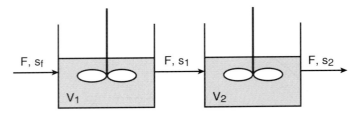

Figure 6.5 Schematic of continuous-flow biodegradation units involving well-mixed vessels in series.

If maintenance is again neglected, the following equations describe the system:

$$\frac{ds_1}{dt} = \frac{F}{V_1}(s_f - s_1) - \frac{1}{Y}\mu_1 b_1 \tag{10}$$

$$\frac{db_1}{dt} = -\frac{F}{V_1}b_1 + \mu_1 b_1 \tag{11}$$

$$\frac{ds_2}{dt} = \frac{F}{V_2}(s_1 - s_2) - \frac{1}{Y}\mu_2 b_2 \tag{12}$$

$$\frac{db_2}{dt} = \frac{F}{V_2}(b_1 - b_2) + \mu_2 b_2 \tag{13}$$

where all symbols are as defined earlier, and subscripts 1 and 2 denote values of the variables (V, volume; s, pollutant concentration; b, biomass) in the first and second reactor, respectively. The specific growth rates in the first and second reactor (μ_1, μ_2) are again given by either the Monod or Andrews expressions.

The analysis shows that this system has the following three, qualitatively different, steady states (SS):

SS1: $s_1 = s_2 = s_f$ and $b_1 = b_2 = 0$

SS2: $s_1 = 0$, $b_1 = 0$; $s_2 < s_f$ $b_2 > 0$

SS3: $s_f > s_1 > s_2$; $b_2 > b_1 > 0$

At SS1, the biomass washes out from the entire system and biodegradation ceases. At SS2, the biomass washes out from the first reactor but establishes itself in the second; SS2 is called the partial washout steady state. At SS3, the biomass establishes itself in the entire reactor configuration. SS1 is clearly unwanted. SS2 is also undesirable because it means that the total system volume is underutilized. If there is a recycle stream from the second to the first reactor, or recycle of biomass to the first reactor after settling of the effluent of the second reactor, SS2 is impossible. This is the only qualitative difference between the system shown in Figure 6.5 and systems of two reactors involving recycle.

An important design parameter for systems of two bioreactors in series is the ratio, ε, of the working volumes of the two vessels defined as $\varepsilon = V_1/V_2$. For the

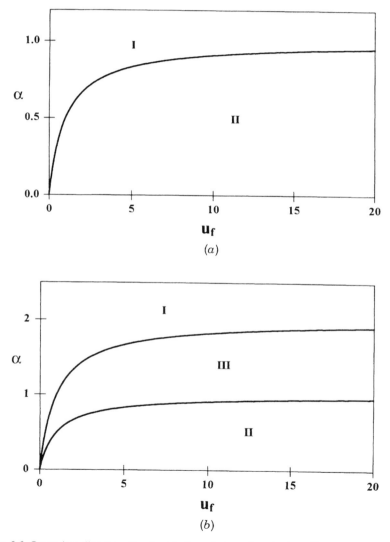

Figure 6.6 Operating diagrams for the biodegradation of a single pollutant using a single population in two well-mixed reactors in series (Fig. 6.5). Diagrams (a) and (b) are for the case where the larger and smaller vessel, respectively, is placed first in the cascade.

system shown in Figure 6.5 the analysis shows that SS2 never arises if $\varepsilon \geq 1$. This implies that if two reactors in series (without recycle) are to be used, the vessels should be either equal in size or the larger vessel should be the one receiving the untreated waste stream.

If biodegradation follows Monod kinetics, the operating diagrams in the $\alpha-u_f$ plane (α, u_f as defined in Section B) are those shown in Figure 6.6a for $\varepsilon \geq 1$ and Figure 6.6b for $\varepsilon < 1$. The boundary between regions I and II in Figure 6.6a, and regions II and III in Figure 6.6b is given by

$$F_{c2} = \frac{V_1 \mu_m s_f}{K + s_f} \tag{14}$$

whereas the boundary between regions I and III in Figure 6.6b is given by

$$F_{c3} = \frac{V_2 \mu_m s_f}{K + s_f} \tag{15}$$

In regions I of Figure 6.6, the system reaches the state of total biomass washout (SS1). In regions II, the culture establishes itself throughout the entire system (SS3) in a survival state that does not exhibit multiplicity (Saghafi, 1988). In region III of Figure 6.6b, the partial washout state (SS2) is reached and again no phenomena of multiplicity are observed. From the foregoing discussion it becomes clear that for the system shown in Figure 6.5, Eq. (14) dictates the maximum volumetric flow rate of the waste stream (F_{c2}), which can be treated without underutilizing the facility. When the feed rate to the first vessel is specified, Eq. (14) can be used to determine the minimum working volume (V_1) for the reactor that is placed first in the cascade. The volume of the second reactor must be less than that of the first reactor.

Under Andrews kinetics, if $\varepsilon \geq 1$, the operating diagram is the same as that shown in Figure 6.3b; whereas if $\varepsilon < 1$, the diagram becomes much more complicated involving six different regions, in some of which various multiplicities occur (Saghafi, 1988).

D. Comparisons Between the One- and Two-Vessel Systems

One should be able to decide whether a single reactor or a cascade of two vessels should be used. The answer to the foregoing question is not unique. It depends on what objective is specified.

One objective could be to avoid washout of the culture as much as possible. In this case comparisons should be made based on the same volumetric capacity of the facility. Thus, the volume (V) of the single reactor must be equal to $V_1 + V_2$ that is, the total working capacity of the two-vessel system. The results under Monod kinetics and without any recycle are shown in Figure 6.7. For the single-vessel system biomass survival occurs in the region under curve 2, whereas for the two-vessel system survival throughout the system (i.e., in both vessels)

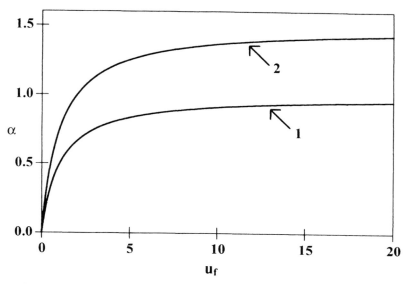

Figure 6.7 Operating diagram for the biodegradation of a single pollutant by a single population under Monod kinetics. Comparison of the configurations shown in Figures 6.2*a* and 6.5 when the total working volume is the same. Curves 2 and 1 are for the case of one and two vessels, respectively.

occurs in the area below curve 1. Curve 1 is defined by Eq. (14) whereas curve 2 is defined by

$$F_{c4} = \frac{(\varepsilon + 1) \, V_1 \mu_m s_f}{\varepsilon(K + s_f)} \tag{16}$$

Comparing expressions (14) and (16), it becomes clear that the single-vessel system allows operation at higher volumetric flow rates without the danger of washing out the biomass. The same conclusion is reached when biomass recycle is used.

An objective that is much more meaningful for waste treatment is to achieve a desired conversion of the pollutant (e.g., 99% destruction, or whatever environmental regulations require) with the minimum reactor volume when F and s_f are specified. To meet this objective, the analysis (Saghafi, 1988) shows that, when no recycle is involved and regardless of Monod or Andrews kinetics, it is better to use two vessels in series with the larger vessel placed first.

As another variant, if biomass recycle is used, the optimal configuration may be one in which two unequal vessels are employed but the smaller vessel is placed first in the cascade (Saghafi, 1988).

It is very important to note that if the kinetics are (erroneously) assumed to be zero-order, the single- or two-vessel configurations will be predicted to be identical in terms of performance. If first-order kinetics are assumed, the two-vessel system turns out to be advantageous, but an error is made again as the prediction is that two equal size vessels should be used. The conclusion is that a good knowledge of the kinetics leads to a correct optimal design.

E. Use of a Cyclically Operated Reactor

Biodegradation of wastes in a single vessel operating with *intermittent* supply of wastewater, and incorporating reaction and settling of the biomass, has been extensively studied (Dennis and Irvine; 1979; Hsu, 1986; Irvine and Busch, 1979; Irvine et al., 1983; Irvine and Richter, 1978; Jones et al., 1990; Palis and Irvine, 1985; Silverstein and Schroeder, 1983). This mode of operation (known as sequencing batch reactor, or SBR) has the advantage of higher reactor productivity when compared to continuous-flow systems, and overcomes the problem of a separate clarifier (settler) for the biomass.

A schematic of SBR operation (in which the flow rates during fill and draw are constant) is shown in Figure 6.8. There are four phases in each cycle: fill, reaction, settle, and draw. For kinetic analysis, the following discussion will neglect the settling phase, which is a physical (rather than kinetic) process.

When a single population is used to biodegrade a single pollutant in a cyclically operated vessel, the process is described by the following mass balances:

$$\frac{dV}{dt} = F - F_e \tag{17}$$

$$\frac{db}{dt} = -\frac{Fb}{V} + \mu b - \mu_c b \tag{18}$$

$$\frac{ds}{dt} = \frac{F}{V}(s_f - s) - \frac{1}{Y}\mu b \tag{19}$$

Comparing Eqs. (18) and (19) with Eqs. (5) and (6) one can see that they are identical, if $\phi = 0$ in Eq. (5). However, there is one fundamental difference. Namely, F and V have constant values in Eqs. (5) and (6), but varying values in Eqs. (18) and (19). Volume variability is described by Eq. (17), which involves two volumetric flow rates; F is the volumetric flow rate of waste supply during the first phase of the cycle (i.e., $0 < t \le t_1$ in Fig. 6.8) and F_e is the volumetric flow rate of the effluent during the third phase of the cycle (i.e., $t_2 \le t \le t_3$ in Fig. 6.8). Except for F and F_e, all symbols in Eqs. (17)–(19) stand for the same quantities defined in Eqs. (5) and (6). Changes in the values of F and F_e, which also result in volume (V) changes, lead to complications for the analysis. Nonetheless, these changes also lead to extremely important results in terms of process dynamics and opportunities for optimization.

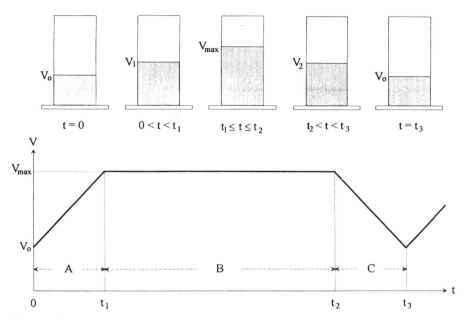

Figure 6.8 Schematic of the volume variation in a cyclically operated reactor. Each cycle comprises three phases: (A) filling, (B) batch, (C) and draw-down. From Wang et al. (1995a).

Note that the minimum value of V, that is, V_0, cannot be zero because the biomass would be lost after the first cycle. The maximum value of V, that is, V_{max}, can be viewed as the equivalent of the constant liquid holdup in a continuous-flow reactor [i.e., V in Eqs. (5) and (6)].

Cyclic operation introduces two new parameters that can be selected for the design of the process. These are the ratio, w, of minimum to maximum volume ($w = V_0/V_{max}$), and the fraction, f_1, of the cycle time allocated to the first phase, that is, to filling the reactor with waste ($f_1 = t_1/t_3$).

When a reactor is operated in a cyclic mode, it cannot reach a true steady state. Instead, after the decay of transients, the process reaches a self-repeating time profile for the concentrations of biomass and reactants (pollutants). These profiles are called limit cycles.

Using phenol as model compound and a pure culture of *Pseudomonas putida* (ATCC 31800), Lenas et al. (1994) investigated the dynamics of biodegradation in a cyclically operated reactor. The kinetics of biodegradation were found to follow the Andrews expression. In this case, the operating parameter space is four dimensional (D–u_f–w–f_1) as opposed to two dimensional (a–u_f), which was the case in Section B. Parameter u_f is the same as before (i.e., $u_f = s_f/K$), whereas D is defined as $D = F/(\mu^* V_{max})$ and is the analog of α used in Sections B and C.

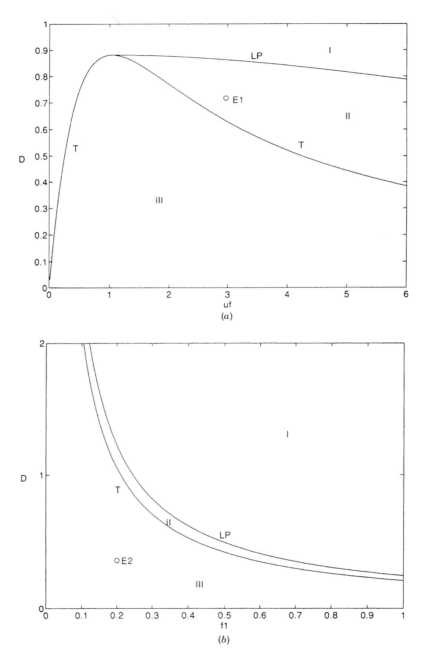

Figure 6.9 Operating diagrams on the (*a*) D–u_f and (*b*) D–f_1 planes for the biodegradation of a single pollutant by a single population in a reactor operating cyclically as shown in Figure 6.8. From Lenas et al. (1994) where other parameters are also specified. Degradation follows Andrews kinetics.

Operating diagrams on the D–u_f and D–f_1 planes are shown in Figure 6.9. It is worth observing that Figure 6.9a is the exact analog of Figure 6.3b, and the three regions shown have the same characteristics. As discussed in Section B, if the reactor is operated in region I, the biomass washes out; in region III there is a single state of culture survival; in region II the biomass will either survive or wash out depending on how the process is started-up. It should be mentioned here that the curves separating region III from regions I and II in the diagrams of Figure 6.9 (denoted with T) are not described by Eq. (8), which was the case for Figure 6.3b. Similarly, the curves (denoted with LP) separating regions I and II in Figure 6.9 are not described by Eq. (9). Curves LP and T of Figure 6.9 can only be constructed numerically.

As mentioned in Section B, the existence of region II is not just a mathematical artifact. In fact, experiments performed under identical operating, but different start-up, conditions (see Lenas et al., 1994) have led to either washout or survival as shown in Figures 6.10 and 6.11, respectively. The operating conditions for these experiments define a single point in the operating parameter space, shown as point $E1$ in Figure 6.9a. On the other hand, experiments performed with identical operating and different start-up conditions in region III of the operating diagram

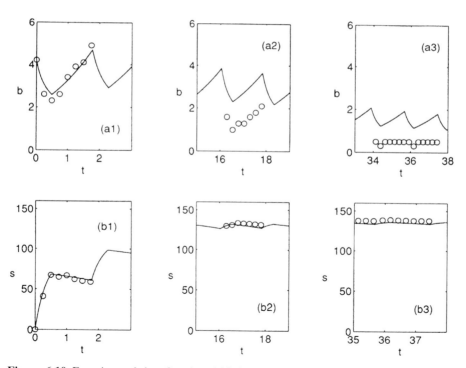

Figure 6.10 Experimental data for phenol biodegradation in a cyclically operated reactor. Operation at point $E1$ of Figure 6.9a. Start-up conditions lead to biomass washout. From Lenas et al. (1994).

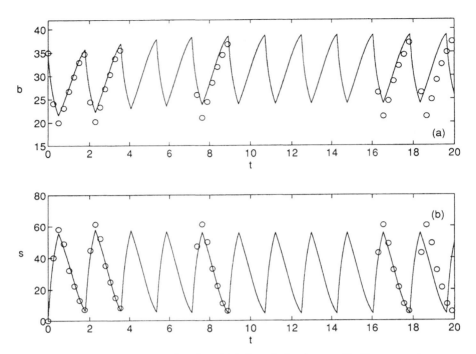

Figure 6.11 Experimental data for phenol biodegradation in a cyclically operated reactor. Operation at point $E1$ of Figure 6.9a. as is also the case for Figure 6.10. Start-up conditions lead to survival of biomass. From Lenas et al. (1994).

(e.g., point $E2$ in Figure 6.9b) always led to the same state of culture survival (Lenas et al., 1994). As discussed in Section B, region II has to be either avoided or used with caution.

For the case of cyclic operation, if one wants to maximize the working capacity, P, of the system and also achieve a required conversion level for the pollutant, Lenas et al. (1994) have shown that there are opportunities in terms of selection of w and f_1. As shown in the diagrams of Figure 6.12, for example, a fast filling phase (low f_1) should be used. However, the optimal value of w is small if u_f is small (Figure 6.12a), or large when u_f is large (Figure 6.12b). Since P is defined as $P = Df_1$, its optimum value dictates the value of F during filling, or the value of V_{max} if F is prespecified [recall that $D = F/(\mu^*V_{max})$]. Results of various computational studies have shown that while the optimal operating parameter values often fall within the safe region III of the operating diagrams, they can also fall in region II of multistability (Lenas et al., 1994).

Again, realistic opportunities for process optimization and cost reduction are lost if oversimplified zero-order or first-order kinetic expressions are used rather than the Monod or Andrews expressions.

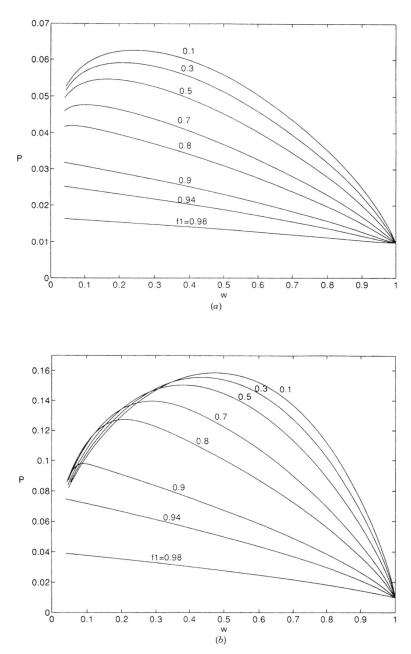

Figure 6.12 Optimal selection of w for various f_1 values for maximizing the volumetric efficiency of a cyclically operated reactor when the pollutant concentration in the untreated waste is (a) low or (b) high. The removal of the pollutant is 98 and 99.7% in (a) and (b), respectively. From Lenas et al. (1994).

COMPETING POPULATIONS–SINGLE POLLUTANT

In the previous sections, the biomass was assumed to consist of either a pure culture or a mixed culture of constant species composition. In biological treatment of wastes, mixed cultures are almost always used, in which interactions among the various species present in the biomass result in variations in the relative species composition. Consequently, the assumption generally made in design of waste treatment facilities of constant species composition can lead to serious errors.

The most common interaction, considered in this section, is microbial competition. There is a vast literature on the dynamics of microbial competitors, which has been discussed and reviewed by various authors (Fredrickson and Stephanopoulos, 1981; Dikshitulu et al., 1993; Baltzis and Wu, 1994). The perspective of the discussion presented here concerns the impact of microbial competition on the design of hazardous waste treatment facilities.

A. Kinetic Studies in Batch Reactors

Consider a mixture of two species, each of which can degrade a given pollutant. If competition for the pollutant is their only mode of interaction, this type of behavior is called "pure and simple" competition. The overall rate of biodegradation in such a case is given by the sum of the individual rates for each species:

$$-r_s = \frac{1}{Y_1} \mu_1 b_1 + \frac{1}{Y_2} \mu_2 b_2 \qquad (20)$$

In a batch reactor,

$$\frac{ds}{dt} = -(-r_s) \qquad (21)$$

$$\frac{db_j}{dt} = \mu_j b_j, \qquad j = 1,2 \qquad (22)$$

where b_1 and b_2 are the concentrations of species 1 and 2, respectively, and Y_j, μ_j ($j = 1, 2$) their corresponding yield coefficient and specific growth rate.

If instead, the mixture of the two species is considered as a single functional population of concentration b (as in the previous sections), the biodegradation rate is expressed as

$$-r_s = \frac{1}{Y} \mu b \qquad (23)$$

Once again, in a batch reactor,

$$\frac{ds}{dt} = -(-r_s) \tag{24}$$

$$\frac{db}{dt} = \mu b \tag{25}$$

where now Y and μ are the apparent yield coefficient and specific growth rate, respectively, of the mixture.

An example of the different consequences of these two approaches was examined by Lewandowski et al. (1989), as described below.

The case of phenol biodegradation by two microbial species was considered. The first (called culture 1) was *Serratia liquefaciens* isolated from a waste treatment facility, whereas the second (called culture 2) was *P. putida* (ATCC 31800). Culture 1 was found to follow Monod kinetics, whereas culture 2 followed Andrews kinetics. The values of the kinetic constants and yield coefficients for each pure culture are shown in Table 6.2. Also given in Table 6.2 are the apparent kinetic constants and yield coefficients for two theoretical mixtures: culture 3 (a 1:2 mixture of pure cultures 1 and 2) and culture 4 (a 2:1 mixture of pure cultures 1 and 2). Culture 3 was determined to follow Andrews kinetics (which is expected since the dominant species in the inoculum follows the same kinetic expression); while, culture 4 was determined to follow Monod kinetics. Therefore, the apparent kinetic expressions for simple mixtures of two species can differ substantially (not only quantitatively, but qualitatively as well) by only a twofold variation in composition. Since individual species compositions in a mixed population can easily vary over time by an order of magnitude, the simple example given here emphasizes the problems inherent in using Eq. (23) to describe the rate of biodegradation by a mixed microbial population.

An important question is what is the error involved when Eq. (23) is used rather than Eq. (20). Three cases are shown in Table 6.3, for which the basis of comparison is the time needed to biodegrade phenol from 180 to 2 mg/L (ppm) in a batch reactor. Cases 1, 2, and 3 are for mixtures of cultures 1 and 2 having initial relative compositions 1:2, 2:1, and 1:4, respectively. The table shows the exact time based

Table 6.2   Real and Apparent Kinetic Constants for Cultures 1–4

	μ^* (h^{-1})	K (mg/L)	K_i (mg/L)	Y
Culture 1	0.41	43.6	∞	0.52
Culture 2	1.40	47.1	51.0	0.29
Culture 3	1.74	93.2	36.8	0.32
Culture 4	0.36	10.0	∞	0.37

From Lewandowski et al. (1989).

Table 6.3   Exact and Approximate Times for Phenol Biodegradation from 180 to 2 mg/L

Relative Composition of S. liquefaciens and P. putida	1:2	2:1	1:4
Exact time	3.3 h	4.0 h	3.0 h
Approximation 1	3.4 h	3.4 h	3.4 h
Approximation 2	3.9 h	3.9 h	3.9 h
Error of approximation 1	4.2%	15.0%	13.3%
Error of approximation 2	18.2%	2.5%	30.0%

From Lewandowski et al. (1989).

on Eqs. (20)–(22), and two approximations based on Eqs. (23)–(25). Approximations 1 and 2 are based on the apparent kinetics of cultures 3 and 4, respectively. The error involved ranges from 2.5 to 30%, which means that the reactor may be oversized or undersized by an equivalent amount.

Thus, laboratory data obtained with one population mixture may lead to significant errors in sizing a large-scale reactor in which the species composition is only slightly different.

B. Biodegradation in Continuous-Flow Reactors

Biodegradation of a pollutant by two competing species in a single continuous-flow reactor with (i.e., $\phi > 0$) or without (i.e., $\phi = 0$) recycle of biomass is described by the following equations:

$$\frac{ds}{dt} = \frac{F}{V}(s_f - s) - \sum_{j=1}^{2} \frac{b_j}{Y_j} \mu_j \tag{26}$$

$$\frac{db_j}{dt} = \frac{F}{V}(\phi - 1)b_j + \mu_j b_j - \mu_{cj}b_j, \qquad j = 1, 2 \tag{27}$$

where all symbols in these equations have been defined previously.

Equations (26) and (27) predict four qualitatively different steady states as follows:

SS1: $b_1 = b_2 = 0,$ $s = s_f$ (total washout)

SS2: $b_1 > 0,$ $b_2 = 0,$ $0 < s < s_f$ (partial washout)

SS3: $b_1 = 0,$ $b_2 > 0,$ $0 < s < s_f$ (partial washout)

SS4: $b_1 > 0,$ $b_2 > 0,$ $0 < s < s_f$ (coexistence)

It is well known (Powell, 1958; Fredrickson and Stephanopoulos, 1981) that SS4 can never really arise. Thus, after starting a process in a continuous-flow reactor with biomass consisting of two competing species, one will eventually end up with a culture consisting of either species 1 or 2, provided that the conditions are selected so that total biomass washout is avoided. Whether SS2 or SS3 is reached for a given value of s_f depends on how the volumetric flow rate, F, is selected.

The inability to maintain a mixed culture of two (in fact, any number of) microbial competitors in a single perfectly mixed continuous-flow reactor has very important consequences. First, if the kinetics have been determined based on batch experiments using the approximate approach (i.e., treating the mixed culture as a single invariant culture), they are definitely not valid for the continuous-flow reactor. In fact, the discrepancies in size calculations will be at least of the order of the errors made in time calculations for case 3 in Table 6.3. The second consequence may be much more important, as discussed in the following.

Consider a case where the waste contains two pollutants, say S and S_1. Assume that the initial culture consists of two species, 1 and 2, both capable of biodegrading and growing on pollutant S for which they thus compete. Assume that pollutant S_1 cannot serve as carbon or energy source for either of the two species, but it can be cometabolized by only one species, say species 1. It should be mentioned here that destruction of pollutants, xenobiotics in particular, through the cometabolic activities of organisms is anything but rare. For example, trichloroethylene (TCE) cannot be degraded alone by any known single or mixed culture. However, there are cultures that can cometabolically degrade and mineralize TCE in the presence of another substrate (e.g., phenol).

If the waste containing pollutants S and S_1 is fed to a continuous-flow bioreactor, the process is still described by Eqs. (26) and (27) as S_1 does not affect the reaction rates. Hence, if total washout is avoided either SS2 or SS3 will be reached. If F is selected so that SS2 is reached, then both pollutants are biodegraded. However, if F is such that SS3 is reached, pollutant S is removed, but pollutant S_1 is not treated at all, and thus the design of the process fails. The important thing here is that if the biomass is treated (as usual) as a single culture, there is no distinction between SS2 and SS3. It is thus impossible to select F in ways ensuring that both pollutants will be degraded. Even if empirically the original selection of the operating condition (F) turns out to be correct, the process may still fail if there is a subsequent operational upset or intentional change in F.

Cometabolism of xenobiotic substances often requires the combined action of various species. Each population performs different transformations, and the overall effect is the mineralization of the original compound. Assume then that in the previous example, both species 1 and 2, which compete for pollutant S, are required for the complete mineralization of S_1. If this is the case, then the process can be successful with a batch reactor but fails when a single continuous-flow reactor is used, since coexistence of the two required species is impossible.

As discussed by various authors (Fredrickson and Stephanopoulos, 1981; Dikshitulu et al., 1993; Baltzis and Wu, 1994), mixed cultures of species competing for a single substrate can be maintained in continuous-flow configurations provided that the reaction environment is not both spatially and temporally homogeneous. Temporal heterogeneities are discussed in the next section where a cyclically operated reactor is considered. Spatial heterogeneities can be created through attachment of microbial cells on surfaces present in a single reactor (Baltzis and Fredrickson, 1983), through the use of cascades of two or more vessels interconnected via recycle streams (Stephanopoulos and Fredrickson, 1979; Kung and Baltzis, 1987; Chang and Baltzis, 1989; Baltzis and Wu, 1994), and through the use of nonmixed environments (Lauffenburger and Calcagno, 1983; Kung and Baltzis, 1992).

C. Biodegradation in a Cyclic Reactor

A cyclically operated reactor leads to a reaction environment that is temporally heterogeneous, meaning that concentrations vary with time. Heterogeneous environments can lead to coexistence of competing species (Stephanopoulos et al., 1979; Davison and Stephanopoulos, 1986; Stephens and Lyberatos, 1987; Pavlou et al., 1990).

The cyclic pattern shown in Figure 6.8 has been experimentally and mathematically studied for the case of two competing species by Dikshitulu et al. (1993). The basic equations describing this system are Eqs. (17), (26), and (27) with $\phi = 0$. As discussed previously, this system does not have equilibrium points (steady states) but rather steady (limit) cycles. There are four qualitatively different limit cycle solutions that this system has, and these are analogs of steady states 1–4 discussed in Section B. Again, total washout, exclusion of either one of the two competitors (partial washout), and coexistence of the two species can arise depending on the operating conditions.

The model system of Dikshitulu et al. (1993) involved two *Pseudomomas* species, both biodegrading phenol following Andrews kinetics. The analysis of the dynamics of the system led to the construction of the operating diagram shown in Figure 6.13. This is a diagram on the $\beta - u_f$ plane. In this case β is defined as $\beta = \mu_1^* V_{max}/(Ff_1)$ and is essentially the inverse of what was previously defined as the dilution rate (D). This diagram is extremely interesting and complex as it involves 15 different regions. In each region, different states are stable or unstable and have different characteristics of instability. Details can be found in the original study (Dikshitulu et al., 1993). Of importance here is what states are stable (thus, physically realizable without application of control) in each of the regions of Figure 6.13. These states are indicated by a plus sign in Table 6.4.

As can be seen from Table 6.4 there are two regions (13 and 14) that guarantee a mixed culture will be maintained. Operation in region 15 can lead to a mixed culture, provided that the process is started up properly since total washout is also possible. Region 7 is also extremely interesting since there are three possible outcomes, and once again process start-up conditions are of great importance and consequence. Experimental verification of the existence of some of the regions of Figure 6.13 can be found in Dikshitulu et al. (1993).

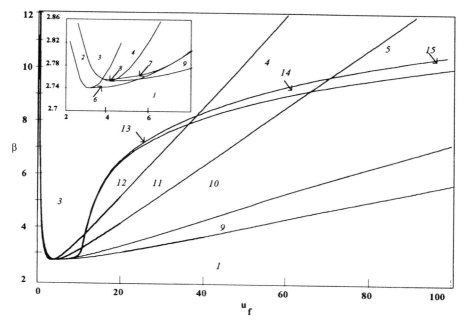

Figure 6.13 Operating diagram on the $\beta-u_f$ plane for the biodegradation of a single pollutant by two competing populations in a cyclically operated bioreactor. Kinetics follow the Andrews expression. The stability of states arising in the various regions is shown in Table 6.4. From Dikshitulu et al. (1993).

Table 6.4   Stability (denoted by +) of Periodic States in the Regions of Figure 6.13

Region	Total Washout	Species 1 only	Species 2 only	Coexistence
1	+			
2			+	
3			+	
4			+	
5	+		+	
6	+		+	
7	+	+	+	
8		+	+	
9	+	+		
10	+	+		
11		+		
12		+		
13				+
14				+
15	+			+

It is important here to once again mention the potential errors inherent in the usual approach that treats a mixed culture as a single one. If this was done in the case of Dikshitulu et al. (1993), then the operating diagram of Figure 6.13 would have been predicted to have three regions only and would have been qualitatively similar to the one shown in Figure 6.9a. Guaranteed survival of biomass would be predicted in a single region (analog of region III in Figure 6.9a), which in actuality would be the combination of regions 2, 3, 4, 8, 11, 12, 13, and 14 of Figure 6.13. However, the biomass composition would be very different within that region, and if a certain mixture of species were required to achieve the treatment objectives, it would be almost a matter of chance to achieve that mixture given the narrow span of regions 13 and 14.

Figure 6.13 also reveals an outcome of the modeling that is counterintuitive. Let us assume that regions 13 and 14 are discovered in trial-and-error laboratory experiments to be desirable operating regimes. However, following standard engineering practice, a "safety factor" is added to the full-scale design by increasing the reactor volume and hence, β. But that puts the full-scale operation into regions 3 or 4, where only one of the two required species survives. As a result, a process that operated successfully at the laboratory scale fails at a commercial scale for reasons that are a mystery to those involved (and therefore attributed to acts of nature).

PRESENCE OF REACTION INTERMEDIATES

All chemical and biochemical processes involve multiple steps. In many cases the reaction intermediates react very fast, and thus they do not affect the overall reaction rates, which can be expressed as functions of the concentrations of the original compound(s). When the reaction intermediates do not react very fast in the reaction sequence, their concentration may affect significantly the kinetics of the overall process. In such cases, the design of the process has to consider the effects of the long-lived intermediates, which is the topic addressed in this section. Two examples are considered, one involving a single (pure) culture and one involving a mixed culture.

A. Use of a Single Culture

Once again, the term *single culture* is meant to imply that we deal with either a pure culture or a mixed culture that has a constant composition. The example concerns denitrification of wastes containing nitrate.

Under anaerobic or anoxic conditions, microorganisms can reduce nitrate to nitrogen through a series of steps that finally lead to formation of nitrogen. One of the intermediates is nitrite, which, as reviewed by Wang et al. (1995b), can be difficult to reduce. These investigators have studied in detail the complex kinetics of nitrate and nitrite reduction with *Pseudomonas denitrificans* (ATCC 13867). Nitrate and nitrite were found to be mutually inhibitory in biodegradation systems, which requires modification of the basic Monod and Andrews kinetic expressions. It should be mentioned here that kinetic interactions can also arise in cases of mixed pollutants without one being an intermediate of the other, for example, benzene and toluene (Chang et al., 1993; Oh et al., 1994).

If s and u denote nitrate and nitrite concentration, respectively, the rate of nitrate reduction according to Wang et al. (1995b) is given by

$$-r_1 = \frac{b}{Y_1} \frac{\mu_1 * s}{K_1 + s + s^2/K_{i1} + K_{21} su} \tag{28}$$

whereas the net rate of nitrite reduction in given by

$$-r_2 = -\left[\frac{\psi}{Y_1} \frac{\mu_1 * s}{K_1 + s + s^2/K_{i1} + K_{21} su} - \frac{1}{Y_2} \frac{\mu_2 * u}{K_2 + u + u^2/K_{i2} + K_{12} us} \right] b \tag{29}$$

where ψ is the stoichiometric ratio of milligrams of nitrite produced per milligram nitrate reduced; and K_{21}, K_{12} are cross-inhibition constants.

Observe that expression (29) consists of two terms; the first indicates nitrite production from nitrate, and the second describes nitrite reduction. The quantity in brackets may be positive (depending on the values of s and u), and thus, $-r_2$ may be negative. A negative rate of nitrite disappearance implies accumulation, which needs to be avoided because it is itself toxic and does not lead to the desired end product (nitrogen gas). Accumulation of nitrite has been shown to occur experimentally when the operating conditions are not properly selected as discussed below.

In a subsequent study, Wang et al. (1995a) used kinetic expressions (28) and (29) to describe denitrification with the cyclically operated reactor discussed in earlier sections. In this case, the process is described by Eq. (17) for the volume variation and the following balances on nitrate, nitrite, and biomass:

$$\frac{ds}{dt} = \frac{F}{V} (s_f - s) - (-r_1) \tag{30}$$

$$\frac{du}{dt} = \frac{F}{V} (u_f - u) - (-r_2) \tag{31}$$

$$\frac{db}{dt} = -\frac{Fb}{V} + \left[\frac{\mu_1 * s}{K_1 + s + s^2/K_{i1} + K_{21} su} + \frac{\mu_2 * u}{K_2 + u + u^2/K_{i2} + K_{12} us} \right] b - \mu_c b \tag{32}$$

where s_f and u_f are the concentrations of nitrate and nitrite, respectively, in the feed stream and all other symbols are as defined previously.

If $u_f = 0$, the waste contains nitrate only. However, there may be situations where the waste contains both nitrate and nitrite. It is also possible that the waste contains

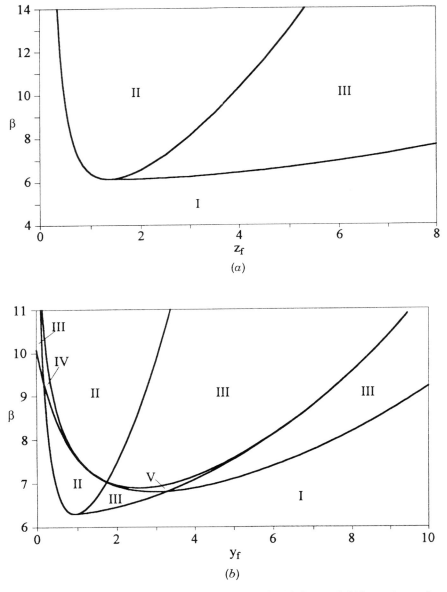

Figure 6.14 Operating diagrams for denitrification on the (*a*) $\beta - z_f$ and (*b*)$\beta - y_f$ planes when the process takes place in a cyclically operated reactor. Diagram (*b*) is characteristic of cases involving a stable intermediate product. States arising in the various regions. and their stability are shown in Table 6.5. From Wang et al. (1995a).

Table 6.5   Stable Periodic States (denoted by +) and Their Multiplicity in the Regions of Figure 6.14

Region	Washout State	Survival State
I	+	
II		+ (1 state)
III	+	+ (1 state)
IV		+ (2 states)
V	+	+ (2 states)

nitrite only, in which case the problem reduces to that of a simple pollutant with no intermediates.

When the waste contains both nitrate and nitrite, there are five operating parameters, namely, β, w, f_1, z_f and y_f: where $\beta = \mu_2 * V_{max}/(F f_1)$; $w = V_0/V_{max}$; $f_1 = t_1/t_3$; $z_f = u_f/K_2$; and $y_f = s_f/K_1$.

Two-dimensional projections of the operating diagrams on the $\beta - z_f$ ($y_f = 0$) and the $\beta - y_f$ ($z_f > 0$) planes are shown in Figure 6.14. The stable states that arise in each one of the regions of these diagrams are shown in Table 6.5.

When nitrate is not present in the feed stream (i.e., $y_f = 0$), the operating diagram for the process, shown in Figure 6.14a, is essentially identical to that shown in Figure 6.9a. Note the slight change in notation: The multistability region is indicated as region II in Figure 6.9a, whereas in Figure 6.14a it is shown as region III. The reason for bringing up this comparison is to show that regions of "unsafe" operation (where washout or survival is uncertain) arise for anaerobic as well as aerobic processes, either in the presence or absence of biomass maintenance requirements.

Figure 6.14b assumes that there is some nitrite present in the feed stream along with nitrate, but its qualitative features are the same even if nitrite is absent from the feed. Observe that Figure 6.14b is much more complex than Figure 6.14a (or 6.9a). There are two new regions (IV and V) where there are two stable survival states under the same operating conditions. The difference between them is the degree of conversion of the pollutants, and once again, the way the process is started-up dictates the final outcome during operation.

Consider region II of Figure 6.14b, where regardless of process start-up conditions biomass survives in a single state (thus, it is the safest region for operation). In reality, this is not a uniform region. There are points in region II leading to complete reduction of both nitrate and nitrite, as well as points leading to complete reduction of nitrate but accumulation of nitrite. This is not a difference from the stability point of view, and thus region II appears as uniform in Figure 6.14b. Experimental verification of the fact that the process may be incomplete (i.e., accumulation of nitrite) or complete (i.e., reduction of both the original and the intermediate pollutant) when operating in region II of Figure 6.14b is shown in Figures 6.15 and 6.16, respectively (Wang et al., 1995a).

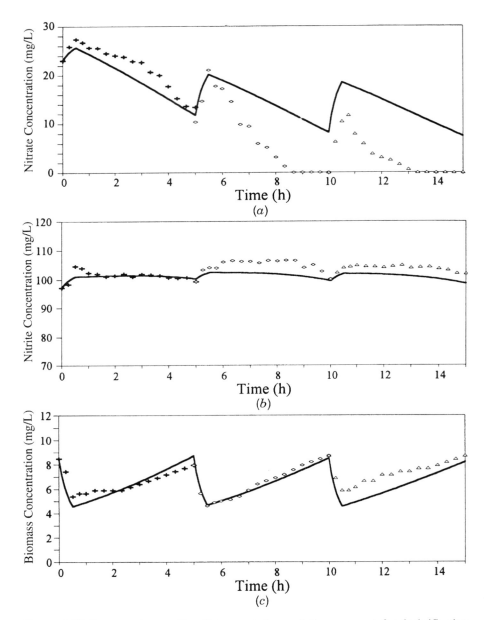

Figure 6.15 Concentration profiles (data: symbols, predictions: curves) for denitrification under operating conditions falling in region II of Figure 6.14*b*. The original compound (nitrate) is fully reduced but the intermediate (nitrite) accumulates in the reactor and its effluent. The process is incomplete. From Wang et al. (1995a).

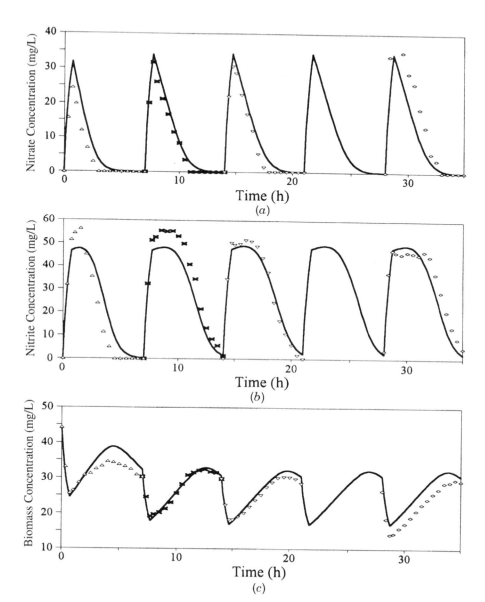

Figure 6.16 Data (symbols) and predicted concentration profiles (curves) for complete treatment of both the original pollutant (nitrate) and the reaction intermediate (nitrite) under operating conditions falling in region II of Figure 6.14b. From Wang et al. (1995a).

From the foregoing discussion, when a process involves a long-lived intermediate, the dynamics may be complicated with extra regions of multistability in the operating diagram, and the concentration of the intermediate must be considered in the design. In fact, intermediates and conditions leading to their complete transformation may entirely decide the overall design parameters for the process.

B. Use of Microbial Consortia

Here, an example is presented where intermediates are formed and noncompeting mixed cultures are needed for complete mineralization of an original pollutant.

Consider a pollutant S_1, which can be metabolized, but not completely, by species 1. If a pure culture of species 1 is used, the pollutant is transformed to compound S_2. Now, another microbial population (species 2), which cannot utilize S_1, can uptake and completely mineralize compound S_2. Clearly, for complete mineralization of the original pollutant S_1, one needs a mixed culture of species 1 and 2. In this case, the two species are not involved in competition. However, they do interact as species 2 depends on species 1. This interaction is called *commensalism*. Species 1 can be also called the *primary degrader*, whereas species 2 can be called the *dependent*. The primary degrader is also the dominant species in the consortium of the two types of organisms.

In this type of system the rates of disappearance of compounds S_1 and S_2 will be, respectively, given by the following expressions:

$$-r_p = \frac{1}{Y_1} \mu_1 b_1 \tag{33}$$

$$-r_1 = -\left[\delta\mu_1 b_1 - \frac{1}{Y_2} \mu_2 b_2 \right] \tag{34}$$

where δ is the stoichiometric ratio of milligrams of S_2 produced per milligram of S_1 transformed.

These are similar to Eqs. (28) and (29). The key difference is that for the case considered here, two types of microbial species (or consortia) are involved, as opposed to a single species (or consortium) for the case of nitrate reduction.

Let us assume that both species follow Monod-type kinetics, and that the kinetic constants in expressions (33) and (34) are those shown in Table 6.6. If the two species were lumped together as a single population (species 3), the rate of reaction (based on the original pollutant only) would be given by

$$-r_3 = \frac{1}{Y_3} \mu_3 b_3 \tag{35}$$

where $b_3 = b_1 + b_2$, Y_3 is an apparent yield coefficient and μ_3 an apparent specific growth rate.

Table 6.6   Kinetic Constants for the Commensal System

Parameter	Species 1	Species 2	Species 3
$\mu_m(h^{-1})$	0.8	0.6	0.71
K (mg/L)	30.0	75.0	46.6
Y	0.5	0.6	0.5

$\delta = 0.2$.

Using the approach of Lewandowski et al. (1989), which was discussed previously, and assuming that batch kinetic experiments were performed with inocula consisting of 80% primary degraders (species 1) and 20% dependents (species 2), the apparent kinetic constants for expression (35) would be those shown in Table 6.6 for species 3.

Now let us undertake the design of this process, to be carried out in a single continuous-flow reactor. The exact description is given by the following mass balances:

$$\frac{ds_1}{dt} = \frac{F}{V}(s_{1f} - s_1) - (-r_p) \tag{36}$$

$$\frac{db_1}{dt} = -\frac{Fb_1}{V} + \mu_1 b_1 \tag{37}$$

$$\frac{ds_2}{dt} = -\frac{Fs_2}{V_2} - (-r_I) \tag{38}$$

$$\frac{db_2}{dt} = -\frac{Fb_2}{V} + \mu_2 b_2 \tag{39}$$

where $-r_p$ and $-r_I$ are given by expressions (33) and (34), respectively; $\mu_j, j = 1, 2$ are given by the Monod expression with the constants shown in Table 6.6; s_{1f} is the concentration of the original pollutant S_1 in the feed stream; and all other symbols are as defined previously.

The analysis shows that this process can reach the following three qualitatively different steady states, none of which exhibits multiplicity:

SS1: $b_1 = b_2 = 0$ (total biomass washout)

SS2: $b_1 > 0,$ $b_2 = 0$ (survival of species 1 only)

SS3: $b_1 > 0,$ $b_2 > 0$ (survival of both species)

The analysis of steady states shows that the operating parameter space, $\alpha - u_f$ with $\alpha = F/(V\mu_{m1})$ and $u_f = s_{1f}/K_1$, is divided in three regions by the following two curves:

$$F_1 = \frac{V\mu_{m1}s_{1f}}{K_1 + s_{1f}} \tag{40}$$

$$F_2 = \frac{V\mu_{m1}\{R - (R^2 - 2\eta\sigma Lu_f)^{1/2}\}}{L} \tag{41}$$

$$\text{where} \quad R = \eta u_f(1 + \sigma) + \eta\sigma + 1 \tag{42}$$

$$L = 2\eta(u_f + 1) + 2 \tag{43}$$

with, $u_f = s_{1f}/K_1$, $\eta = \delta K_1 Y_1/K_2$, and $\sigma = \mu_{m2}/\mu_{m1}$.

The boundaries defined by Eqs. (40) and (41) are, respectively, shown as curves 1 and 3 in the operating diagram of Figure 6.17. If the operating conditions (F, s_{1f}) are selected in a way defining a point above curve 1 in Figure 6.17, the system reaches SS1. For points between curves 1 and 3, SS2 is reached; whereas for points below curve 3, SS3 is reached.

Since for the process considered here the presence of both species is required, Eq. (41) defines the maximum volumetric flow rate (F_2) for a given reactor volume (V).

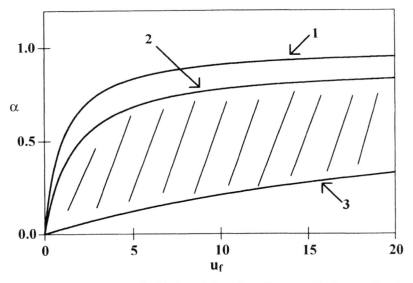

Figure 6.17 Operating diagram for biodegradation of a pollutant and its intermediate by two populations. The shaded area is the "error region" when the mixed culture is treated as a single population.

If the approach of treating the two species as a single culture was used, there are only two outcomes, washout and biomass survival, and the maximum volumetric flow rate that can be used in order to ensure biomass survival is given by Eq. (7) when s_{1f} is substituted for s_f. Using the values of μ_m and K for species 3 (mixture) from Table 6.6, the boundary defined by Eq. (7) is the one shown as curve 2 in Figure 6.17.

The shaded area of Figure 6.17 is essentially the error of the approximation. Operation in that area would lead to conversion of the original pollutant S_1 to compound S_2, but not to its complete mineralization, which is the process objective. This error is not just quantitative. It is also qualitative and leads once again to an incorrect process design.

CONCLUSIONS

Biodegradation of wastes of industrial origin can be kinetically quite complex. The complexity results from the inherent kinetics of biodegradation (such as Monod or Andrews), and from population shifts in mixed cultures, and is not simply a mathematical artifact. Erroneous simplifications can (and often do) lead to significant design and operational deficiencies.

The material presented in this chapter suggests that if the black box is opened, operation of waste treatment units could be better understood and a number of opportunities for optimal design (and thus economic benefits) could arise. At the very least, one should be extremely cautious in making extrapolations of data obtained using unspecified mixtures of microorganisms (such as activated sludge) that are traditionally presumed to be invariant with time.

REFERENCES

Andrews, J. F. (1968). A Mathematical Model for the Continuous Culture of Microorganisms Utilizing Inhibitory Substrates, *Biotech. Bioeng.*, **10**:707–723.

Baltzis B. C. and A. G. Fredrickson (1983). Competition of Two Microbial Populations for a Single Resource in a Chemostat When One of Them Exhibits Wall Attachment, *Biotech. Bioeng.*, **25**:2419–2439.

Baltzis B. C. and M. Wu (1994). Steady-state Coexistence of Three Pure and Simple Competitors in a Four-Membered Reactor Network, *Math. Biosci.*, **123**:147–165.

Chang, S. W. and B. C. Baltzis (1989). Impossibility of Coexistence of Three Pure and Simple Competitors in Configurations of Three Interconnected Chemostats, *Biotech. Bioeng.*, **33**:460–470.

Chang, M. K., T. C. Voice, and C. S. Criddle (1993). Kinetics of Competitive Inhibition and Cometabolism in the Biodegradation of Benzene, Toluene, and *p*-Xylene by Two *Pseudomonas* Isolates, *Biotech. Bioeng.*, **41**:1057–1065.

Davison, B. H. and G. Stephanopoulos (1986). Effect of pH Oscillations on a Competing Mixed Culture, *Biotech. Bioeng.*, **28**:1127–1137.

Dennis, R. W. and R. L. Irvine (1979). Effect of Fill:React Ratio on Sequencing Batch Biological Reactors, *J. Water Pollut. Contr. Fed.*, **51**:255–263.

Dikshitulu, S. (1993). Competition between Two Microbial Populations in a Sequencing Fed-Batch Reactor and Its Implications for Waste Treatment Applications, Ph.D. Thesis, New Jersey Institute of Technology, Newark, NJ.

Dikshitulu, S., B. C. Baltzis, G. A. Lewandowski, and S. Pavlou (1993). Competition between Two Microbial Populations in a Sequencing Fed-Batch Reactor: Theory, Experimental Verification, and Implications for Waste Treatment Applications, *Biotech. Bioeng.*, **42**:643–656.

Fredrickson, A. G. and G. N. Stephanopoulos (1981). Microbial Competition, *Science*, **213**:972–979.

Herbert, D. (1958). The Continuous Culture of Microorganisms: Some Theoretical Aspects, in *Continuous Cultivation of Microorganisms: A Symposium*, I. Malek, Ed., Publishing House of the Czechoslovak Academy of Sciences, Prague, pp. 45–52.

Hsu, E. H. (1986). Treatment of a Petrochemical Wastewater in Sequencing Batch Reactors, *Environ. Prog.*, **5**:71–81.

Irvine, R. L. and A. W. Busch (1979). Sequencing Batch Biological Reactors—An Overview. *J. Wat. Pollut. Contr. Fed.*, **51**:235–243.

Irvine, R. L. and R. O. Richter (1978). Comparative Evaluation of Sequencing Batch Reactors, *J. Environ. Eng. Div. ASCE.*, **104**:503–514.

Irvine, R. L., L. H. Ketchum, R. E. Breyfogle, and E. F. Barth (1983). Municipal Application of Sequencing Batch Treatment, *J. Wat. Pollut. Contr. Fed.*, **55**:484–488.

Jones, W. L., P. A. Wilderer, and E. D. Schroeder (1990). Operation of a Three-Stage SBR System for Nitrogen Removal from Wastewater, *J. Wat. Pollut. Contr. Fed.*, **62**:268–274.

Kung, C. M. and B. C. Baltzis (1987). Operating Parameters' Effects on the Outcome of Pure and Simple Competition between Two Populations in Configurations of Two Interconnected Chemostats, *Biotech. Bioeng.*, **30**:1006–1018.

Kung, C. M. and B. C. Baltzis (1992). The Growth of Pure and Simple Microbial Competitors in a Moving Distributed Medium, *Math Biosci.*, **111**:295–313.

Lauffenburger, D. and B. Calcagno (1983). Competition between Two Microbial Populations in a Nonmixed Environment: Effect of Cell Random Motility, *Biotech. Bioeng.*, **25**:2103–2125.

Lenas, P., B. C. Baltzis, G. A. Lewandowski, and Y. F. Ko. (1994). Biodegradation of Wastes in a Cyclically Operated Reactor: Theory, Experimental Verification and Optimization Studies, *Chem. Eng. Sci.*, **49**:4547–4561.

Lewandowski, G. A., B. C. Baltzis, C. M. Kung, and M. E. Frank (1989). An Approach to Biocatalyst Modelling of Mixed Populations Using Pure Culture Kinetic Data, in *Biotechnology Applications in Hazardous Waste Treatment*, G. A. Lewandowski, P. M. Armenante, and B. C. Baltzis, Eds., Engineering Foundation, New York, pp. 95–110.

Monod, J. (1942). *Recherches sur la croissance des cultures bactériennes*, Hermann et Cie., Paris.

Oh, Y. S., Z. Shareefdeen, B. C. Baltzis, and R. Bartha (1994). Interactions between Benzene, Toluene and *p*-Xylene (BTX) During Their Biodegradation, *Biotech. Bioeng.*, **44**:533–538.

Palis, J. C. and R. L. Irvine (1985). Nitrogen Removal in a Low-Loaded Single Tank Sequencing Batch Reactor, *J. Wat. Pollut. Contr. Fed.*, **57**:82–85.

Pavlou, S., I. G. Kevrekidis, and G. Lyberatos (1990). On the Coexistence of Competing Microbial Species in a Chemostat under Cycling, *Biotech. Bioeng.*, **35**:224–232.

Powell, E. O. (1958). Criteria for Growth of Contaminants and Mutants in Continuous Culture, *J. Gen. Microbiol.*, **18**:259–268.

Saghafi, F. (1988). Optimal Design Aspects of a Unit Treating Hazardous Wastes via Biodegradation: A Theoretical Approach, M. S. Thesis, New Jersey Institute of Tehnology, Newark, NJ.

Silverstein, J. A. and E. D. Schroeder (1983). Performance of SBR Activated Sludge Processes with Nitrification/Denitrification, *J. Wat. Pollut. Contr. Fed.*, **55**:377–384.

Stephanopoulos, G. and A. G. Fredrickson (1979). Effect of Spatial Inhomogeneities on the Coexistence of Competing Microbial Populations, *Biotech. Bioeng.*, **21**:1491–1498.

Stephanopoulos, G., A. G. Fredrickson, and R. Aris (1979). The Growth of Competing Microbial Populations in a CSTR with Periodically Varying Inputs, *AIChE J.*, **25**:863–872.

Stephens, M. L. and G. Lyberatos (1987). Effect of Cycling on Final Mixed Culture Fate, *Biotech. Bioeng.*, **29**:672–678.

Varuntanya, C. P. (1986). The Use of Pure Cultures as a Means of Understanding the Performance of a Mixed Culture in the Biodegradation of Phenol. Ph.D. Thesis, New Jersey Institute of Technology, Newark, NJ.

Wang, J. H., B. C. Baltzis, and G. A. Lewandowski (1995a). Reduction of Nitrate and Nitrite in a Cyclically Operated Continuous Biological Reactor, *Biotech. Bioeng.*, **46**:159–171.

Wang, J. H., B. C. Baltzis, and G. A. Lewandowski (1995b). Fundamental Denitrification Kinetic Studies with *Pseudomonas denitrificans*, *Biotech. Bioeng.*, **47**:26–41.

CHAPTER 7

Hydrogeologic Factors Affecting Biodegradation Processes

Kelton D. Barr

Delta Environmental Consultants Inc, 2770 Cleveland Ave, Roseville, Minnesota 55113

Of all the factors influencing the microbial activity in a subsurface environment, the hydrogeologic factors are perhaps the most important. These control the movement of fluids through the geologic matrix that constitutes the subsurface environment. In turn, these factors control the flow and transport of all rate-limiting compounds needed for bioactivity, that is, organic carbon, electron acceptors, and inorganic nutrients. In order to overcome these bioactivity rate limitations and achieve greater rates of biodegradation, the flow and transport processes must be manipulated.

However, any manipulation of these processes to enhance biodegradation rates must be viewed in the context of the other ongoing hydrogeologic processes and the associated geochemical processes. Many of the problems encountered in early in situ bioremediation projects can be attributed to the failure to account for the transport of other dissolved inorganics, the geochemical processes prevailing inside and outside of contaminated zones, and the subsequent dissolution and precipitation processes.

The purpose of this chapter is to describe the major hydrogeologic processes that affect the nature and rates of biodegradation acting on organic contamination. Flow and transport parameters and groundwater and soil air flow processes will be described, and some of the more useful equations, approaches, and references will be provided. This chapter should provide the reader with the conceptual framework and necessary fundamentals for the application of hydrogeology to biodegradation processes.

Biological Treatment of Hazardous Wastes, Edited by Gordon A. Lewandowski and Louis J. DeFilippi
ISBN 0-471-04861-5   ©1998 John Wiley & Sons, Inc.

FLOW PARAMETERS

Porosity

Many of the hydrogeologic factors affecting biodegradation processes have to do with the void spaces within the geologic materials and the surfaces of the geologic materials within these voids. Within most rock units, these voids are primarily fractures and constitute less than 1% of the bulk volume of the unit. Within unconsolidated sedimentary units and soils (i.e., those geologic units whose individual mineral particles are not cemented or fused together), these voids are much more prevalent and can constitute up to over half of the bulk volume. It is within these voids that the subsurface microbial populations reside, fluid flow occurs, and biodegradation takes place.

Most biodegradation processes of interest involve shallow unconsolidated sedimentary deposits or, more simply, soils. These are complex assemblages of rock particles that are the result of the geologic processes that deposited the parent materials and the geomorphologic and weathering processes that have subsequently acted on and altered the soils. These soils also consist of primary mineral particles, secondary minerals, amorphous mineral and inorganic components, water, and organic components, including the microbial communities.

Several generalizations about these soils' physical properties can be made. Those soils deposited by moving water (e.g., stream, beach, and deltaic deposits) tend to be composed of more rounded soil particles (also called grains or clasts) that are deposited in layers or strata, each relatively well sorted (i.e., most clasts' diameters being approximately the same). Porosities of these soils generally range from 25 to 50%. Soils deposited in still waters (e.g., lacustrine and marine deposits) tend to be composed of fine-grained soil particles or clasts that are also well sorted (also referred to as poorly graded). Those soils created by other processes (e.g., glacial tills, bedrock weathering) tend to be composed of more angular clasts and of amorphous mineral components that are poorly stratified and poorly sorted (i.e., well graded). Porosities of these soils generally range from 40 to 70%. Soils tend to the lower end of these ranges if they have undergone consolidation. Also, the smaller the soil particle, the more irregular the shape generally is and the poorer the packing. This generally results in greater porosity and surface area.

Most soils are at least somewhat graded, and their particle constituents consist of a range of particle diameters and shapes, with the smaller particles found in the void spaces between larger particles. These void spaces can be further diminished by secondary deposition of chemical precipitates and amorphous materials and in some cases by the buildup of biomass on the soil particle surfaces. The remaining void space is occupied by fluids, most typically water and air, but in the proximity of a contaminant spill, the void space can also be occupied by nonaqueous liquids.

Figure 7.1 summarizes the relationship of porosity to particle diameter. This relationship shown in the figure was originally based on the D_{90} (i.e., 90% of the

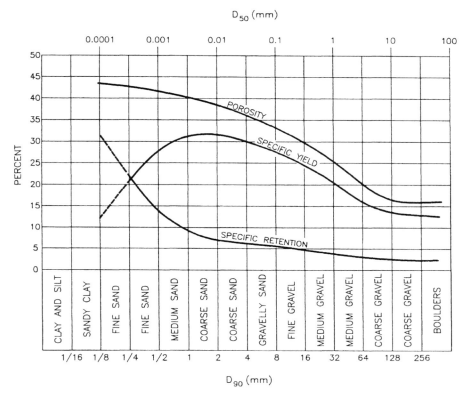

Figure 7.1 Relationships between porosity, specific retention, and specific yield (from Johnson, 1967, and deMarsily, 1986).

sample's particles having smaller diameters) by Johnson (1967) and has been modified (de Marsily, 1986) to apply to the mean grain diameter (D_{50}) as well. The porosity ranges of various soils and rocks are summarized on Table 7.1.

Generally, soils in both the vadose and saturated zones are water wetted and have a layer of "adhesive water" immediately surrounding the soil particles (Polubarinova-Kochina, 1962; de Marsily, 1986). This is due to the generally negative residual electrostatic charges on the particle surfaces due in turn to incomplete or broken crystal structures of the mineral components of the soil particles interacting with the polar water molecules. These water molecules will be oriented perpendicular to the crystal surfaces and will form an adsorbed layer tens of molecules thick (about 0.1 μm) and a transition zone 0.1–0.5 μm thick where the water molecules are still dominated by hydrogen bonding (Hem, 1970). This is the primary basis for the macroscopic properties of viscosity and surface tension, two properties that are major factors in governing the saturated and unsaturated flow of water.

Table 7.1 Selected Physical Properties of Different Types of Geologic Materials

	Porosity (%)[a]	Porosity (%)[b]	Porosity (%)[a]	Porosity (%)[d]	Porosity (%)[c]	Hydraulic Conductivity (cm/s)[b]	(cm/s)[a]	(cm/s)[c]	(cm/s)[f]	Specific Yield[g] Minimum (%)	Maximum (%)	Average (%)	Surface Area[b,h] (m²/g)	Cation Exchange Capacity[b,h] (meq/100 g)
Clay	44–53	41–60	40–70	33–71	40–55	10^{-10}–10^{-4}	10^{-10}–10^{-4}	$10^{-5.3}$–$10^{-3.3}$	$<10^{-7}$	0	19	8	150–250	5–60
Kaolinite													5–39	3–15
Illite													100–200	10–40
Smectite													700–800	80–150
Sandy clay				20–64										
Silt	15–48	20–45	35–50	29–52		10^{-7}–10^{-3}	10^{-7}–10^{-3}	10^{-5}–$10^{-3.3}$	10^{-6}–10^{-5}	3	12	7	1–3	9–27
Sand		25–45	25–50		30–45	$10^{-5.5}$–10^{0}	$10^{-3.5}$–10^{0}			3	19	9	2	2–7
Fine sand				29–50				10^{-2}	10^{-3}–10^{-2}	10	28	21		
Medium sand								10^{-1}		15	32	26		
Coarse sand				33–44				10^{0}	10^{-2}–10^{-1}	20	35	27		
Gravelly sand				12–46					10^{-1}–10^{0}	20	35	25		
Gravel		25–35	25–40			10^{-1}–10^{2}	10^{-1}–10^{2}							
Fine gravel									$>10^{0}$	21	35	25		
Medium gravel										13	26	23		
Coarse gravel										12	26	22		
Peat					60–80	10^{-4}–10^{-2}			10^{-7}–10^{-5}					>200
Fibric peat							$10^{-2.7}$–$10^{-2.2}$							
Sapric peat							10^{-4}–$10^{-3.7}$							
Loam		20–40			35–50								50–100	8–22
Sandy loam														
Sandstone		3.5–38	5–30				10^{-8}–$10^{-3.5}$						10–40	2–7
Limestone		0.5–12.5	0–20				$10^{-7.2}$–$10^{-3.5}$							
Shale		0.5–7.5	0–10				10^{-11}–10^{-7}							
Granite		0.02–1.8					10^{-12}–10^{-9}							
Unfractured			0–5											
Fractured			0–10											

[a] deMarsily (1986).
[b] Dragun (1988).
[c] Freeze and Cherry (1979).
[d] Hough (1969).
[e] Polubarinova–Kochina (1962).
[f] Verruijt (1970).
[g] Johnson (1967).
[h] Lambe and Whitman (1969).

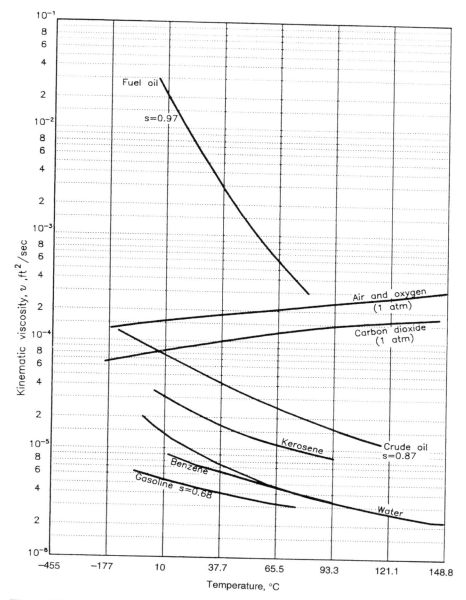

Figure 7.2 Kinematic viscosity (v) of selected fluids (Note: multiply viscosity values by 0.0929 to convert to units of m²/s; s = specific gravity at 70°F) (from Olson, 1973).

Viscosity is the property that determines the degree of shear stress required to maintain a velocity gradient, that is, maintaining flow through the middle of a pore while the water next to the soil particles remains motionless. Viscosity is a function of temperature and pressure. While the pressure at a subsurface location changes very little, the temperature at shallow depths can vary considerably. This will cause the thickness of adsorbed water to change. The relationship of kinematic viscosity v (i.e., the ratio μ/ρ of dynamic viscosity μ to density ρ for a fluid) to temperature is summarized in Figure 7.2 for several liquid and gaseous fluids. These relationships will be discussed further in the following section.

When the viscosity and the thickness of the adsorbed layer of water are taken into account, a number of pores will have insufficient diameter for groundwater to flow through them. These pores become nonfunctional for through-flow, or "dead ends;" consequently, the specific retention is higher for finer soils, as is shown in Figure 7.1. In order for flow to occur, there needs to be a network of interconnected, functional pores. These interconnected pore pathways can be quite round-about, resulting in a large disparity between the apparent flow path l_{app} (ignoring the matrix) and the actual flow path l_{act} (considering the matrix) between two points. This ratio l_{app}/l_{act} is termed *tortuosity*. Diffusional tortuosity can range from 0.7 for sands to 0.1 for clays, and the range can be more disparate for flow (de Marsily, 1986).

In well-sorted, coarse-grained soils, such as sands and gravels, most of the pores are on the order of $10^2 - 10^3$ μm in diameter, and most of the water is unadsorbed and able to flow under the influence of gravity and potentiometric forces (also called *gravitational water* or *funicular water*). However, in poorly sorted or fine-grained soils, many of the pores are on the order of $1-10$ μm in diameter, and much of the water, sometimes even most of the water, is adsorbed and immobile.

The degree of pore interconnectedness and its ability to yield water and allow for groundwater flow are all represented by *specific yield*. This is a measurable property defined as the ratio of the volume of water that a saturated rock or soil (termed sample here) will yield by gravity-based drainage to the total volume of the sample (Johnson, 1967). Conversely, the ratio of the volume of water retained after gravity-based drainage of a sample to the total sample volume is termed *specific retention*; this includes both adsorbed water and dead-end water and is equivalent to the *field capacity* of unsaturated soils.

These are important factors for biodegradation processes. For vadose zone soils, the locations of the retained water are the most favorable for the sustained establishment of biomass. For saturated zone soils, the biofilms that are established will be located in the adsorbed water. In both of these zones, the delivery of oxygen and potentially other growth supplements will be by diffusion from the fluids migrating past the adsorbed water. The thickness of the adsorbed water or the distance from these migrating fluids determine the delivery rates of these supplements and in turn determines the type and viability of microbial activity.

The relationship between particle sizes and both specific yield and specific retention is summarized on Figure 7.1. As can be seen, fine-grained soils can have much of the pore water retained as adsorbed water. Also the average and range of specific yield values compiled by Johnson (1967) are summarized in Table 7.1.

Permeability

In almost all circumstances groundwater flow in the saturated zone is laminar, that is, viscous forces predominate over inertial forces (Domenico, 1972). This is important because Darcy's law, the fundamental equation used for groundwater flow, is valid only for laminar flow. The criterion for assessing the applicability of Darcy's law is the Reynolds number, Re. Although originally developed for pipe flow, the Reynolds number can be defined for groundwater flow (Strack, 1989) as:

$$\text{Re} = D \cdot q/v \tag{1}$$

where Re is dimensionless, D is the average clast diameter (m), q is the groundwater specific discharge (m/s), and v is the kinematic viscosity of water (m^2/s), using SI units. Re has been experimentally determined to be equal to or less than 1 for laminar flow, although some researchers reportedly extend the Reynolds number range up to 10 (Chauveteau and Thirriot, 1967; Polubarinova-Kochina, 1962; Freeze and Cherry, 1979).

These parameters are also involved in determining the conductivity of a soil for its saturating fluid. The basic equation can be stated as:

$$K_f = k \cdot g/v_f \tag{2}$$

where K_f = fluid conductivity of the soil (m/s)
   k = intrinsic permeability of the soil (m^2)
   g = acceleration due to gravity (m/s^2)
   v_f = kinematic viscosity of the fluid (m^2/s), in SI units

K_f is termed the *hydraulic conductivity* for groundwater flow in the saturated zone and is termed *pneumatic conductivity* for air flow in the vadose zone. Typical ranges of hydraulic conductivity values are summarized in Table 7.1. As can be seen, as the finer-grained members of the soil become more coarse, the conductivity values are higher.

The intrinsic permeability is a property of the porous medium that includes its porosity interconnectedness, pore sizes, and tortuosity. It is primarily determined by measuring the fluid conductivity of the soil in either the laboratory or the field. Because the intrinsic permeability is a property of the soil only, the pneumatic conductivity of a soil can be determined from its hydraulic conductivity, and vice versa, by the relationship

$$K_B = K_A \, (v_A/v_B) \tag{3}$$

where A and B denote the different fluids.

This relationship assumes comparable distribution of adsorbed water within the soil pores, and this is generally true if an unsaturated soil's moisture content is at field capacity. However, if the soil has experienced drying and the soil moisture is lower than field capacity, the effective pore size will be larger, and the pneumatic conductivity will be greater than would be calculated by Eq. (3). The permeabilities of dual-fluid systems are described in more detail by Brooks and Corey (1966), Corey (1967), Falta et al. (1989), and Stylianou and DeVantier (1995). Another complication in directly relating air and groundwater flow through porous media is slipflow, also called the Klinkenberg effect, where nonzero velocities occur along pore walls. These occur at low air flow rates and can be important in low-permeability materials (Massmann, 1989; Thorstenson and Pollock, 1989). These interrelationships of fluid conductivities are important because they allow the field researcher to take full advantage of the permeability information collected at a site for the design of manipulation systems for soil air, groundwater, and other organic fluids.

TRANSPORT PARAMETERS

The physicochemical processes in general decrease dissolved hydrocarbons concentrations by redistributing the hydrocarbon mass within the subsurface, over a larger volume. However, by decreasing their concentrations and by redistributing mass, these processes can render the petroleum hydrocarbons constituents more bioavailable and hasten their attenuation by biodegradation. Several major physicochemical processes are discussed in this section. For further overview of these processes, the reader is referred to Domenico and Schwartz (1990).

Solubility

Because most inorganic compounds tend to dissolve in water via dissociation reactions that produce dissolved ions and because of the polar nature of water molecules, the dissolved loading of groundwater is generally composed principally of inorganic compounds. Consequently, this section will describe the dissolution processes important for inorganic compounds, including nutrients. Because of their generally nonpolar nature, most organic compounds are less soluble and interact with water and the porous media following the processes described in the next section.

Groundwater is continually evolving toward equilibrium with its load of dissolved material and the geologic material through which it flows. The longer the flow path and the slower the rate of movement, the longer the residence time in the subsurface and the higher is the mineralization or total inorganic content of the groundwater. Chebotarev (1955) outlined the sequence in dominant water quality types that are associated with successively longer residence times:

$$(\text{Short residence times})$$
$$\{HCO_3^-\} \rightarrow \{HCO_3^- + Cl^-\} \rightarrow \{Cl^- + HCO_3^-\} \rightarrow \{Cl^- + SO_4^{2-}\} \text{ or}$$

$$(\text{Long residence times})$$
$$\{SO_4^{2-} + Cl^-\} \rightarrow \{Cl^-\}$$

where the dominant species are listed first in each bracketed pair. Schoeller (1959) summarized a similar relationship between total ionic concentration and dominant water quality type. Both Chebotarev's and Schoeller's sequences show that with increasing mineralization, water tends toward the composition of seawater.

In most terrains, groundwater receives its solutes mainly from the topsoil zone (Schoeller, 1959). The concentration of solutes in natural waters depends on the degree of evapotranspiration and, inversely, on the degree of precipitation. The smaller the proportion of infiltrating water that reaches the water table with respect to the total amount initially infiltrating (including rainfalls that only partially penetrate), the more concentrated will the chemical content be in that recharge. Consequently, more highly mineralized waters are more commonly found in the more arid zones.

The principal agent by which groundwater decomposes geologic material is carbonic acid, H_2CO_3, and over 99% of Earth's carbon exists in carbonate minerals, which are predominantly calcite, $CaCO_3$, and dolomite, $CaMg(CO_3)_2$ (Freeze and Cherry, 1979). There are two sources for carbonate, and its two protonated forms, bicarbonate and carbonic acid. One is the dissolution of calcite (or dolomite):

$$CaCO_{3(s)} \rightleftharpoons Ca^{2+} + CO_3^{2-}$$

The other source is carbon dioxide, either from the atmosphere or from biological processes in the soil and groundwater:

$$CO_{2(aq)} + H_2O \rightleftharpoons H_2CO_3$$

The amount of $CO_{2(aq)}$ in the subsurface is in turn dependent on the distribution constant:

$$K_D = [CO_{2(aq)}]/[CO_{2(g)}]$$

If the water is in equilibrium with the atmosphere ($p_{co/2}$, the partial pressure contributed by carbon dioxide, is $10^{-3.5}$ atm), then

$$[CO_{2(aq)}]_{atm} = 1.1 \times 10^{-5} M$$

However, in the topsoil zone in groundwater recharge areas $[CO_{2(aq)}]$ can be much higher, since CO_2 is about 200 times more soluble than oxygen (Hutchinson, 1957). However, the interactions with other carbonate species and other bioprocesses keep CO_2 concentrations lower than the solubility limit, with the groundwater $p_{co/2}$ in the vicinity of 10^{-3}–10^{-2} atm (Schoeller, 1959). Consequently, the concentration of carbonic acid is dependent on the concentration of carbon dioxide in the soil, which is in turn a function of the biochemical processes of the organic portion of the topsoil.

Because H_2CO_3 is a polyprotic acid, its dissociation reactions produce hydrogen ions i.e., protons) by the reactions:

$$H_2CO_3 \rightleftharpoons H^+ + HCO_3^-$$
$$HCO_3 \rightleftharpoons H^+ + CO_3^{2-}$$

so that the transformations of carbonate species both causes and results from changes in acidity, as measured by pH. Another, less common polyprotic acid involved in soil processes is phosphoric acid, H_3PO_4, which produces phosphatic compounds and protons by its dissociation reactions:

$$H_3PO_4 \rightleftharpoons H_2PO_4^- + H^+$$

$$H_2PO_4^- \rightleftharpoons HPO_4^{2-} + H^+$$

$$HPO_4^{2-} \rightleftharpoons PO_4^{3-} + H^+$$

In most terrains the concentrations of these different phosphate and carbonate species are largely determined by pH and carbon dioxide concentrations, both influenced by biological processes in the subsurface (Stumm and Morgan, 1970; Freeze and Cherry, 1979). These subsurface biological processes are predominantly microbial respiration and transformation of organic materials that involve the transfer of electrons, that is, oxidation and reduction processes. Both of these processes involve and affect inorganic compounds. These effects are principally

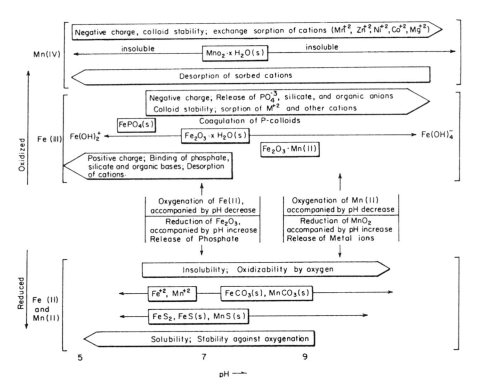

Figure 7.3 Major features of chemistry of iron, manganese, and phosphorus (from Stumm and Morgan, 1970).

indicated by the *reduction-oxidation potential* or *redox potential*, denoted by Eh (also denoted E_h in biochemical texts). Changes in Eh are particularly pronounced at the pore level in the immediate proximity to the active microbial consortia.

The relative reducing strength of electrons, as indicated by the redox potential, and concentrations of protons, as measured by pH, are important in determining the solubility limits and the stability of a particular form of many of the dissolved compounds. This is an important factor for determining the stability of bioavailable forms of nutrients and metals important for bioactivities. This is illustrated in Figure 7.3 where the changes in the chemistry of iron, manganese, and phosphate caused by changes in the redox potential and pH are descriptively summarized. For a comprehensive discussion of this, as well as for a collection of Eh–pH diagrams for a large number of inorganic compounds, the reader is refered to the classic works on the geochemistry and solution chemistry of inorganic compounds by Garrels and Christ (1965) and by Stumm and Morgan (1970).

Sorption

Because many organic contaminants' molecules are to some degree nonpolar, there is a finite limit to their solubility in a volume of more polar molecules such as water. This results in low ($<10^{-3}$ mg/L) to moderate (10^3 mg/L) aqueous solubility limits for these constituents. The solubility limits of many organic compounds are summarized in Verschueren (1983). Because of their limited solubilities, organic compounds generally tend to adsorb, or partition, to other forms of organic carbon in the subsurface. This can cause their migration in the dissolved plume as micelles, that is, "packs" of organic molecules with the more hydrophilic compounds or the more hydrophilic portions of each molecule on the outer surface of the micelle. It also leads to adsorption of these constituents to the organic coatings of the soil particles or to their adsorbed water (Perlinger and Eisenreich, 1991).

The tendency to adsorb to the subsurface material is different for each of the constituents, so that the constituents that sorb strongly tend to migrate down gradient more slowly than those that sorb less strongly This leads to a "chromatographic" effect within a migrating dissolved plume with the more mobile constituents arriving at downgradient locations before the more retarded constituents.

This adsorption property is represented in transport equations by the *retardation factor*, which is defined as:

$$R_f = 1 + (\rho_b K_d)/n \tag{4}$$

where n = porosity (dimensionless fraction; cf. Fig. 7.1)
  ρ_b = soil bulk density (kg/m^3)
  K_c = distribution coefficient (m^3/kg), in SI units

The distribution coefficient, K_d, is derived from the relationship between aqueous concentration and degree of adsorption as defined by a linear Freundlich isotherm. Several studies have determined a relationship in turn between the distribution coefficient and the mass fraction of organic carbon:

$$K_d = K_{oc}f_{oc} \tag{5}$$

where f_{oc} = mass fraction of organic carbon (kg oc/kg soil, dry weight)
       K_{oc} = partition coefficient between organic carbon and water

These studies have also found a good statistical correlation between the partition coefficient and the octanol–water partition coefficient, K_{ow} and between a compound's solubility and K_{ow}. Examples of these correlations are:

$$\log(K_{oc}) = -0.21 + \log(K_{ow}) \tag{6a}$$

$$\log(K_{oc}) = 0.44 - 0.54 \cdot \log(S_m) \tag{6b}$$

where S_m is the aqueous solubility (mole fraction), which was developed by Karickhoff et al. (1979) for 10 aromatic or polynuclear aromatic compounds, and

$$\log(K_{oc}) = 0.49 + 0.72 \log(K_{ow}) \tag{7}$$

which was developed by Schwarzenbach and Westall (1981) for 13 aromatic, alkylbenzene, and chlorinated aliphatic compounds (Hassett et al., 1983). Other relationships, along with the chemical compounds to which they are applicable, are summarized in Dragun (1988) and Lyman (1990).

While these empirical expressions are useful for general predictions of sorption equilibrium, there is accumulating evidence that there are significant time effects that can cause considerable variability in the partitioning of organic compounds. While much of the partitioning that is attributed to sorption is fast enough to be considered instantaneous sorption and to be at equilibrium, there are slower sorption processes that involve diffusion mechanisms into the organic material and the smaller pores of the porous media; these smaller pores make up more than 95% of the organic material surface area and sorption sites and have pore diameters less than 0.5 nm (de Jonge and Mittelmeijer-Hazeleger, 1996). These processes are important to consider, for the sites of this slower sorption restrict the bioavailability of the sorbed material. These slower sorption processes are described by Pignatello and Xing (1996).

Relationships between aqueous solubility and the octanol–water partition coefficient are well established. Examples of the relationship are shown in Figure 7.4. Consequently, the parameters that typically need to be obtained in order to assess the sorption and retardation of organic constituents are the soil bulk density (ρ_B), the octanol–water partition coefficient (K_{ow}), and the mass fraction of soil organic carbon (f_{oc}).

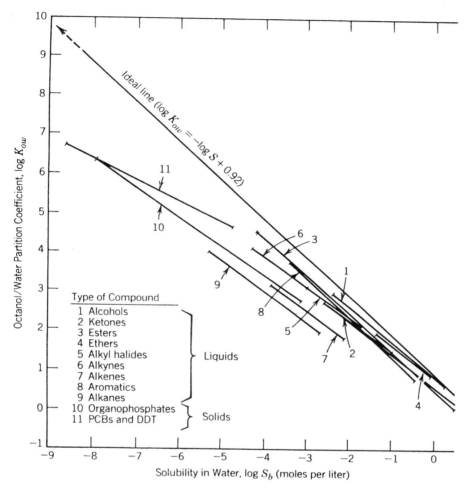

Figure 7.4 Relationship between solubility in water and octanol–water partition coefficients for various organic chemicals (from Chiou et al., 1982).

Mass Transfer

In a multifluid environment, such as a contaminated vadose zone, the transfer of compounds among soil air, soil moisture, and nonaqueous organic fluids can be an important factor in determining the compounds' bioavailability. Two laws are useful in describing these mass-transfer processes. The first is Raoult's law, which describes the transfer equilibrium at the nonaqueous fluid–vapor interface. This is described as an equilibrium partial pressure of an organic compound in air in contact with an organic liquid:

$$P_i = \chi_i p^\circ_i \tag{8}$$

where P_i = partial pressure of the organic compound vapor in the gaseous phase
(atm)
χ_i = mole fraction of the organic compound
p°_i = vapor pressure of the pure organic compound (atm)

This law is valid for describing the volatilization of pure organic solvents, but it provides good estimates for components of mixtures of organic compounds, such as gasoline (Domenico and Schwartz, 1990; Ellerd and Massmann, 1995).

The mass-transfer equilibrium at the water vapor interface is described by Henry's law, which states that the partial pressure of a solute in aqueous solution is proportional to its aqueous phase concentration:

$$P_i = K_{Hi} C_{aqi} \tag{9}$$

where P_i = equilibrium vapor pressure of solute i (atm)
C_{aqi} = equilibrium aqueous concentration of solute i (mol/m^3)
K_{Hi} = Henry's law constant for solute i (atm $\times$ m^3/mol), in SI units

This relationship is also useful for determining mass transfer between soil gases and groundwater. Values for vapor pressures, solubility, and Henry's law constants can be found in references such as Verschueren (1983) and Domenico and Schwartz (1990).

Dispersion

Because dissolved constituents migrate through subsurface materials that are not homogeneous, there are inherently different constituent migration rates in different portions of the plume. This can result in an apparent lateral spreading from the plume's flanks and an apparent spreading of the plume's leading edge. This apparent spreading is represented by transverse and longitudinal dispersivity coefficients, respectively. Empirical field studies of dispersion revealing relationships with flow velocity and scale of plume have been compiled and summarized by Gelhar et al. (1985) and Gelhar et al. (1992). Figure 7.5a shows the relationship between longitudinal dispersivity and plume length in the saturated zone, as compiled by Gelhar et al. (1992); it indicates that the apparent longitudinal dispersivity determined in the field can be several orders of magnitude greater that the dispersivity measured in the laboratory. Gelhar et al. (1985) also compiled a relationship between longitudinal dispersivity and scale of observation in the unsaturated zone which is shown in Figure 7.5b.

While these empirical relationships can be useful, these studies are known to be hampered by limitations in the field data (e.g., large spacing between observation wells, poor definition of the initial subsurface volume occupied by injected tracer) and limited information on field conditions (e.g., definition of the permeability

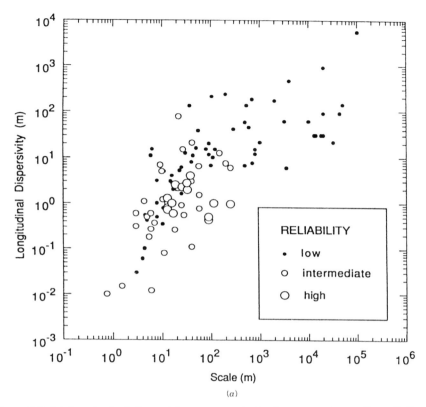

Figure 7.5 (*a*) Longitudinal dispersivity versus scale for the saturated zone (from Gelhar et al., 1985).

distribution, temporal fluctuations in the flow system; cf. Domenico and Schwartz, 1990). Recent quantitative studies are finding lateral dispersivity values more in line with laboratory values (e.g. van der Kamp, 1994). Rehfeldt and Gelhar (1992) have concluded that the major factor for determining longitudinal dispersivity is aquifer heterogeneities, and the major factor for determining transverse dispersivity is temporal changes in groundwater flow directions. Consequently, the benefit of groundwater intermixing by dispersion for overcoming limitation in electron acceptors or nutrients is likely to be minimal, unless it is artificially enhanced by pumping.

Diffusion

The migration of dissolved constituents because of concentration gradients is known as diffusion. This is commonly described by Fick's law:

$$J = -D \, \mathrm{grad}(C) \tag{10}$$

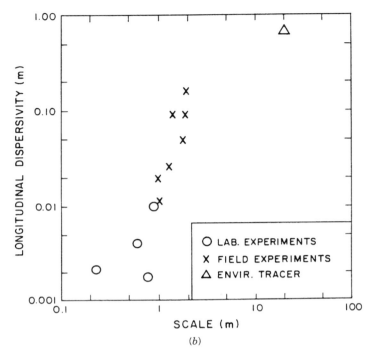

Figure 7.5 (*b*) Longitudinal dispersivity versus scale for the the unsaturated zone (from Gelhar et al., 1985).

where J = mass flux (mol/cm^2/s)
       D = effective diffusion coefficient (cm^2/s)
       C = concentration (mol/cm^3)
       grad = gradient operator (cm^{-1}), in SI units

This equation applies both to the dissolved constituents in groundwater and to the volatile constituents in the soil air in the vadose zone. Values for the effective diffusion coefficient of selected constituents in air and water can be found in Cussler (1984) and Domenico and Schwartz (1990). Order-of-magnitude values for diffusion coefficients for most compounds are 10^{-2} cm^2/s in air and 10^{-5} cm^2/s in water. Equations for estimating air and water diffusion coefficients for a range of constituents can be found in Tucker and Nelken (1990).

While seemingly making only a modest contribution to the redistribution in permeable settings, diffusion can be a dominant transport mechanism in low-permeability settings, in both the vadose and saturated zones (Gillham and Cherry, 1983; Johnson et al., 1989) and plays a major role in limiting sorption kinetics (Pignatello and Xing 1996). The diffusion process is also potentially important in the replenishment of dissolved oxygen in the near-water-table portion of the saturated zone by influx from the vadose zone (Barr, 1993; Barr et al., 1995). For these reasons, it can be important to determine diffusion coefficients for the constituents of interest and to monitor oxygen (O_2), carbon dioxide (CO_2), and methane (CH_4) in the soil air in the base of the the vadose zone.

Microbial Transport

Porosity, and minimum pore diameter in particular, are important parameters for determining the migration potential for bacteria and viruses in soils (Romero, 1970). Dragun (1988) applied the following guideline for assessing whether microbial migration was potentially possible:

$$D_{S15}/D_{M85} > 29 \qquad (11)$$

where D_{S15} = particle diameter of soil, where 15% of the soil is finer
D_{M85} = migrating microbial particle, where 85% of such particles are finer

The ranges in particle diameters for soils and biota are shown on Figure 7.6. As can be seen, microbial migration through clays, silts, and clayey sands would not occur by this relationship, and this is consistent with empirical experience. However, it does not preclude microbial migration through macropores or secondary porosity features in these materials, such as cracks, root tubes, solution openings, and the like.

The electrostatic charge on soil particles is another factor impeding microbial migration. Both soil particles and microorganisms generally have a negative surface

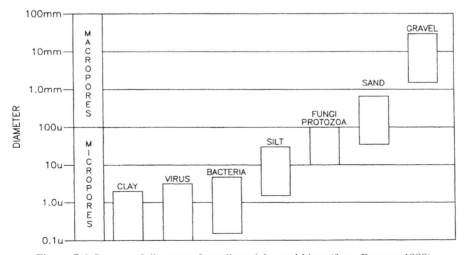

Figure 7.6 Ranges of diameters for soil particles and biota (from Dragun, 1988).

charge. Where pore diameters are large enough, microbial migration can still be inhibited by positively charged soil particles, such as iron oxides; by relatively high dissolved cation concentrations in the soil water (Dragun, 1988); or by microbial properties such as size, motility, and surface hydrophobicity (Rutter and Vincent, 1988; Camper et al., 1993). Another empirical observation is that microbial migration is inhibited in the vadose zone by the mechanisms of unsaturated flow conditions (Lance and Gerba, 1984). The movement and survival of enteric bacteria and viruses are discussed by Gerba and Bitton (1984). Other aspects involving microbial migration are discussed elsewhere in this book.

FLOW PROCESSES

The flow, or advection, processes that move fluids through the subsurface both enables the spread of contamination as well as the initiation of the attenuation processes. For the biodegradation processes within the subsurface, these flow processes are the major means for the redistribution of substrates, electron acceptors, nutrients, and biotransformation products. As long as these redistributions occur at rates greater than diffusion alone, these flow processes largely determine the types, locations, magnitudes, and rates of bioactivity within the subsurface.

Soil Air Flow

Any subsurface contamination problem and its associated biodegradation processes are located within the context of a naturally occurring groundwater flow system. If located within the vadose zone, groundwater flow is dominated by episodes of downward gravitational flow and its redistribution in other directions by capillary forces. This limits the pathways of migration for both dissolved organic carbon compounds as well as the nutrients and electron acceptors. Consequently, it is typically difficult to greatly increase the general biomass and the abundance of the microbial consortia capable of biodegrading contaminants of concern. However, because air is the saturating fluid within the vadose zone, there is the possibility for its flow processes, both naturally occurring and artificially manipulated, to introduce and distribute the electron acceptor oxygen, resulting in potentially significant biodegradation activity.

Under natural conditions the air exchange in the vadose zone is largely the result of two processes. First, the fluctuations of the water table causes a like volume of air to be pushed out or drawn into the vadose zone through the unpaved land surface. The effects of this was investigated by Rainwater et al. (1989) who studied the effects of cyclic water table fluctuations on the biodegradation rates in a capillary fringe zone contaminated with residual concentrations of diesel fuel. The study found that the alternating wetting and aeration of the cyclic fluctuations of the capillary fringe provided a more conducive environment for hydrocarbon biodegradation in that zone. After 9 weeks, the zone with cyclic fluctuations had almost twice as much biodegradation of the diesel fuel compared to a duplicate setting with a static capillary fringe, reducing the diesel fuel volume in the soil by 25%.

The second process affecting the air exchange in the vadose zone is the barometric pressure fluctuation of the atmosphere. These fluctuations cause a compression and decompression of the air column, resulting in air exchange through the unpaved soil surface. The effects of barometric pressure changes on the movement of air in the vadose zone has been described by Massman and Farrier (1992) who found that "fresh" air can intrude several meters into the vadose zone during a typical barometric pressure cycle, and tens of meters during a storm event. However, these air exchanges occur only at the ground surface, and their effects are progressively diminished with depth from the land surface. Consequently, the movement of oxygen in the lowermost portion of the vadose zone is generally dominated by diffusion.

Under natural conditions, the interfacial mass transfer of oxygen from the vadose zone into the saturated zone maintains saturated oxygen concentrations at the capillary fringe. With subsurface contamination in the vadose zone, the biochemical oxygen demand from the geochemical and biological processes initiated by the presence of the contamination can deplete the oxygen in this zone, resulting in oxygen diffusion into the zone.

Another process that can effect air movement within the vadose zone are density-driven gas flows. Vapors comprised of a number of organic compounds, chiefly chlorinated solvents, were found by Falta et al. (1989) to sink through the vadose zone air to the capillary fringe and then migrate radially outward above the capillary fringe. These flow may also occur in response to very small pressure gradients, due to their low viscosity.

With soil venting such as soil vapor extraction or bioventing, these conditions are drastically changed. The movement of air is dominated by the lateral movement induced by venting, and elevated oxygen concentrations can be found throughout the vadose zone, including directly above the capillary fringe. This results in a significant increase in the interfacial mass transfer of oxygen into the saturated zone by one or two orders of magnitude and in turn promotes increased rates of aerobic biodegradation within the contaminant plume in the saturated zone under the portion of the vadose zone influenced by venting operations (Barr et al., 1995). Methods of estimated mass transfer of oxygen across the water table interface can be found in Tucker and Nelken (1990).

Groundwater Flow

Most biodegradation reactions in the saturated zone involve the microbial oxidation of organic compounds using electron acceptors such as oxygen, sulfate, nitrate, iron, and manganese. These electron acceptors undergo reduction at rates mediated and catalyzed by the microbial consortia and their associated enzymes. In addition, the growth of the microbial consortia requires the availability of nutrients for biomass production.

All of these substrates, needed either continuously or intermittently for metabolism or growth in the saturated zone, are to some degree supplied to the biomass by groundwater advection. Furthermore, any subsurface contamination that reaches the water table will have its associated biodegradation processes proceeding within the context of a naturally occurring groundwater flow system. Movement of all dissolved substances is predominantly by advection with redistribution on the pore scale by diffusion between moving and stagnant water.

In order to enhance biodegradation processes within the saturated zone, the groundwater flow patterns must be manipulated. This manipulation must introduce these growth-limiting substrates into the saturated zone, transport them to the area of contamination so that they move through the contaminated area at an appropriate rate, and remove the degradation products that might subsequently impair further biodegradation of the contamination. Because this manipulation is superimposed upon the natural groundwater flow system prevailing in the area of contamination, an effective manipulation of groundwater must fully take into account the features and details of the flow regime. These features are occurring at several scales.

At the large scale, there are several types of flow systems that can be prevailing within the area of interest that range from regional to local in scale. Regional flow systems have relatively long residence times with recharge and discharge area separated by large distances, often several counties apart. Nevertheless, Tóth (1963) found that within a regional flow system almost all water is involved in local flow systems, 90% of which never penetrates deeper than 250–300 ft, and that at a regional discharge point most of the groundwater originates within the adjoining local groundwater flow system. Consequently, where regional flow systems are present, local flow systems are typically superimposed on them. These local flow systems typically have relatively short residence times with recharge and discharge areas generally adjoining, so that vertical flow directions can change direction over relatively short lateral distances.

Under different conditions where the regional slope of the groundwater system is low or where the local topographic relief is relatively high, groundwater flow occurs only in local flow systems that can be deeply penetrating (Tóth, 1963; Freeze and Witherspoon, 1966, 1967, 1968). As a consequence, the groundwater quality may reflect a longer residence time, have relatively high dissolved inorganics concentrations, and can interact with growth-limiting substrates introduced, even with a local groundwater flow system.

On an intermediate scale, surface and subsurface hydrogeologic features can greatly influence the three-dimensional configuration of groundwater flow. This has been systematically studied by Winter (1976, 1978) who found that subsurface factors such as the presence and locations of buried high-permeability zones, permeability contrasts, and anisotropy greatly influence the depth and areal extent of local flow systems. Surface water factors such as water depths, presence and thickness of fine-grained bottom sediments, and location of surface water bodies relative to buried high-permeability zones were also found to influence the depth and areal extent of local flow systems. These features can greatly influence the distribution of vertical groundwater flow. Failure to take the influence of such features into account can result in an incomplete delineation of the contamination and ineffective manipulation of groundwater for enhanced bioremediation.

On a small scale, the specific flow path within a groundwater flow system is greatly affected by the site-specific heterogeneity of geologic units and subunits. When a site is investigated in detail, permeabilities can vary over several orders of magnitude, even within an apparently homogeneous, high-permeability setting such as an outwash sand deposit (Sudicky et al., 1983; Sudicky, 1986). The presence of an underlying higher permeability unit can cause significant downward flow toward the underlying unit, without an apparent downward gradient, as has been described by Strack (1984, 1989). Furthermore, groundwater flow paths can have a downward trajectory simply from ongoing infiltration into the top of an aquifer. The position of a streamline within an aquifer is defined (Strack, 1989, p. 322) as

$$Z/H = x_0/x \qquad (12)$$

where Z = height of streamline above the aquifer base at a distance x downgradient of the groundwater divide

h = saturated thickness of the aquifer at distance x

x_0 = distance downgradient from the groundwater divide that the streamline entered the aquifer

The manipulation of groundwater flow must take these small-scale effects into account to avoid incomplete or ineffective delivery of growth-limiting substrates to the contaminated area.

Acknowledgments

William A. Newman, Rich G. Soule, and Louis J. DeFilippi are gratefully acknowledged for their technical assistance, their review of the manuscript, and their insightful suggestions and comments.

REFERENCES

Barr, K. D. (1993). Enhanced Groundwater Remediation by Bioventing and Its Simulation by Biomodeling, *Proceedings of the U.S. Air Force Environmental Restoration Technology Transfer Symposium*, San Antonio, TX, 3.63–3.72, Jan. 26–28.

Barr, K. D., B. D. O'Flanagan, W. A. Newman, J. K. Julik, and D. W. Wetzstein (1995). In-Situ Treatment of Landfill Leachate and Wastes Using Bioventing Technology, *Proceedings, Superfund XVI Conference, Washington, DC, 882–891 November 6–8, pp. 882–891.*

Brooks, R. H. and A. T. Corey (1966). *Properties of Porous Media Affecting Fluid Flow*, ASCE J. Irrigat. Drainage Div., **92**:IR2.

Camper, A. K., J. T. Hayes, P. J. Sturman, W. L. Jones, and A. B. Cunningham (1993). Effects of Motility and Adsorption Rate Coefficient on Transport of Bacteria through Saturated Porous Media, *Appl. Environ. Microbiol.*, **59**(10):3455–3462.

Chauveteau, G. and C. Thirriot (1967). Régimes découlement en milieu poreux et limite de la loi de Darcy, *La Houille Blanche*, **1**(22):1–8.

Chebotarev, I. I. (1955): Metamorphism of Natural Waters in the Crust of Weathering, Part I, *Geochim. Cosmochim. Acta.*, **8**:1–48.

Chiou, C. T., D. W. Schmedding, and M. Manes, (1982): Partitioning of Organic Compounds in Octanol-Water Systems, *Environ. Sci. Tech.*, **16**:4–10.

Corey, A. T. (1967). *Mechanics of Heterogeneous Fluids in Porous Media*, Water Resources Publications, Fort Collins, CO.

Cussler, E. L. (1984). *Diffusion: Mass Transfer in Fluid Systems*, Cambridge University Press, New York,

de Jonge, H. and M. C. Mittelmeijer-Hazeleger (1996). Adsorption of CO_2 and N_2 on Soil Organic Matter: Nature of Porosity, Surface Area, and Diffusion Mechanisms, *Environ. Sci. Tech.*, **30**(2):408–413.

de Marsily, G. (1986). *Quantitative Hydrogeology*, Academic Press, Orlando, FL.

Domenico, P. A. (1972). *Concepts and Models in Groundwater Hydrology*, McGraw-Hill, New York.

Domenico, P. A. and F. W. Schwartz (1990). *Physical and Chemical Hydrogeology*, Wiley, New York.

Dragun, J. (1988). *The Soil Chemistry of Hazardous Materials*, Hazardous Materials Control Research Institute, Silver Springs, MD.

Ellerd, M. G. and J. W. Massmann (1995). *Analytical Tools for Designing Subsurface Gas Extraction and Control Systems*, Department of Engineering Professional Development, University of Wisconsin—Madison, Madison, WI.

Falta, R.W., I. Javendel, K. Pruess, and P. A. Witherspoon (1989). Density-Driven Flow of Gas in the Unsaturated Zone Due to the Evaporation of Volatile Organic Compounds, *Wat. Resour. Res.*, **25**(10):2159–2169.

Freeze, R. A. and J. A. Cherry (1979). *Groundwater*, Prentice-Hall, Englewood Cliffs, NJ,

Freeze, R. A. and P. A. Witherspoon (1996). Theoretical Analysis of Regional Ground Water Flow: 1. Analytical and Numerical Solutions to the Mathematical Model, *Wat. Resour. Res.*, **2**(4):641–656.

Freeze, R. A. and P. A. Witherspoon (1967). Theoretical Analysis of Regional Ground Water Flow: 2. Effect of Water-Table Configuration and Subsurface Permeability Variation, *Wat. Resour. Res.*, **3**(2): 623–634.

Freeze, R. A. and P. A. Witherspoon (1968). Theoretical Analysis of Regional Ground Water Flow: 3. Quantitative Interpretations, *Wat. Resour. Res.*, **4**(3):581–590.

Garrels, R. M. and C. L. Christ (1965). *Solutions, Minerals and Equilibria*, Harper & Row, New York,

Gelhar, L. W., A. Mantoglou, C. Welty, and K. R. Rehfeldt (1985). *A Review of Field-Scale Physical Solute Transport Processes in Saturated and Unsaturated Porous Media*, Electric Power Research Institute Report EA-4190, Palo Alto, CA.

Gelhar, L. W., C. Welty, and K. R. Rehfeldt (1992). A Critical Review of Data on Field-Scale Dispersion in Aquifers, *Wat. Resour. Res.*, **28**(7): 1955–1974.

Gerba, C. P. and G. Bitton (1984). Microbial Pollutants: Their Survival and Transport Pattern to Groundwater, *Groundwater Pollution Microbiology*, in G. Bitton and C. P. Gerba, Eds., Wiley, New York, Chapter 4.

Gillham, R. W. and J. A. Cherry (1983). Predictability of Solute Transport in Diffusion-Controlled Hydrogeologic Regimes, in *Proceedings of the Symposium on Low-Level Waste Disposal: Facility Design, Construction and Operating Practices*, Sept. 28–29, 1982, Washington, DC., U. S. Nuclear Regulatory Commission NUREG/CP-0028, CONF-820911, **3**:379–410.

Hassett, J. J., W. L. Banwart, and R. A. Griffin (1983). Correlation of Compound Properties with Sorption Characteristics of Nonpolar Compounds by Soils and Sediments: Concepts and Limitations, in *Environment and Solid Wastes: Characterization, Treatment and Disposal*, Francis, C. W. and S. I. Auerbach, Eds., Butterworth, Stoneham, MA, pp. 161–178.

Hem, J. D. (1970). *Study and Interpretation of the Chemical Characteristics of Natural Water*, U. S. Geological Survey Water-Supply, Paper 1473.

Hough, B. K. (1969). *Basic Soils Engineering*, Ronald Press, New York,

Hutchinson, G. E. (1957). *A Treatise on Limnology*, Wiley, New York,

Johnson, A. I. (1967). *Specific Yield—Compilation of Specific Yields for Various Materials*, U. S. Geological Survey Water-Supply, Paper 1662-D.

Johnson, R. L., J. A. Cherry, and J. F. Pankow (1989). Diffusive Contaminant Transport in Natural Clay: A Field Example and Implications for Clay-Lined Waste Disposal Sites, *Environ. Sci. Tech.*, **23**(3):340–349.

Karickhoff, S. W., D. S. Brown, and T. A. Scott (1979). Sorption of Hydrophobic Pollutants on Natural Sediments, *Wat. Res.*, **13**:241–248.

Lambe, T. W. and R. V. Whitman (1969). *Soil Mechanics*, Wiley, New York,

Lance, J. C. and C. P. Gerba (1984). Virus Movement in Soil During Saturated and Unsaturated Flow, *Appl. Environ. Microbiol.*, **47**:335–337.

Lyman, W. J. (1990). Adsorption Coefficient for Soils and Sediments, in *Handbook of Chemical Property Estimation Methods*, Lyman, W. J., W. F. Reehl, and D. H. Rosenblatt, Eds., American Chemical Society, Washington, DC, Chapter 4.

Massman, J. W. (1989). Applying Groundwater Flow Models in Vapor Extraction System Design, *J. Environ. Eng.*, **115**(1):129–149.

Massman, J. and D. F. Farrier (1992). Effects of Atmospheric Pressures on Gas Transport in the Vadose Zone, *Wat. Resour. Res.*, **28**(3):777–791.

Olson, R. M. (1973). *Essentials of Engineering Fluid Mechanics*, 3rd ed., Intext Press, New York,

Perlinger, J. A. and S. J. Eisenreich (1991). *Innovative Approach to Determine Retardation Factors of Organic Contaminants in Groundwater Systems*, Report to U.S. Geological Survey, NTIS Document No. PB91–195792.

Pignatello, J. J. and B. Xing (1996). Mechanisms of Slow Sorption of Organic Chemicals to Natural Particles, *Environ. Sci. Tech.*, **30**(1):1–11.

Polubarinova-Kochina, P. Ya. (1962). *Theory of Groundwater Movement* (trans. by R. J. M. deWiest), Princeton University Press, Princeton, NJ.

Rainwater, K., M. P. Mayfield, C. Heintx-Wyatt, and B. J. Claborn (1989). Laboratory Studies of the Effects of Cyclic Vertical Water Table Movement on In Situ Biodegradation of Diesel Fuel, *Proceedings of the NWWA/API Conference on Petroleum Hydrocarbons and Organic Chemicals in Ground Water—Prevention, Detection, and Restoration*, Houston, TX, November 15–17, pp. 673–686.

Rehfeldt, K. R. and L. W. Gelhar (1992). Stochastic Analysis of Dispersion in Unsteady Flow in Heterogeneous Aquifers, *Wat. Resour. Res.*, **28**(8):2085–2099.

Romero, J. C. (1970). The Movement of Bacteria and Viruses Through Porous Media, *Ground Water*, **8**:37–48.

Rutter, P. R. and B. Vincent (1988). Attachment Mechanisms in the Surface Growth of Microorganisms, in *Physiological Models in Microbiology, Vol. II.*, M. J. Bazin and J. I. Prosser, Eds., CRC Press, Boca Raton, FL, pp. 87–107.

Schwarzenbach, R. P. and J. Westall (1981). Transport of Nonpolar Organic Compounds from Surface Water to Groundwater. Laboratory Studies, *Environ. Sci. Tech.*, **15**:1300–1367.

Schoeller, H. (1959). Geochemistry of Ground Water, in *Arid Zone Hydrology—Recent Developments*, UNESCO, Paris, France, pp. 54–83.

Strack, O. D. L. (1984). Three-Dimensional Streamlines in Dupuit-Forchheimer Models, *Wat. Resour. Res.* **20**(7):812–822.

Strack, O. D. L. (1989). *Groundwater Mechanics*, Prentice-Hall, Englewood Cliffs, NJ.

Stumm, W. and J. J. Morgan (1970). *Aquatic Chemistry, An Introduction Emphasizing Chemical Equilibria in Natural Waters*, Wiley-Interscience, New York.

Stylianou, C. and B. A. DeVantier (1995). Relative Air Permeability as Function of Saturation in Soil Venting, *J. Environ. Eng.*, **121**(4):337–347.

Sudicky, E. A. (1986). A Natural Gradient Experiment on Solute Transport in a Sand Aquifer: Spatial Variability of Hydraulic Conductivity and its Role in the Dispersion Process, *Wat. Resour. Res.*, **22**(13):2069–2082.

Sudicky, E. A., J. A. Cherry, and E. O. Frind (1983). Migration of Contaminants in Groundwater at a Landfill: A Case Study, *J. Hydrology*, **63**(1/2):81–108.

Thorstenson, D. C. and D. W. Pollock (1989). Gas Transport in Unsaturated Porous Media: The Adequacy of Fick's Law, *Rev. Geophys.*, **27**(1):61–78.

Tóth, J. (1963). A Theoretical Analysis of Groundwater Flow in Small Drainage Basins, *J. Geophys. Res.*, **63**(16):4795–4812.

Tucker, W. A. and L. H. Nelken (1990). Diffusion Coefficients in Air and Water, in *Handbook of Chemical Property Estimation Methods*, W. J. Lyman, W. F. Reehl, and D. H. Rosenblatt, Eds., American Chemical Society, Washington, DC.

van der Kamp, G., L. D. Luba, J. A. Cherry, and H. Maathuis (1994). Field Study of a Long and Very Narrow Contaminant Plume, *Ground Water*, **32**(6):1008–1016.

Verruijt, A. (1970). *Theory of Groundwater Flow*, Macmillan, London.

Verschueren, K. (1983). *Handbook of Environmental Data on Organic Chemicals*, Van Nostrand Reinhold, New York,

Winter, T. C. (1976). *Numerical Simulation Analysis of the Interaction of Lakes and Ground Water*, U.S. Geological Survey Professional, Paper 1001, U.S. Government Printing Office, Washington, DC.

Winter, T. C. (1978). Numerical Simulation of Steady State Three-Dimensional Groundwater Flow Near Lakes, *Wat. Resour. Res.*, **14**(2):245–254.

CHAPTER 8

Assessment of the Potential for Clogging and Its Mitigation During In Situ Bioremediation

Peter R. Jaffé and Stewart W. Taylor

Department of Civil Engineering and Operations Research, Princeton University, Princeton New Jersay 08544 (P.R.J); Bechtel International, Inc., Oakridge, Tennessee (S.W.T)

The proper design and operation of in situ biological remediation schemes requires careful attention to the alteration of soil permeability due to clogging. Clogging is a common occurrence in the vicinity of injection wells and at the soil surface during artificial groundwater recharge. This clogging can be physical, chemical, or biological.

Physical clogging occurs when suspended solids are deposited in soil pores that then restrict the passage of additional particles through these pores. When the diameter of the suspended particles is small compared to the pore diameter, the particles will penetrate deeper into the soil formation. Hence, as the size of the suspended solids increases, clogging occurs closer to the point of injection and for the same mass of solids injected will also occur faster. Physical clogging has been studied extensively and will not be addressed in this chapter.

Chemical clogging can occur during the in situ bioremediation of aquifers, especially when hydrogen peroxide is utilized as an oxidizing agent. Hydrocarbon-degrading bacteria can adapt to concentrations of hydrogen peroxide equivalent to 100 mg/L of oxygen or more. Hydrogen peroxide will decompose rapidly in the presence of Fe^{2+}, leading to the formation of oxygen bubbles that block the soil pores in the vicinity of the injection well. To avoid this process, phosphate needs to be added to the injection water to precipitate the iron. This may lead to chemical clogging in very hard waters due to the possible precipitation of magnesium and calcium phosphates. Hence, careful consideration of the groundwater geochemistry has to be given to avoid clogging of the formation due to the addition of hydrogen peroxide.

Biological Treatment of Hazardous Wastes, Edited by Gordon A. Lewandowski and Louis J. DeFilippi
ISBN 0-471-04861-5   ©1998 John Wiley & Sons, Inc.

Biological clogging is caused by a decrease of the effective pore size due to biological growth. This chapter will concentrate mainly on this type of clogging. The two field applications where biological clogging is of most concern are during the artificial recharge of groundwaters with treated wastewater and during the in situ bioremediation of contaminated aquifers.

Bacteria, especially aerobic ones, can attach to solid surfaces by means of exopolysaccharides, which continue to protect the attached biomass as it grows. This exopolysacharide matrix consists largely of anionic polymers that can trap organic and inorganic nutrients by ion exchange, providing a source of nutrients for the bacteria even in environments where nutrients exist at very low concentrations (Costerton et al., 1981). This attached matrix of bacteria and exopolysaccharides is called a biofilm. Because biofilms can develop under very low nutrient conditions, biological clogging has been observed even during the injection of tertiary treated wastewater for the artificial recharge of groundwater.

The reduction in the permeability of soils due to biological growth has been studied by many investigators in soil column experiments. Gupta and Swartzendruber (1962) clearly illustrated the link between biological growth and a reduction in soil permeability. They observed a significant permeability reduction in the top 15 cm of their soil columns during prolonged flow at ambient temperatures. This reduction in permeability correlated well with bacterial numbers, as long as they were in excess of 400,000 per gram of sand. No decrease in permeability was observed when the conditions were adverse to microbial growth, such as temperatures just above freezing or addition of a 0.1% phenol solution. In similar experiments, Mitchell and Nevo (1964) observed that clogging of a sandy porous media after a prolonged percolation of water containing organic matter correlated stronger with the accumulation of polysaccharides than with bacterial numbers, suggesting that conditions that favor the accumulation of polysaccharides may result in more severe clogging. The importance of these polysaccharides in soil clogging was demonstrated further by Shaw et al. (1985). They showed that after the permeability in a soil core was reduced due to the development of a biofilm, addition of a biocide that killed the bacteria in the biofilm did not have an immediate effect on the permeability. Whereas, the addition of an oxidizing agent, such as 5% sodium hypochlorite, that killed the bacteria and also dissolved the biofilm's exopolysaccharide matrix, restored the permeability of the cores. This is one of the reasons chlorine, at a concentration of about 2 mg/L, has been used successfully as an oxidizing agent to avoid clogging at a number of injection wells (Ehrlich et al., 1979, Vecchioli et al., 1980, Clementz et al., 1982).

Several investigators have linked the biofilm accumulation (in terms of an average biofilm thickness) to changes in the permeability and porosity. Under different operating conditions they also observed that the permeability approached asymptotically a minimum value (Frankenberger et al., 1979; Taylor and Jaffé, 1990a; Cunningham et al., 1991). For constant-head column experiments and homogenous porous media, Cunningham et al. (1991) noted that the biofilm reached a maximum thickness in 5 days, resulting in a porosity reduction between 50 and 96%, and a permeability decrease between 92 and 98%. Their observations indicated that a minimum permeability between 3×10^8 and $7 \times 10^8 \text{cm}^2$ was reached after the biofilm developed to its maximum thickness. In constant-flow column experiments, Taylor and Jaffé (1990a) observed a minimum hydraulic conductivity of $8 \times 10^{-3} \text{cm/min}$. The absolute value for such a minimum permeability is expected to vary depending on the field circumstances. In constant-head experiments, which more closely simulate a groundwater recharge process, as the permeability decreases and less water is infiltrating through the soil, less nutrients are available for further biomass growth. The mechanism leading toward a minimum permeability during constant-flow experiments will be examined further in this chapter.

Because biofilm growth is limited by nutrient availability, more growth will occur in regions with a higher permeability. This increased biofilm growth will in turn reduce the permeability of these regions, having the overall effect that an initially heterogeneous permeability field will become more homogeneous during bacterial clogging. This was observed by several investigators who evaluated the possibility of stimulating microbial growth to selectively plug the more permeable regions of a formation in order to increase the efficiency of oil recovery during water injection (Macleod et al., 1988, Raiders et al. 1989). This selective clogging is related to the heterogeneities at the pore scale. Consider two porous media with the same overall permeability, one being homogeneous (narrow range of pore size) while the other is heterogeneous (wide range of pore sizes). If biological growth occurred in both of these media, we would expect a relatively uniform growth in the homogeneous one, while in the heterogeneous medium we would expect to see a larger biomass accumulation in the larger pores. Since the larger pores are the main flow conduits of the heterogeneous medium, clogging of these large pores will result in a more pronounced decrease in permeability. Hence, we would expect a larger decrease in the overall permeability in the heterogeneous media than in the homogeneous media. Whereas most experimental results linking the biomass growth to changes in permeability are for homogeneous media, there is a clear need to link biomass growth to changes in permeability for heterogeneous media. This will be discussed in more detail later in this chapter.

Due to the filtration properties of porous media, and the affinity of bacteria for surfaces, much of the biomass introduced in solution will become immobilized in the pore space and contribute to permeability reduction. The role of bacterial filtration and adsorption on permeability reduction and biomass transport has been described by Taylor and Jaffé (1990b).

In addition to biological clogging due to the accumulation of biomass, the production and accumulation of biogases has been observed to result in clogging. Clogging during denitrification was studied by Soares et al. (1988, 1989) in sand column experiments. They observed that once denitrification was well established, accumulation of nitrogen gas in the top 15 cm of the column occurred rapidly, with a consequent reduction in the permeability. Removals on the order of 100 ppm of nitrate resulted in a decrease in permeability of more than an order of magnitude in only 100 h of operation.

The establishment of an active denitrification process in column experiments to which secondary wastewater effluent was applied intermittently for up to 3 years has also been observed to contribute toward clogging (Rice, 1974). In these experiments, physical clogging, near the surface, due to the deposition of suspended sediments was the main cause of clogging. Biological clogging became noticeable only after longer infiltration periods, which were possible if the suspended sediments were low (less than 10 mg/L). During the drying cycle the permeability of the media was restored to values similar to those at the beginning of the infiltration experiments. After operating the columns intermittently for 3 years, a reduction of 50–60% in the permeability was observed. This reduction was attributed to the accumulation of insoluble gases resulting from the denitrification process.

CLOGGING DURING GROUNDWATER RECHARGE

Groundwater recharge is becoming common practice in arid areas, and its objective is to prevent further lowering of groundwater levels. This is achieved by either injecting surface water, or treated wastewater, through injection wells, or allowing the water to infiltrate directly through permeable soil from the surface. In most cases the infiltrating water contains some suspended sediments as well as nutrients, which may lead to physical as well as biological clogging. Column infiltration studies (Ripley and Saleem, 1973) have shown that biological clogging occurs mainly in the upper few centimeters, while physical clogging is dependent on the suspended particles and may occur throughout the soil column.

In field applications of secondary wastewater, Bouwer et al. (1974) observed a 50% reduction in the permeability over 20- to 30-day infiltration periods. Drying periods of 10 days in length resulted, in most cases, in a complete restoration of the permeability. An interesting finding of these field studies was that vegetated surfaces resulted in higher infiltration rates, due to the removal of suspended sediments. Again, clogging was observed to occur close to the soil surface, and no reduction in permeability could be observed in the underlying aquifer.

From these observations it is clear that a constraint for the artificial recharge of groundwater by secondary wastewater is the reduction of the infiltration rate over time due to soil clogging. This clogging is mainly a physical process due to the filtration of suspended sediments, and to a lesser degree a biological process. Biological clogging becomes more important when the organic matter content of the wastewater is high, and the flooding time is long so that sufficient biomass can be formed. For even longer flooding times, denitrification can lead to the accumulation of nitrogen bubbles that also cause clogging. During a drying cycle organic suspended solids, as well as biomass, can be oxidized, restoring the soil permeability. A drying cycle may also allow trapped gas to escape to the atmosphere.

BIOLOGICAL CLOGGING DURING IN SITU BIOREMEDIATION: EXPERIMENTAL AND THEORETICAL RESULTS

To better understand and quantify the permeability reduction caused by enhanced biological growth at a nutrient injection well, a bench-scale experimental investigation was conducted by Taylor and Jaffé (1990a). A theoretical interpretation of these experimental results was provided by Taylor et al. (1990). This section summarizes and highlights the significant findings of these and other investigations.

Experimental Setup

The experimental investigation was conducted using packed-column reactors measuring 5.08 cm in diameter and 52 cm in length. Columns were packed with a uniform test sand having a mean grain diameter of 0.7 mm, a porosity of 0.35, and a saturated hydraulic conductivity of 2.5×10^{-1} cm/s. Piezometers were located along the longitudinal axis of each column at 1 to 4 cm intervals for the measurement of piezometric head. A sample port was integrated with each piezometer to allow acquisition of a water sample from the interior of the column for analysis of substrate concentration. Columns were operated in an up-flow mode at continuous flow rates controlled by a peristaltic pump, resulting in Darcy velocities ranging from 3.2 to 9.5 m/day, which were representative of the forced gradient conditions imposed at an injection well.

Columns were seeded with bacteria derived from a sewage treatment plant and acclimated to the growth substrate (methanol). A solution comprised of a mineral salts medium and substrate was then continuously supplied to each column over the duration of the experiment. The mineral salts medium provided all essential and trace nutrients other than carbon. Dissolved oxygen levels were maintained at sufficient levels to prevent oxygen limitation and ensure that biomass growth was substrate limited.

Piezometric head and substrate concentration were monitored over space and over the duration of each experiment, which lasted 284–356 days. Piezometric head differentials across sample ports, along with the imposed Darcy velocities, were used to calculate the corresponding changes in hydraulic conductivity over space and time. Column experiments were terminated when changes in piezometric head and substrate concentrations were small with respect to time. Upon termination of the experiment, the columns were dismantled and their contents analyzed to determine the bulk biomass density (biomass per unit volume of porous medium) as a function of location in the column.

Experimental Results

Observed permeability reductions for two column experiments are plotted in Figures 8.1 and 8.2 as a function of axial position and time. Permeability reduction is defined as the permeability with in situ biomass divided by initial permeability of

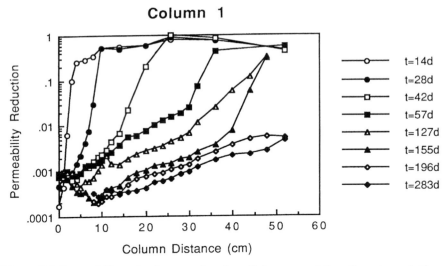

Figure 8.1 Permeability reduction as a function of distance and time for column 1 (from Taylor and Jaffé, 1990a).

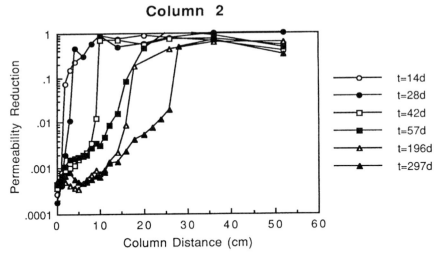

Figure 8.2 Permeability reduction as a function of distance and time for column 2 (from Taylor and Jaffé, 1990a).

the medium without biomass. A value of unity corresponds to no permeability reduction, while a value less than unity represents the fractional reduction in permeability due to biomass growth. Initial results show that permeability is reduced by roughly 3 orders of magnitude near the column inlet, a region of high substrate utilization and biomass growth. Over time, the region of reduced permeability propagates into the column and the permeability reduction is apparently bounded

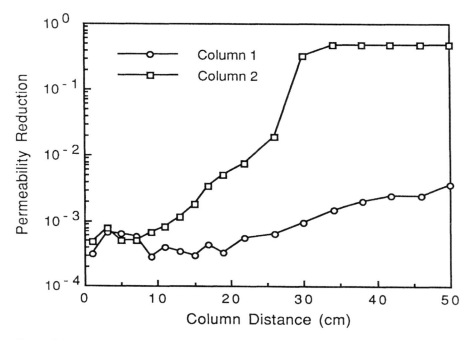

Figure 8.3 Permeability reduction as a function of distance at $t = 284$ days for column 1 and $t = 365$ days for column 2 (from Taylor and Jaffé, 1990a).

between 10^{-4} and 10^{-3}. Profiles obtained near the end of the experiment show that the region of reduced permeability encompasses the entire length of column 1, whereas the profile of column 2 shows the region of reduced permeability extends over the first 28 cm, or roughly half of the column.

The permeability reduction profiles observed at the end of each column experiment have also been plotted in Figure 8.3. The column 1 profile shows the permeability reduction to be nearly constant over the first 20 cm of the column, the average value being about 4.3×10^{-4}. Similarly, the column 2 profile shows the permeability reduction to be constant over the first 8 cm of the column, having an average value of 5.7×10^{-4}.

Substrate concentrations monitored over the duration of the column experiments are plotted at selected times in Figures 8.4 and 8.5. These data show that substrate is depleted within the first 5–10 cm of the column. The data also show the substrate penetrating more deeply into the column over time, suggesting either changes in the kinetics of substrate utilization or increased rates of advective and dispersive transport. Given that the Darcy velocity was maintained constant for a given experiment, increased advective and dispersive transport rates suggest a reduction in effective porosity.

The spatial distributions of the bulk biomass density, expressed in terms of bacterial organic carbon (BOC), measured at the end of each experiment are plotted

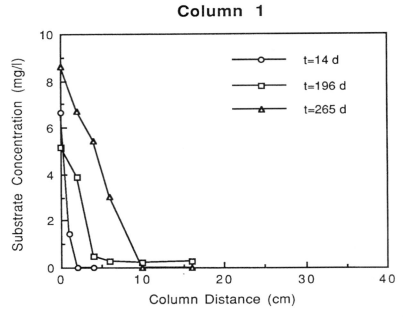

Figure 8.4 Substrate concentration as a function of distance and time for column 1 (from Taylor and Jaffé, 1990a).

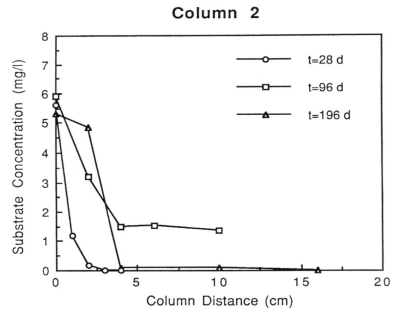

Figure 8.5 Substrate concentration as a function of distance and time for column 2 (from Taylor and Jaffé, 1990a).

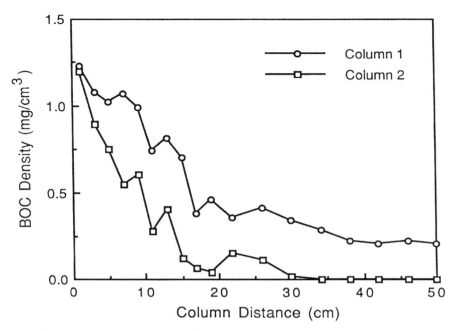

Figure 8.6 Bulk biomass density as a function of distance at t = 284 days for column 1 and t = 365 days for column 2, where biomass is expressed in terms of BOC (from Taylor and Jaffé, 1990a).

in Figure 8.6. Column 1 data show the biomass density to be a monotonically decreasing function of column distance and indicate that biomass has penetrated the entire column length. Column 2 data behave similarly, although biomass has penetrated only 28–32 cm into the column. The relationship between permeability reduction and bulk biomass density is illustrated in Figure 8.7. Permeability reduction is seen to decrease with increasing bulk biomass density for densities less than about 0.4 mg/cm^3. At higher densities, permeability reduction is independent of the bulk biomass density. This region of independence corresponds to the lower limit of permeability reduction described earlier.

Discussion of Experimental Results

Results from these experiments clearly show that in situ biological growth can cause substantial reductions in the permeability of a porous medium. These results also suggest the existence of a lower limit beyond which no further permeability reduction can occur. In the experiments documented by Taylor and Jaffé (1990a), the magnitude of this limit was found to be about 5×10^{-4}, which corresponds to a hydraulic conductivity of about 1×10^{-4} cm/s. One of two conceptual models may account for the limit on permeability reduction resulting from in situ biomass

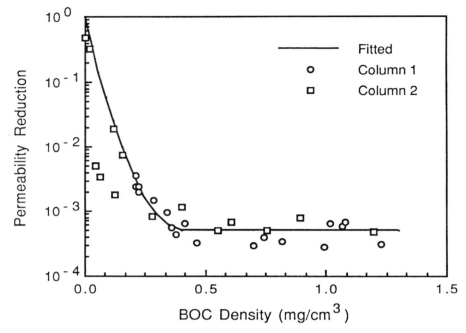

Figure 8.7 Permeability reduction as a function of bulk biomass density, where biomass is expressed in terms of BOC (from Taylor and Jaffé, 1990a).

growth, and have been referred to as the *closed-pore* model and the *open-pore* model.

The underlying premise of the closed-pore model is that biomass has grown and completely filled the pore space of the medium and that water is flowing through the biomass matrix itself. If this were the case, measured permeabilities would reflect the intrinsic permeability of the biomass matrix; permeability would therefore be constant for regions of the medium for which biomass has accumulated in the pore space. This model would explain the nearly constant permeability observed in the first 8–20 cm of the column experiments described by Taylor and Jaffé (1990a). Furthermore, the average linear (pore) velocity would be nearly unaffected by the presence of biomass, since water comprises about 99% of the biomass matrix. Dispersivity, a function of pore space geometry, would also appear unaffected as the pore space geometry for flow would be defined by the solid (mineral) phase in this conceptual model.

The open-pore model assumes that fluid shear stresses, increased by the reduction in permeability, detach bacteria from the pore biofilm and thus maintain an open cross section. A mass balance for the biofilm includes terms for biomass growth and decay, as well as terms for detachment due to fluid shear and reattachment due to the physicochemical process of filtration. As biomass accumulates due to growth, the biofilm thickness would increase and the pore space would decrease, resulting in increased pore water velocities and shear stresses at the biofilm–water interface. At steady state, there would exist a critical shear stress, related to the pore water velocity and the kinetics of biomass growth and decay, for which the net rate of removal due to shearing and filtration would balance the net biomass growth rate. Permeability, a function of pore space, would then be constant for regions of a medium for which this steady-state condition exists and would explain the nearly constant values of biomass-affected permeability observed by Taylor and Jaffé (1990a) in their column experiments. In this conceptual model, the biofilm effectively acts as a second, impermeable solid phase. Biofilm growth would alter the pore space geometry, which would result in a reduction in porosity and affect the dispersivity of the medium as well.

While the decrease in effective porosity, as evidenced by the increased advective and dispersive transport of substrate, provides indirect evidence supporting the concept of an open-pore model, later experiments conducted by Cunningham et al. (1991) provide a more definitive basis for this conceptual model. They constructed experimental porous media reactors that allowed optical, in situ measurements of the microscale biomass distribution, including measurement of the biofilm thickness. Media tested consisted of glass spheres and sand of various sizes. These reactors were operated under constant gradient conditions, as opposed to the constant flux conditions imposed by Taylor and Jaffé (1990a), which resulted in a flow rate decrease over time with biomass growth. Within 2–3 days after inoculation, a uniform biofilm of detectable thickness could be observed on the exposed edges of the reactor media. Biofilm development and thickness, and the biofilm-affected permeability were monitored over time. The reactors reached steady state after a period of 5–10 days, after which no appreciable change in biofilm thickness or permeability occurred. At steady-state, biofilms with thickness ranging from 10 to 60 μm were observed that had displaced a substantial part of the original pore space, with porosity being reduced by as much as 50%. Permeability reductions ranged between 1 and 5% of the original value at steady state. Similar to Taylor and Jaffé (1990a), Cunningham et al. (1991) observed a minimum permeability that persisted after biofilm thickness reached a maximum value and noted that such results indicate substantial interaction among mass transport, hydrodynamics, and biomass accumulation at the biofilm–liquid interface in porous media.

Vandevivere and Baveye (1992a,b) and Baveye et al. (1992) have cited evidence that bacteria do not form continuous biofilms, but rather isolated bacterial aggregates on the sand particles used in their column studies. They also observed similar reductions in permeability as Taylor and Jaffé (1990a) and Cunningham et al. (1991) and argue that the permeability reduction is caused by the accumulation of bacterial aggregates at pore constrictions, where they form plugs.

Rittmann (1993) used the normalized substrate loading, J, as a criterion for establishing whether bacterial colonization of the solid porous media surface is a continuous biofilm or discrete bacterial aggregates. This parameter is defined as the ratio of actual substrate flux into the biofilm to the minimum flux giving a steady-state biofilm that is "deep." The term "deep" means that diffusional resistance and biofilm thickness are great enough that the substrate concentration within the biofilm eventually approaches zero. Once a biofilm is deep, the addition of more biomass does not increase the flux because the added biomass is exposed to zero substrate concentration. Rittmann (1993) analyzed a number of porous media experiments for which the continuity of colonization was known. The results showed that biofilms are continuous for J greater than 1.0 but appear to become discontinuous for values less than about 0.25. He further noted that distinguishing between continuous and discontinuous biofilms may be of the greatest importance when permeability reduction is being modeled.

Physical Models for Permeability Reduction

The empirical results reported by Taylor and Jaffé (1990a), Cunningham et al. (1991), and Vandevivere and Baveye (1992a,b) provide insight into the process of permeability reduction via in situ biological growth, but such results are media specific and cannot be generalized to media having different pore size or grain size distributions. A physically based model is required to generalize the relationship between permeability reduction and biofilm thickness or biovolume, accounting for the physical characteristics of the porous medium. Physically based models for permeability reduction have been developed by Taylor et al. (1990) and Cunningham et al. (1991) for the case of continuous biofilms, which are described in more detail later. These models have been developed around two geometric representations of porous media: *sphere-type* models, which represent the solid phase as various stable and regular packings of uniform spheres, and *cut-and-random-rejoin-type* models, which consider the random nature and interconnectedness of the pores. Both types of models assume that the porous medium is rigid, flow is saturated, and the biofilm has negligible permeability. No physically based models have yet been developed for discontinuous biofilms, particularly when the permeability reduction is conceptualized as being caused by the accumulation of bacterial aggregates at pore constrictions, where they form plugs.

Taylor et al. (1990) used the sphere-type model to describe the porosity and permeability reduction due to a continuous biofilm of uniform thickness for cubic, orthorhombic, tetragonal-spheroidal, and rhombohedral packing arrangements. Cunningham et al. (1991) report results for the cubic packing arrangement. The conceptual basis for this type of model breaks down when the biofilm thickness is large relative to the grain diameter. This occurs when the biofilm fills the narrow passageways between adjacent spheres and isolates the enclosed pore space. Conceptually, no flow could occur because the pore space is no longer continuous. The application of sphere-type models is therefore limited to relatively thin biofilms on homogeneous, unconsolidated media that can be represented as nearly spherical particles.

Taylor et al. (1990) also used the cut-and-random-rejoin-type model to describe the reduction in porosity and permeability due to a continuous biofilm of uniform thickness. The cut-and-random-rejoin type uses three parameters to describe the

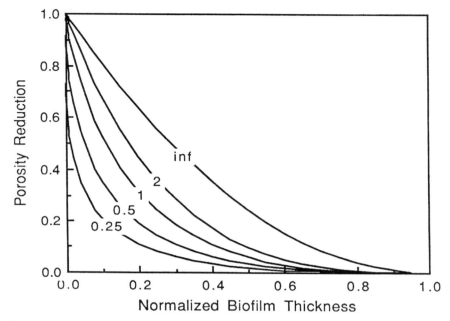

Figure 8.8 Porosity reduction as a function of normalized biofilm thickness and pore size distribution index for the cut-and-random-rejoin-type model ($r_{min} = 0$) (from Taylor et al., 1990a).

pore size distribution: the maximum pore radius, r_{max}; the minimum pore radius, r_{min}; and the pore size distribution index, λ. The physical significance of r_{max} and r_{min} are self-evident, while λ is a measure of the uniformity of pore sizes that can theoretically range from zero (soils with a wide range of pore sizes) to infinity (only one pore size). Figure 8.8 illustrates the porosity reduction as a function of the normalized biofilm thickness and λ for the cut-and-random-rejoin-type model. It is seen that porosity reduction is more pronounced in media having a wide range of pore sizes (λ small) for a given biofilm thickness. Figure 8.9 shows the permeability reduction as a function of the porosity reduction and λ for the cut-and-random-rejoin-type model, using the Mualem-based model for hydraulic conductivity. For a given porosity reduction, the permeability reduction is less significant when pore sizes are widely distributed than when the pore sizes are homogeneous. This behavior is consistent with the physical reality that clogging of relatively small pores has less effect on permeability than clogging of large pores.

Both sphere-type and cut-and-random-rejoin-type models predict quantitatively similar reductions in porosity and permeability for relatively uniform pore sizes (e.g. $\lambda = 2$). However, the sphere-type model is too cumbersome to represent the continuum of pore sizes that characterize natural porous media. Due to the conceptual flaw that occurs for relatively thick biofilms, sphere-type models also cannot predict permeability reductions of the magnitudes often reported in the

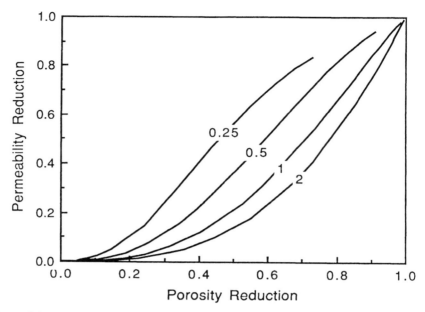

Figure 8.9 Mualem-based permeability reduction as a function of porosity reduction and pore size distribution index for the cut-and-random-rejoin-type model ($r_{min} = 0$) (from Taylor et al., 1990a).

literature. Cut-and-random-rejoin-type models, on the other hand, impose no limitation on the pore size distribution. While Taylor et al. (1990) used a power function to represent the pore size distribution, any pore size distribution function can in fact be accommodated. Furthermore, cut-and-random-rejoin-type models exhibit the proper asymptotic behavior as the pore space fills with biomass, that is, permeability approaches zero as the biofilm-affected porosity approaches zero and the biofilm thickness approaches r_{max}. This attribute allows cut-and-random-rejoin-type models to predict the three-order-of-magnitude permeability reductions often reported in the literature.

DESIGN CONSIDERATIONS/MITIGATION

The bench-scale experiments described above show that high bacterial growth rates, which might occur as a consequence of introducing nutrients into the subsurface, can reduce permeability to 0.1% of the original value. Permeability reductions of these magnitudes could result in the biofouling of a nutrient injection well or infiltration gallery to the point where the facilities are no longer usable, requiring difficult and costly rejuvenation.

Design and operating strategies to mitigate biofouling during in situ bioremediation are discussed below.

Design Considerations

Since the process of permeability reduction is a coupling of biological, chemical, and physical processes, all aspects must be considered in the design of in situ bioremediation systems on a site-specific basis. These design considerations may include the metabolic pathway leading to the biodegradation of the contaminants of concern, additional metabolic pathways and bacterial growth, and the physical properties of the porous medium. Additional factors to be considered are the proliferation of denitrifying bacteria upon introduction of nitrate and the proliferation of iron bacteria (obligate aerobes) upon introduction of oxygen.

The physical properties of the aquifer can significantly influence the biofouling propensity of nutrient injection wells, as demonstrated by Taylor and Jaffé (1991). They developed a mathematical model describing the biofouling of a nutrient injection well from which both electron acceptor and donor are introduced into an aquifer from a single well. For example, the introduction of oxygen and methane into an aquifer to promote the cometabolic biodegradation of trichloroethylene. This model considered the transport and utilization of an electron acceptor and donor, the transport of biomass (including attachment and detachment), the growth and decay of both suspended and fixed biomass, and the changes in porosity, permeability, and pore scale dispersivity of the aquifer with in situ biomass growth. The model was used to simulate the head buildup as a function of time at a nutrient injection well pumped at a constant rate.

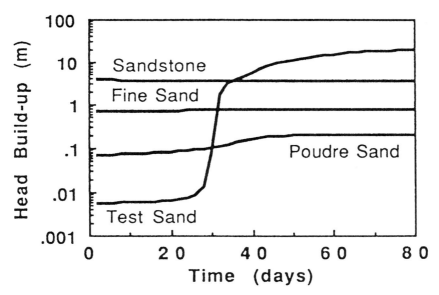

Figure 8.10 Head buildup at a nutrient injection well as a function of time for C-190 test sand ($k = 2.9310^{-6}$), Poudre River sand ($k = 2.2610^{-7}$), fine sand ($k=2.8510^{-8}$), and Berea sandstone ($k = 4.8110^{-9}$). Injection rate is constant for all cases (from Taylor and Jaffé, 1991).

Results are displayed in Figure 8.10 for three types of sand and for Berea sandstone. These results show that relative head buildup may increase significantly with time and asymptotically approach a steady-state value. Based on the steady-state head buildup, it is seen that a medium's susceptibility to biofouling increases with increasing permeability of the initial ("clean") medium. While this result appears to be counterintuitive, it may be explained by considering the biofilm mass balance. For media with a low permeability (e.g., Berea sandstone), the fluid shear acting on the biofilm is large for a given injection rate. The biomass detachment rate from the biofilm, which increases with increasing shear stress, dominates the mass balance and keeps biofilm growth in check. For highly permeable media (e.g., test sand), shear stresses are initially low for the same injection rate, allowing biomass growth to dominate the mass balance. This allows the biofilm to grow and occupy a significant fraction of the pore space, resulting in more head buildup and a greater permeability reduction relative to (initially) less permeable media. Ultimately, the increase in shear stress (and resulting biomass detachment rate) balances the growth rate, resulting in a steady-state biofilm thickness and permeability distribution.

Taylor and Jaffé (1991) also used the same model to identify the relationship between a medium's biofouling propensity, as measured by the steady-state head buildup at the injection well, and the medium's pore structure, which is completely defined by the porosity ϕ, λ, and r_{max} for the cut-and-random-rejoin-type model. The relationships between the head buildup at the injection well and these three parameters are illustrated in Figures 8.11, 8.12, and 8.13. These results show that a porous medium having a large porosity, a wide range of pore sizes, and a large maximum pore radius is most susceptible to clogging due to in situ biomass growth. The head buildup is most sensitive to λ, and it appears that for values of $\lambda < 1$

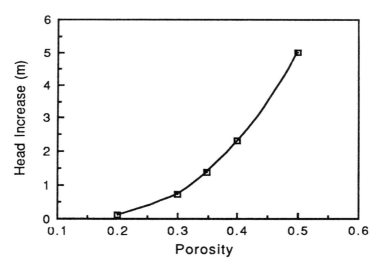

Figure 8.11 Head increase at a nutrient injection well as a function of the (clean) porosity (λ = 2 and r_{max} = 0.016 cm) (from Taylor and Jaffé, 1991).

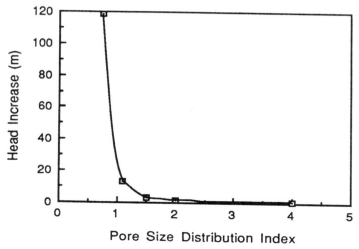

Figure 8.12 Head increase at a nutrient injection well as a function of the pore size distribution index ($\phi = 0.35$ and $r_{max} = 0.016\,cm$) (from Taylor and Jaffé, 1991).

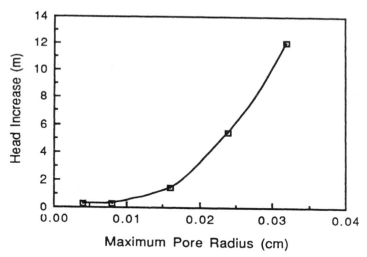

Figure 8.13 Head increase at a nutrient injection well as a function of the maximum pore radius ($\phi = 0.35$ and $\lambda = 2$) (from Taylor and Jaffé, 1991).

biofouling would preclude the simultaneous introduction of both the electron acceptor and donor through a single nutrient injection well. Other results presented by Taylor and Jaffé (1991) show that the use of a threshold permeability alone as a physical criterion for determining the feasibility of in situ biological treatment can be misleading with respect to biofouling. In general, the values ϕ, λ, and r_{max} must be considered together when making this determination.

Mitigation of Biofouling During In Situ Bioremediation

Injection wells and infiltration galleries are most susceptible to biofouling when all constituents required for bacterial growth are introduced simultaneously in a single injection well or infiltration gallery. This could occur when nutrients (nitrogen, phosphorous) and electron acceptor (oxygen) are introduced for the aerobic biodegradation of petroleum hydrocarbons, or when electron acceptor (oxygen) and electron donor (methane) and trace nutrients are introduced for the cometabolic biodegradation of chlorinated hydrocarbons. A number of strategies may be used to mitigate biofouling in these instances.

Clogging in infiltration galleries (above the saturated zone) and in surface soils can be easily mitigated by allowing for a drying period of several days. A method that is appropriate for saturated conditions, and that is limited to aerobic processes, involves the use of hydrogen peroxide as an alternate source of oxygen. Hydrogen peroxide is toxic to bacteria at concentrations greater than about 500 mg/L. When supplied to the subsurface at toxic concentrations, bacterial growth near the injection well or infiltration gallery would be suppressed. Away from such structures, concentrations dilute to nontoxic levels due to hydrodynamic dispersion and reaction. In addition to its relatively high cost, there are disadvantages to using hydrogen peroxide, which have been discussed earlier. Similarly, and applicable also to anaerobic processes, is the injection of chlorine at a concentration of about 2 mg/L, which has been used successfully to suppress clogging at injection wells. These strategies are attractive if the biological activity in the vicinity of the injection well is relatively unimportant with respect to the overall cleanup goal.

Another method involves introducing growth-enhancing constituents in separate wells. Biofouling would be suppressed because only one constituent would be present at any given well. Hydrodynamic mixing of the constituents, and associated biological growth, would occur away from injection wells. The obvious disadvantage to this strategy is the increased capital and operating costs associated with the increased number of nutrient injection wells. Furthermore, by using different injection wells, the proper stoichiometric ratio of these growth-enhancing constituents will not be obtained throughout the whole zone of interest, resulting in a significant loss in efficacy of some or all of these constituents.

A third method is based on the sequential introduction of the growth-enhancing constituents over time. This method has been field tested at the pilot scale (Semprini et al., 1988) and mechanistically modeled (Taylor and Jaffé, 1991) for the sequential injection of an electron acceptor and electron donor into an aquifer. When sequentially injected over time, these constituents are not simultaneously available near the injection well so that bacterial growth near the well screen is suppressed. As these constituents are transported downgradient of the well, hydrodynamic dispersion mixes the constituents at some distance from the well, supplying the bacteria with a nearly continuous and simultaneous supply of the appropriate constituents. Results of model simulations reported by Taylor and Jaffé (1991) showed that this strategy can be very effective in reducing the head buildup at an injection well. Using the same nutrient injection well model as described previously,

Table 8.1   Simulated Head Increases at a Nutrient Injection Well for Simultaneously Introduced and Sequentially Pulsed Electron Acceptor and Donor

Pulse Period (days)	Head Increase (m)
Simultaneous injection	10.10
10	3.50
5	2.66
2.5	2.26
1	1.86
0.5	1.70
0.25	1.68

Injection rate is constant for all cases. The simlation time was 75 days.

the electron acceptor and electron donor concentrations at the injection well were each varied sinusoidally, but the pulses were lagged over time to be out of phase. Results are reported in Table 8.1. These results show that the pulsing strategy can successfully limit the head increase at an injection well to a manageable amount (of the order of 1 m), whereas the simultaneous injection of electron acceptor and donor yields head increases of unacceptable magnitudes (of the order of 10 m after a 75-day injection period).

The mitigation of clogging due to the production and accumulation of a biogas phase (i.e., nitrogen or methane) has not been addressed in a rigorous manner in the literature. An obvious mitigation strategy would be the periodic injection of nontoxic surfactant solutions, which if combined with higher flushing rates should remove a significant part of the gas phase from the vicinity of the injection well. Care has to be taken to avoid the formation of an emulsion, which would be counterproductive.

SUMMARY

Clogging of soils during bioremediation can occur due to physical, chemical, and biological processes. In this chapter we have focused on the clogging mechanism due to biological growth. Significant conclusions that were drawn, based on experimental studies and mathematical simulations are:

- Biological clogging is most likely to occur in regions where the concentration of the electron acceptor and donor are both high, such as in the vicinity of injection wells, or near the surface of infiltration galleries.
- Soils that have a wide pore size distribution have a higher propensity for a significant reduction in their permeability due to biological clogging than soils that are homogeneous, with a narrow pore size distribution.

- During biological growth, the permeability will decrease until it reaches a minimum. This minimum permeability is a function of the soil properties as well as the operating conditions.

Effective strategies to minimize clogging in the vicinity of an injection well are:

- Out-of-phase injection into the same well of the electron donor and electron acceptor
- Injection of the electron donor and electron acceptor into separate wells
- Discontinuous operation, which allows for the periodic decay of biomass

REFERENCES

Baveye, P., P. Vandevivere, and D. deLozada (1992). Comment on Biofilm Growth and the Related Changes in the Physical Properties of a Porous Medium, 1, Experimental Investigation *Wat. Resour. Res.*, **28**:1481–1482.

Bouwer, H., R. C. Rice, and E. D. Escarceca (1974). High-Rate Land Treatment 1: Infiltration and Hydraulic Aspects of the Flushing Meadows Project, *J. WPCF*, **46**(5):834–834.

Clementz, D. M., D. E. Patterson, R. J. Aseltine, and R. E. Young (1982). Stimulation of Water Injection Wells in the Los Angeles Basin Using Sodium Hypochlorite and Mineral Acids. *J. Pet. Tech.*, **34**:2087–2096.

Costerton, J. W., R. T. Irvin, and K. J. Cheng (1981). The Bacterial Clycocalyx in Nature and Disease, *Ann. Rev. Microbiol.*, **35**:299–324.

Cunningham, A. B., W. G. Characklis, F. Abedeen, and D. Crawford (1991). Influence of Biofilm Accumulation on Porous Media Hydrodynamics, *Environ. Sci. Tech.*, **25**(7):1305–1311.

Ehrlich, G. G., H. F. H. Ku, J. Vecchioli, and T. A. Ehlke (1979). Microbiological Effects of Recharging the Magothy Aquifer, Bay Park, New York, with Tertiary Treated Sewage, U.S. Geol. Surv. Prof. Pap. 751-E.

Frankenberger, W. T., Jr., F. R. Troeh, and L. C. Dumenil (1979). Bacterial Effects on Hydraulic Conductivity in Soils, *Soil Sci. Soc. Am. J.*, **43**:333–338.

Gupta, R. P., and D. Swartzendruber (1962). Flow-Associated Reduction in the Hydraulic Conductivity of Quarz Sand, *Soil Sci. Soc. Proceed.*, **26**:6–10.

Macleod, F. A., H. M. Lappin-Scott, and J. W. Costerton (1988). Plugging of a Model Rock System by Using Starved Bacteria, *Appl. Environ. Microbiol.*, **54**:1365–1372.

Mitchell, R., and Z. Nevo (1964). Effect of Bacterial Polysaccharide Accumulation on Infiltration of Water through Sand, *Appl. Microbiol.*, **12**(3):219–223.

Raiders, R. A., M. J. McInerney, D. E. Revus, H. M. Torbati, R. M. Knapp, and G. E. Jenneman (1989). Selectivity and Depth of Microbial Plugging in Berea Sandstone Cores, *J. Ind. Microbiol.*, **1**:195–203.

Rice, R. (1974). Soil Clogging During Infiltration of Secondary Effluent, *J. WPCF*, **46**(4): 708–716.

Ripley, D. P. and Z. A. Saleem (1973). Clogging in Simulated Glacial Aquifers Due to Artificial Recharge, *Wat. Resour. Res.*, **9**(4):1047–1057.

Rittmann, B. E. (1993). The Significance of Biofilms in Porous Media, *Wat. Resour. Res.*, **29**:2195–2202.

Semprini, L., P. V., Roberts, G. D., Hopkins, and P. L. McCarty (1988). Field Evaluation of Aquifer Restoration by Enhanced Biotranformation, *Proc., Int. Conf. on Physiochemical and Biological Detoxification of Hazardous Wastes*, Technomic, Lancaster, PA.

Shaw J. C., B. Bramhill, N. C. Wardlaw, and J. W. Costerton (1985). Bacterial Fouling in a Model Core System, *Appl. Environ. Microbiol.*, **49**(3):693–701.

Soares, M. I. M., S. Belkin, and A. Abelovich (1988). Biological Groundwater Denitrification: Laboratory Studies, *Wat. Sci. Tech.*, **20**(3):189–195.

Soares, M. I. M., S. Belkin, and A. Abelovich (1989). Clogging of Microbial Denitrification Sand Columns: Gas Bubbles of Biomass Accumulation, *Z. Wasser-Abwasser-Forch.*, **22**:20–24.

Taylor, S. W. and P. R. Jaffé (1990a). Biofilm Growth and the Related Changes in the Physical Properties of a Porous Medium. 1. Experimental Investigation, *Wat. Resour. Res.*, **26**:2153–2159.

Taylor, S. W., and P. R. Jaffé (1990b). Substrate and Biomass Transport in a Porous Medium. *Wat. Resour. Res.*, **26**:2181–2194.

Taylor, S. W., P. C. D. Milly, and P. R. Jaffé (1990). Biofilm Growth and the Related Changes in the Physical Properties of a Porous Medium, 2. Permeability, *Wat. Resour. Res.*, **26**:2161–2169.

Taylor, S. W., and P. R. Jaffé (1991). Enhanced In-situ Biodegradation and Aquifer Permeability Reduction, *J. Environ. Eng.*, **117**:25–46.

Vandevivere, P. and P. Baveye (1992a). Saturated Hydraulic Conductivity Reduction Caused by Aerobic Bacteria in Sand Columns, *J. Soil. Sci. Soc. Am.*, **56**:1–13.

Vandevivere, P. and P. Baveye (1992b). Effect of Bacterial Extracellular Polymers on the Saturated Hydraulic Conductivity of Sand Columns, *Appl. Environ. Microbiol.*, **58**:1690–1698.

Vecchioli, J., H. F. H. Ku, and D. J. Sulam (1980). Hydraulic Effects of Recharging the Magothy Aquifer, Bay Park, New York, with Tertiary Treated Sewage, U. S. Geol. Surv. Prof. Pap. 751–F.

CHAPTER 9

Design Considerations for In Situ Bioremediation of Organic Contaminants

Edward J. Bouwer, Neal D. Durant, Liza P. Wilson, and Wei-xian Zhang

Department of Geography and Environmental Engineering, Johns Hopkins University 3400 North Charles Street, Baltimore, Maryland 21218 (E.J.B., N.D.D); American Association for the Advancement of Sciences Fellow, National Center for Environmental Assessment, U.S. Environmental Protection Agency, 401 M Street S. W. (8620), Washington, D. C. 20460 (L.P.W.); Department of Civil and Environmental Engineering, Lehigh University, Bethlehem, Pennsylvania 18015 (W.X.Z.)

The contamination of groundwater and soils with organic compounds is widespread (National Research Council, 1994). The extensive production and use of organic compounds makes them among the most prevalent pollutants in the subsurface at waste disposal sites. Because many of these synthetic compounds are known or potential threats to public health and the environment, there is an urgent need to understand their fate in the environment and develop cost-effective methods for their control.

The cost for the remediation of known sites in the United States is estimated well above one trillion dollars (Bredehoeft, 1994). Billions of dollars are being spent each year to decontaminate soil and groundwater aquifers in the United States alone. Remediation of groundwater and soil at contaminated sites is often attempted using groundwater pumping so that contaminants can dissolve and be pumped to the surface for treatment (Mackay and Cherry, 1989). It is difficult to extract hydrophobic organic contaminants from the subsurface, and such pump-and-treat systems are generally inefficient and slow. Surveys of pump-and-treat remediation (Travis and Doty, 1990; National Research Council, 1993) in the United States suggest it could take tens and even hundreds of years to reach health-based cleanup objectives by this method. As a result, there is substantial interest in exploring in situ biological treatment as an alternative approach for site remediation.

Biological Treatment of Hazardous Wastes, Edited by Gordon A. Lewandowski and Louis J. DeFilippi
ISBN 0-471-04861-5 ©1998 John Wiley & Sons, Inc.

Exploiting the metabolic capabilities of indigenous or introduced micro-organisms in regions of subsurface contamination offers the prospect of converting dissolved and sorbed contaminants to harmless products. This cleanup approach, termed *in situ bioremediation*, involves direct contact between microorganisms and the dissolved and sorbed contaminants for biotransformation (Bouwer, 1992; Bouwer and Zehnder, 1993).

As discussed in other chapters, biotransformation in the subsurface environment is a very complex process. There are many engineering challenges associated with the design and implementation of in situ bioremediation (Bouwer et al., 1994). The combination of the intricacies of microbial processes and the physical challenge of monitoring both microorganisms and contaminants in the subsurface makes bioremediation difficult to understand. The inherent complexity involved in performing bioremediation in situ means that special attention must be given to evaluating the success of a project. There are many site-specific factors that control the success of in situ bioremediation. This chapter addresses some important issues concerning the design and implementation of in situ bioremediation. These include (1) intrinsic and engineered in situ bioremediation, (2) the need for adapted microorganisms that can be the indigenous bacteria or introduced bacteria, (3) the need for a proper environment and suitable chemicals (nutrients, electron donor, and electron acceptor) for the microorganisms to function, and (4) the importance of mass transfer in the contact between bacteria and the contaminant.

MICROBIAL METABOLISM OF ORGANIC CONTAMINANTS

The metabolic capabilities of subsurface microorganisms are quite diverse. For growth of organisms, the presence of electron donors and acceptors, a carbon source, and nutrients is essential. Besides natural compounds, many contaminants can provide these growth requirements. Most organic contaminants can typically be categorized as either aliphatic or aromatic compounds that contain different functional groups, such as $-OH$, $-Cl$, $-NH_2$, $-NO_2$, and $-SO_3$. As electron donors these chemicals are oxidized during microbial metabolism to yield energy for growth; in the best case they are mineralized. Some of the breakdown intermediates may be assimilated as a carbon source for microbial growth. Functional groups (e.g. $-NH_2$, $-NO_2$, and $-SO_3$) may either be used as nutrients or cleaved from the carbon skeleton when the compound is reduced or oxidized. Oxidation can take place aerobically (in the presence of oxygen) or anaerobically (in the absence of oxygen). Oxygen serves two distinct functions. It can act as terminal acceptor of electrons released during the oxidation of the electron donors or it can react directly with the organic molecule. As an electron acceptor, oxygen can be replaced by other oxidized inorganic compounds such as nitrate, metal ions [e.g. Fe (III), or Mn(IV)], sulfate, or carbon dioxide, although the energy gains to the microorganism are then smaller. These alternate electron acceptors are reactants in anaerobic microbial processes, although they cannot substitute for the function of oxygen as a direct reactant (Schink, 1988).

Classes of organic contaminants known to be biotransformed by subsurface microorganisms are listed in Table 9.1 along with an indication of the required electron acceptor. In some instances, the compounds are the primary energy and carbon supply for the microorganisms (upper portion of Table 9.1). For other compounds, the biotransformation occurs as co-substrate utilization where enzymes involved in the metabolism of one substrate are also able to degrade the contaminant

Table 9.1 Some Important Classes of Organic Contaminants Susceptible to In Situ Bioremediation

Chemical Class	Frequency of Occurrence	Favorable Electron Acceptor(s)	Notes
Primary Metabolism			
Monoaromatic hydrocarbons	Very common	Oxygen[a], anaerobic[b]	
Polyaromatic hydrocarbons	Common	Oxygen[a], anaerobic[b]	Difficult to degrade if >4–5 rings
Aliphatic hydrocarbons	Common	Oxygen[a]	
Phenols	Infrequent	Oxygen[a], anaerobic[b]	
Nitroaromatics	Common	Oxygen[a], anaerobic[b]	
Alcohols, ketones, esters	Common	Oxygen[a], anaerobic[a]	
Some chlorinated solvents (CH_2Cl_2, CH_3CH_2Cl, $CH_2 = CHCl$)	Common	Oxygen[a]	Biodegradable under a narrow range of conditions
Less halogenated aromatics	Common	Oxygen[a]	
Less chlorinated polychlorinated biphenyls	Infrequent	Oxygen[a]	Biodegradable under a narrow range of conditions
Co-substrate Metabolism			
Highly halogenated aliphatics	Very common	Oxygen,[a] anaerobic[a]	Aerobic co-metabolism by methanotrophs in special cases; anaerobic cometabolism by many bacteria
Less halogenated aliphatics	Very common	Oxygen,[a] anaerobic[a]	Aerobic cometabolism by methanotrophs; anaerobic cometabolism by many bacteria
Highly halogenated aromatics	Common	Aerobic,[a] anaerobic[a]	

[a] Many observations, confirmed in laboratory and supported by field evidence.
[b] Demonstrated, effectiveness in field uncertain.

(lower portion of Table 9.1). Halogenated compounds are among the most prevalent organic contaminants at waste sites (Plumb, 1991). The aerobic co-oxidation of chlorinated solvents by methanotrophic bacteria while using methane as a co-substrate (Table 9.1) is under intensive investigation for in situ bioremediation applications. In the absence of molecular oxygen (anaerobic conditions), halogenated organic compounds may be reductively dehalogenated during their degradation or be used as terminal electron acceptor for growth of microorganisms (Mohn and Tiedje 1991; Holliger et al., 1992). In reductive dehalogenation, the halogenated compound becomes an electron acceptor, and in this process, a halogen is removed and is replaced with a hydrogen atom. Detailed information on biotransformation of halogenated compounds is presented elsewhere (Holliger et al., 1990; Vogel et al., 1987; Norris et al., 1994). For similar processes that occur in a fixed-film reactor, please refer to Figure 2.2 in Chapter 2.

TREATMENT APPROACHES

The most important principle of bioremediation is that microorganisms (mainly bacteria) can be used to destroy hazardous contaminants or transform them to less harmful forms. The microorganisms act against the contaminants only when they have access to a variety of materials—compounds to help them generate energy and nutrients to build more cells. In some cases the natural conditions at the contaminated site provide sufficient quantities of essential materials that bioremediation can occur without human intervention—a process called *natural* or *intrinsic bioremediation*. In most cases, bioremediation requires construction of engineered systems to supply microbe-stimulating materials—a process called *engineered bioremediation*. Engineered bioremediation relies on accelerating the desired biodegradation reactions by encouraging the growth of more organisms, as well as by optimizing the environment in which the organisms must carry out the detoxification reactions. Consequently, bioremediation can be considered as a continuum, extending from completely natural at one end to fully engineered at the other. The common and distinct issues for these two broad types of in situ bioremediation are discussed in the following two sections.

Intrinsic Bioremediation

Intrinsic bioremediation relies on the innate capabilities of naturally occurring microbial populations to convert environmental pollutants to harmless forms. Intrinsic bioremediation occurs at many sites, sometimes at a rate significant enough to attenuate plume migration. There is increasing interest in relying on intrinsic bioremediation for control of all or some of the contamination at waste sites (National Research Council, 1993). Intrinsic bioremediation is attractive economically because it is relatively passive, requiring only a demonstration (via extensive site characterization) that natural biological processes are destroying contaminants in situ, followed by long-term monitoring. Examples of sites where intrinsic bioremediation has been shown to play a significant role in attenuating organic contaminants in groundwater have been presented by Madsen et al. (1991), Godsy et al. (1992), Klecka et al. (1990), Cozarelli et al. (1990), Davis et al. (1994), Major et al. (1991) and Mondello et al. (Chapter 11).

Before intrinsic bioremediation can be considered as a legitimate cleanup method for a given site, the capacity of the indigenous microorganisms to metabolize the contaminants must be documented through field and laboratory testing. Furthermore, the effectiveness of intrinsic bioremediation must be proven with a site-monitoring program to confirm the progress of contaminant biodegradation. Chemical analyses of contaminants, terminal electron acceptors, and/or other reactants and products indicative of biodegradation processes should be performed. Consequently, employing intrinsic bioremediation is in contrast to no-action alternatives.

For intrinsic bioremediation to be effective as a stand-alone approach to aquifer or vadose zone restoration, the naturally occurring hydrogeochemical conditions at the site must allow the rate of biodegradation to exceed the rate of contaminant migration. Site conditions that favor intrinsic bioremediation include consistent and known groundwater flow throughout the year to ensure that contamination is not spreading with the flowing groundwater, presence of carbonate minerals to buffer acidity produced during biodegradation, adequate supply of electron acceptors and nutrients to meet the microbial growth requirements, and an absence of compounds that might be toxic to microorganisms (e.g., Hg, cyanide). Intrinsic bioremediation is especially applicable to sites where contamination is in the vadose zone and oxygen is more readily available. Some of the factors that need to be considered and methods that can be used in establishing the importance or role of intrinsic bioremediation at a site are discussed in the following sections.

Environmental and Chemical Factors Controlling Subsurface Metabolic Activity The presence of a microbial population in the subsurface capable of biodegrading a contaminant does not ensure that bioremediation will occur. The site's environment and chemistry must be favorable for microbial growth. Often environment factors, such as pH, concentration, temperature, and nutrient availability, determine whether or not biotransformation takes place. Relatively few microbial species can grow in acidic (pH < 4) or basic (pH > 10) environments. Aerobic naphthalene biodegradation has been shown to be substantially slower at pH 5 than pH 6.5 (Hambrick et al., 1980). In addition, decreases in pH have been observed down gradient of contaminant sources at several sites (Cozarelli et al., 1990; Godsy et al., 1992) resulting from organic acid accumulation associated with the incomplete anaerobic degradation of aromatic compounds. Degradation of nonaromatic contaminants may yield similar results. Thus, low pH may be both an indicator of previous biodegradation and a signal of inhibitory conditions. The rates of many microbial reactions studied for wastewater biodegradation (Metcalf and Eddy, 1991) typically double for each 10°C rise in temperature [described by the van't Hoff–Arrhenius relationship (Schwarzenbach et al., 1993)]. Colder temperatures often slow biodegradation rates, which could lead to long operating times for bioremediation.

Many organic contaminants become toxic to microorganisms at high concentrations. Consequently, at sites containing substantial amounts of nonaqueous phase liquids (NAPL), intrinsic bioremediation cannot be the primary means of source control and containment. Also, the greater electron acceptor and nutrient requirements for biodegradation of high concentrations of organics add to the complexity of achieving bioremediation. Biodegradation of waste mixtures containing metals, such as Hg, Pb, cyanide, and As, at toxic concentrations can also be problematic. At low concentrations, often in the microgram per liter to nanogram per liter ($\mu g/L$ to ng/L) range, insufficient energy and carbon may be available for growth and maintenance. Consequently, biodegradation of organic contaminants may be inhibited at both high and low concentrations. However, if concentrations are in the $\mu g/L$ to ng/L range, many compounds may be at or below cleanup goals, and thus, further remediation may not be necessary.

Groundwater Evidence of Intrinsic Bioremediation   Characterizing and monitoring the groundwater chemistry at a site is one of the first steps toward detecting natural in situ bioremediation. In addition to measuring the disappearance of the organic pollutants near and down gradient of the contaminant source, it is necessary to also measure for the presence and absence of electron acceptors. A conceptual model of intrinsic bioremediation at a site can be constructed by overlaying isopleth (concentration contour) maps of contaminant distributions against maps of electron acceptor distributions within an aquifer. Because H_2 concentrations and redox conditions are closely related, Lovley et al. (1994) have proposed that measuring H_2 concentrations in groundwater is also an effective means for delineating the terminal electron accepting condition at a given location. Groundwater analysis can also be used to detect organic intermediates of contaminant biodegradation (e.g., organic acids), as well as biodegradation end products (e.g., CH_4, CO_2, Fe^{2+}, NO_2^-, and/or H_2S).

Because concentrations of contaminants in groundwater diminish down gradient of the source as a result of hydrodynamic dispersion, sorption, and volatilization (unconfined aquifers only), it is essential to distinguish those groundwater quality variations that are caused by chemical and physical forces from those that are caused by bacteria. To eliminate the possibility of sorption, demonstrations of intrinsic bioremediation often focus on the less hydrophobic, biodegradable compounds in a plume. Estimates of the magnitude of sorption can also be obtained by measuring the organic carbon content of target formation samples, and applying a sorption coefficient from the literature (Mackay et al., 1992). The magnitude of dispersion is often estimated by tracking the movement of a conservative tracer (a nonsorbing, nonbiodegradable, nonreactive component in the groundwater) relative to the contaminant source. In some intrinsic bioremediation demonstrations, it is possible to distinguish biodegradation from dispersion by comparing along a flow path the concentration of the biodegradable pollutant against that of an inorganic or organic conservative tracer, such as Ca^{2+} and trimethylbenzene (Baedecker et al., 1993; Wiedemeier et al., 1995). In plumes where none of the existing constituents act conservatively, tracers such as Br^- and Cl^- can be introduced to track dispersion versus attenuation (Thierrin et al., 1995). Dispersion coefficients can also be predicted with the aid of solute transport codes that incorporate data on the aquifer hydrogeologic properties and sorption (Bedient et al., 1994).

The literature contains extensive information on the biodegradability of dozens of the more soluble (bioavailable) organic compounds commonly detected at hazardous waste and leaking underground storage tank sites (Lee et al., 1988; Young and Cerniglia, 1995). Any demonstration of intrinsic bioremediation should include a review of the biodegradation rates reported in the literature, as well as a review of the electron acceptor conditions known to favor biodegradation of the site compounds. Subsurface environments contaminated with organic compounds are often anaerobic due to bacterial consumption of oxygen and relatively slow rates of oxygen replenishment. Thus, reducing conditions exist near source, and aerobic or microaerobic conditions are typically only present at the edges of organic plumes. A review of past research should provide an indication of which compounds will persist in anaerobic environments and which will biodegrade.

Microbial Evidence of Intrinsic Bioremediation   A second line of evidence for demonstrating the significance of in situ bioremediation at contaminated sites is to directly confirm that the subsurface bacteria in contact with the organic contaminant plume are capable of synthesizing the enzymes necessary to degrade the pollutant of interest. This approach may not be necessary for all demonstrations of intrinsic bioremediation but can be quite valuable for documenting the potential success of bioremediation. The ability of subsurface bacteria to degrade a given mix of pollutants in groundwater is dependent on the type and concentration of compounds, the available electron acceptors, and the duration that bacteria have been exposed to the contamination. While many organic pollutants are known to biodegrade, bacteria often require an adaptation period before they can manufacture the enzymes necessary to biodegrade these compounds (Spain and van Veld, 1983; Wilson et al., 1985). Few data are available on the time span required for microbial adaptation in situ. In the laboratory, observed biodegradation rates typically exceed those observed in the field by 4–10 times (Sturman et al., 1995). Given this outcome, we might expect laboratory acclimation periods to also significantly underestimate actual field adaptation times (Wilson et al., 1985). Regardless of the duration of field adaptation period, evidence of in situ microbial acclimation has been reported at the vast majority of contaminated sites where intrinsic bioremediation has been investigated, and the number of these reports continues to rapidly grow.

A variety of laboratory techniques can demonstrate that subsurface bacteria are acclimated to the pollutants present at a site (Madsen, 1991). These techniques require the sampling of subsurface bacteria representative of the contaminated site. Research suggests that the majority of subsurface microorganisms are associated with sediment and soil particles (Wilson et al., 1983). Well water is generally a poor indicator of aquifer microbial activity because well water microflora are characteristically more diverse than aquifer populations (Balkwill and Ghiorse, 1985). Consequently, most studies on subsurface microbiology have focused on microflora adhered to aquifer material (Wilson et al., 1983; Balkwill and Ghiorse, 1985; Phelps et al., 1989; and Madsen et al., 1991). Recovering subsurface sediments aseptically (without introducing surface bacteria) can be challenging because drilling fluids may seep into cores. Tracers such as rhodamine-WT and bromide are sometimes added to drilling mud as a means for detecting extraneous fluids in aquifer samples collected for microbiological purposes (Barranco et al., 1994). Detailed approaches (including anaerobic techniques) for aseptic sampling of aquifer sediments are presented by Wilson et al. (1983), Phelps et al. (1989), and Russell et al. (1992).

Two of the more common techniques for determining the ability of indigenous bacteria to degrade site contaminants are heterotrophic plate counts and microcosm studies. Plate counts are typically prepared such that the carbon source in the microbiological agar represents one of the primary constituents in the contaminant plume from which the bacteria were sampled. If colonies grow on this agar when incubated under the redox conditions present in the subsurface, then the bacteria are likely capable of in situ biodegradation. Plate counts are limited, however, because the inherent selectivity of agar media and incubation conditions capture only a fraction of the viable cells present in any application of this method (Hickman and Novak, 1989).

A more accurate approach for determining in situ biodegradation potential is measuring the disappearance of site pollutants in laboratory microcosms (Dobbins et al., 1987; Hickman and Novak, 1989) containing aquifer material and groundwater incubated at temperatures resembling those found in the site subsurface. Using microcosms to measure biodegradation of model compounds under site conditions can be used to approximate in situ biodegradation rates. It should be recognized, however, that if compounds are assayed individually, the synergistic and/or antagonistic effects of other compounds mixed in the groundwater plume may not be adequately represented. Predictions of in situ biodegradation rates should recognize that biodegradation rates of certain compounds are enhanced by the presence of certain compounds and inhibited by the presence of others (Alvarez and Vogel, 1991; Chang et al., 1993; Bouchez et al., 1995). In addition, if the microcosm study involves only a small number of samples, it may fail to capture the true metabolic activity present in an aquifer. As an example, Wilson et al. (1987) observed that a laboratory microcosm study underestimated the amount of in situ biodegradation actually occurring at the Canadian Forces Base, in Ontario, Canada.

One drawback of microcosm studies is that they can be expensive and labor intensive. A more rapid measure of the biodegradative potential is to measure the content of adenosine triphosphate (ATP) in the sediment (Webster et al., 1985; Wilson et al. 1986). Because ATP is an integral component in bacterial biochemistry, the presence of ATP can be a sensitive indicator of metabolic activity. In a study of sediments from an aquifer contaminated with wood-creosoting wastes, Wilson et al. (1986) observed a direct correlation between sediment ATP content and biodegradation of toluene.

Many techniques for directly measuring in situ microbiological activity are still in the development stage. One novel technique is the use of bioluminescent reporter bacteria for in situ monitoring of catabolic gene expression and aromatic hydrocarbon biodegradation. By inserting light-producing genes into naphthalene degrading bacteria, King et al. (1990) demonstrated that bioluminescence can serve as a promising tool for evaluating the potential for naphthalene biodegradation in contaminated soils. The amount of bioluminescence can be used as a quantitative measure of a pollutant's bioavailability for biodegradation. Application of bioluminescent reporter bacteria to contaminated sediments and groundwater may become useful for determining the likelihood that in situ biodegradation is occurring based on the extent of bioavailability.

Another promising technique for evaluating biodegradation potential and quantifying biodegradation rates is the use of in situ microcosms (ISM), first introduced by Gilham et al. (1990). The method entails isolating a portion of the contaminated aquifer with a test cylinder installed at the bottom of a borehole. Biotic and abiotic transformation within the ISM are measured by extracting the pore water from the isolated zone. In general, biodegradation rates measured with ISM systems have been consistent with those measured in laboratory studies of sediments collected from the same aquifer zone (Hunt et al., 1995; Nielsen et al., 1996), but the ISM approach may be more favorable then laboratory studies because the former provides an in situ measurement under site conditions.

Engineered In Situ Bioremediation

Although the potential for intrinsic bioremediation is substantial at many sites, it can often be unacceptably slow as a sole remediation strategy due to poorly adapted microorganisms, the limited availability of electron acceptors and nutrients, cold temperatures, high concentrations of contaminants (NAPL), and mass-transfer limitations in the subsurface. When site conditions are not favorable for natural biotransformation, bioremediation requires construction of engineered systems to supply microbe-stimulating materials. Engineered bioremediation relies on accelerating the desired biodegradation reactions by encouraging the growth of more organisms, as well as by optimizing the environment in which the organisms must carry out the detoxification reactions.

Frequently, the necessary stimuli for microbial growth in the subsurface are oxygen and other electron acceptors (such as nitrate or sulfate) and nutrients (such as nitrogen and phosphorus). Typical engineered bioremediation systems therefore perfuse electron acceptors and nutrients through the contaminated region, as shown in Figure 9.1. Engineering in situ bioremediation near the land surface can be achieved by using infiltration galleries that allow water amended with nutrients and electron acceptors to percolate through the soil. When contamination is deeper, in situ bioremediation systems inject the amended water through wells. As shown in Figure 9.1, some in situ bioremediation systems use extraction and injection wells in combination to control the flow of electron acceptors and nutrients and to hydraulically isolate the contaminated area. Another approach to engineered in situ bioremediation is extraction of contaminated groundwater combined with above-ground bioreactor treatment and subsequent reinjection of

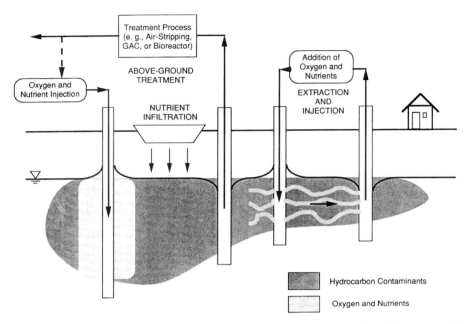

Figure 9.1 Subsurface treatment using above-ground reactors, injection of oxygen and nutrients, or dynamic system that involves injection and extraction wells.

nutrient-spiked effluent (Fig. 9.1) (Lee et al., 1988; Bouwer, 1992; National Research Council, 1993).

Biodegradation in the vadose zone can be stimulated in a process called bioventing. Bioventing involves inducing air movement through the vadose zone. However, the main purpose of bioventing is not to extract volatile contaminants but to enhance aerobic biodegradation of contaminants by supplying oxygen to soil microbes. Airflow rates are kept at the minimum rate required to deliver oxygen. Inorganic nutrients may also be added if necessary (Lesson et al., 1993; Downey et al., 1995).

The water circulation systems and air injection systems described require continuous energy input for pumping fluids. These systems thus require continuous site management and maintenance, which are generally expensive processes. Nonpumping approaches to engineered in situ bioremediation that do not require a continuous energy input are being developed. Such approaches under investigation use hydraulic barriers to direct groundwater flow through a biologically reactive medium (Fig. 9.2). As natural or induced hydraulic gradients move the contaminated groundwater through the biologically active zone, contaminant biodegradation occurs, leaving uncontaminated water to emerge from the downstream side (Pankow et al., 1993; Devlin and Barker, 1994).

Biological reaction zones can be created by adding growth-sustaining chemicals (e.g., electron acceptors and/or nutrients) to enhance contaminant biodegradation. For example, some researchers have tested biological removal of nitrate from

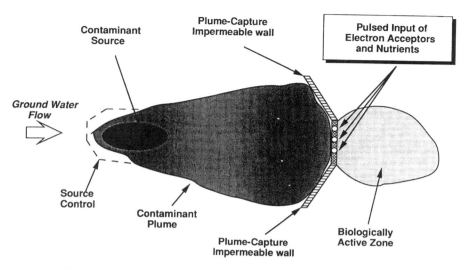

Figure 9.2 Process diagram for plume-capture bioremediation wall.

groundwater by periodically injecting a readily biodegradable organic compound to stimulate denitrifying bacteria, which convert nitrate to nitrogen gas (Gillham and Burris, 1992). Other researchers have used denitrifying bacteria to oxidize aromatic compounds while using nitrate as an electron acceptor (Hutchins et al., 1991). In tests at a commercial site in North Carolina, researchers have used a series of wells packed with briquettes that slowly release oxygen to stimulate aerobic bacteria that degrade gasoline components (Kao and Borden, 1994). Researchers at Lawrence Livermore National Laboratory are testing ways to inject methanotrophs and methane to create an in situ microbial filter for passive decontamination of chlorinated solvents (Taylor et al., 1992). While the concept of an in situ bioremediation "wall" shows promise, it is in the developmental stage and has not yet been demonstrated on a large field scale.

Biotransformation Stoichiometry   Organic contaminants are normally bio-transformed because the organisms can use the contaminants for their own growth. Other chemicals, such as nutrients and electron acceptors, are required during this metabolism. If any of these chemicals essential to cell growth are in short supply relative to the contaminant, then there is likely to be slow or limited contaminant removal. Thus, the bioremediation system must be designed and operated to supply the proper concentrations and ratios of these nutrients and electron acceptors if the subsurface does not provide them. The addition of nutrients and electron acceptors often stimulates microbial growth in the subsurface. The reaction stoichiometry helps to define the chemical needs of a microbial system. Some examples of stoichiometric reactions for important biotransformations being investigated for bioremediation systems are given in Table 9.2. The stoichiometric relationships can

Table 9.2   Stoichiometric Relationships for Possible Bioremediation Reactions[a]

Reaction 1: Aerobic biotransformation of naphthalene

$$C_{10}H_8 + 3.912O_2 + 1.62HCO_3^- + 1.62NH_4^+ \rightarrow 1.62C_5H_7O_2N^b + 3.53CO_2 + 2.38H_2O$$

Reaction 2: Biotransformation of naphthalene with nitrate

$$C_{10}H_8 + 5.66NO_3^- + 5.66H^+ \rightarrow 0.86C_5H_7O_2N + 2.4N_2 + 5.71CO_2 + 3.81H_2O$$

Reaction 3: Methane utilization by methanotrophs

$$CH_4 + 1.04O_2 + 0.137NO_3^- + 0.137H^+ \rightarrow 0.137C_5H_7O_2N + 0.315CO_2 + 1.59H_2O$$

Reaction 4: Reductive dechlorination of trichloroethylene to dichloroethylene

$$C_2HCl_3 + 0.47CH_3COO^- + 0.089NH_4^+ + 0.619H_2O \rightarrow 0.089C_5H_7O_2N + C_2H_2Cl_2 + HCl + 0.117CO_2 + 0.385HCO_3^-$$

[a] Stoichiometric reactions were obtained by making energy balances according to the procedures detailed by McCarty (1971). All compounds were considered in aqueous phase except CO_2, N_2, and CH_4 were taken as gaseous. Free energy of formation values for the organic compounds were obtained from *Handbook of Organic Chemistry* (Dean, 1987).
[b] $C_5H_7O_2N$ = empirical formula for biomass.

help select the appropriate solution of nutrients and electron acceptors to flush throughout the zone of contamination. However, in situ bioremediation is site specific, and the appropriate solution will vary depending upon the chemistry of the subsurface environment. In addition to considering microbial stoichiometric requirements, formulation of bioremediation additives also needs to recognize potential abiotic reactions that could consume chemical additives (e.g., O_2, PO_4^{3-}) intended for bacteria (Durant et al., 1997).

The hypothetical chemical needs for in situ bioremediation of naphthalene (common in coal tars, wood preservatives, creosotes, and petroleum residues) at residual saturation in the subsurface are illustrated in Figure 9.3. The reaction stoichiometries for the data in Figure 9.3 are derived from reactions 1 and 2 given in Table 9.2. Retention capacities for NAPL in soil have been observed to range from 15 to 75 L/m^3 (Pankow and Cherry, 1996). Using the lower bound to illustrate the best case, each cubic meter of soil would contain 15L or 17.3 kg of residual naphthalene (Fig. 9.3). The aerobic biodegradation of this 17.3 kg naphthalene requires 16.9 kg oxygen, 3.1 kg nitrogen, and 0.52 kg phosphorus using the stoichiometry given in Table 9.2. During the aerobic oxidation of the naphthalene, 24.7 kg of biomass would be formed per cubic meter of soil. If the aerobic growth occurs at once, it forms a larger mass than that of the original naphthalene. In practice, large amounts of biomass accumulation does not occur because the growth rate is slowed by the limited availability of the contaminant, electron acceptor, and/ or nutrients. The low solubility of oxygen in water greatly limits the aerobic biotransformation of naphthalene in the contaminated soil. In order to deliver the amount of oxygen required using water at air saturation (approx. 8 mg oxygen/L at 25°C), 8450 pore volumes of this water must be passed through the contaminated zone. Use of pure oxygen to saturate the water with about 40 mg oxygen/L reduces the pore volumes of water required to 1690. Hydrogen peroxide and nitrate are two

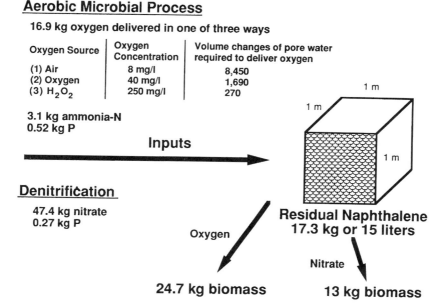

Aerobic Microbial Process

16.9 kg oxygen delivered in one of three ways

Oxygen Source	Oxygen Concentration	Volume changes of pore water required to deliver oxygen
(1) Air	8 mg/l	8,450
(2) Oxygen	40 mg/l	1,690
(3) H_2O_2	250 mg/l	270

3.1 kg ammonia-N
0.52 kg P

Inputs

Denitrification

47.4 kg nitrate
0.27 kg P

1 m

1 m

1 m

Residual Naphthalene
17.3 kg or 15 liters

Oxygen

Nitrate

24.7 kg biomass   **13 kg biomass**

Figure 9.3 Example chemical requirements and products of naphthalene bioremediation for one cubic meter of soil under aerobic and denitrifying conditions.

possible electron acceptors. These two chemical additives have a much higher water solubility than oxygen and are easier to deliver in the subsurface. The mass balances with hydrogen peroxide and nitrate are included in Figure 9.3.

At some sites, where aquifer conditions are reduced and significant concentrations of Fe(II) and Mn(II) are present, use of certain oxidants, especially O_2 and H_2O_2, for biostimulation may lead to the precipitation of metal oxides that could lead to plugging of the formation and injection well. A second approach for increasing the oxidant capacity is to add nitrate. This will stimulate denitrifying bacteria that can possibly oxidize organic contaminants while using nitrate as the electron acceptor as summarized in Table 9.1 and described in Chapter 12. Nitrate is advantageous because it is inexpensive, is very soluble ($\sim 660\,g/L$), will not precipitate with reduced iron, and is nontoxic to microorganisms. Many genera of denitrifying bacteria are commonly found in the subsurface, such as *Arthrobacter, Bacillus, Pseudomonas, Agrobacterium Alcaligenes,* and *Flavobacterium* (Payne, 1981). Thus, there is a good chance that denitrifying bacteria will already be present in the contaminated soil. Biotransformation rates with nitrate are generally similar to those with oxygen, although some compounds appear to resist biodegradation in the absence of oxygen. Nitrate could be added in sufficient quantity so that the organic contaminant becomes the rate-limiting compound for microbial growth. This will greatly reduce the number of pore volumes of water that will need to be recirculated to decompose the organic contaminant and will shorten the time for cleanup. Nitrate will also be easier to distribute throughout the contaminated soil in contrast with dissolved oxygen that will be depleted by the microorganisms within a short travel distance from an injection well. For the example in Figure 9.3 with naphthalene, $47.4 \, kg$ nitrate/m^3 soil would be required to supply the needed electron acceptor (reaction 2, Table 9.2). Since nitrate is a contaminant of concern in drinking water supplies (current maximum contaminant level is $10 \, mg/L \ NO_3$—N), controlled stoichiometric addition of nitrate will be needed to prevent nitrate pollution during a bioremediation operation.

The influence of nutrient and electron acceptor additions on the performance of in situ bioremediation is being investigated by the authors at a former manufactured gas plant (MGP) in Baltimore. The objective of this effort is to examine ways to apply microbial processes for remediation. The site has extensive coal tar contamination. Dissolved oxygen in the site groundwater ranges from below detection to $2 \, mg/L$, indicating microaerophilic or anoxic conditions throughout the site. Sediment samples obtained aseptically from boreholes drilled on-site were incubated in the laboratory to determine if the indigenous bacteria are capable of mineralizing the principal aromatic compounds in the groundwater plume (benzene, naphthalene, and phenanthrene) (Durant et al., 1995).

The study concluded that aerobic conditions were most favorable for bio-transformation of the aromatic compounds. Of the 49 samples assayed under aerobic conditions, 13 samples exhibited capacity to mineralize (observed by trapping $^{14}CO_2$) significant ($p \leq 0.05$) quantities of naphthalene (8 ± 3 to $43\pm7\%$) and/or phenanthrene (3 ± 1 to $31 \pm 3\%$) during 4 weeks of incubation at 22°C. Aerobic naphthalene mineralization was most common as illustrated in Figure 9.4. In general, all the aromatic substrates resisted degradation under anaerobic conditions (microcosms incubated in an anaerobic glovebox). Phenanthrene was not observed to degrade under anaerobic conditions. Significant benzene mineralization (6 ± 2 to $24 \pm 1\%$) was observed in 3 of 11 samples assayed for benzene, but no mineralization of this compound was observed in the 7 anaerobic samples tested. Biodegradation of acetic acid is an indication of the presence of metabolically active microorganisms because it is easily biodegraded under a variety of conditions. Aerobic mineralization (5 ± 2 to $70 \pm 2\%$) of acetic acid was detected in 20 of the 49 aerobic samples, and anaerobic mineralization (42 ± 17 to $60 \pm 3\%$) in 4 of the 12 samples tested. Although these bacteria did not always successfully degrade the aromatic compounds, these data suggest the proliferation of metabolically active bacteria throughout the site.

Supplementing the natural sediment microcosms with nutrients (N and P) and dissolved oxygen often enhanced the extent of mineralization (Durant et al., 1997). In sediments from one borehole at the MGP site, the amount of naphthalene mineralized increased at least threefold in the presence of stoichiometric quantities of nitrogen (as ammonia), phosphorus (as phosphate), and dissolved oxygen in equilibrium with pure oxygen (Fig. 9.4).

In general, however, subsurface sediments from the MGP site exhibited significant variability in their response to nutrient and electron acceptor amend-ments (Durant et al., 1997). A 6-week sediment–water microcosm study was

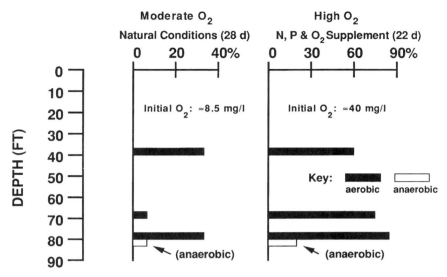

Figure 9.4 Depth profile diagram of naphthalene mineralization in sediment–water microcosms prepared with sediments from a single borehole. Only statistically significant mineralization ($p \leq 0.05$) is shown. The anaerobic results correspond to samples collected at a depth of 76–78 ft.

conducted with 10 of the MGP sediments to determine how naphthalene mineralization varied under five separate conditions (either no nutrients, elevated NO_3^-, elevated PO_4^{3-}, elevated NO_3^- and PO_4^{3-}, or elevated O_2 but no nutrients). Of the sediments exhibiting significant naphthalene mineralization, microcosms receiving water with elevated O_2 (21 mg/L) typically mineralized at least twice as much naphthalene as other conditions (Fig. 9.5). In sediments from borehole F (38 and 43 ft), however, microcosms amended with elevated nitrate (25 mg/L) mineralized naphthalene more rapidly than microcosms amended with elevated O_2. These data indicate that mixed electron acceptor conditions (applying both O_2 and NO_3^-) might be appropriate for enhancing biodegradation at some sites (Wilson and Bouwer, 1997).

In general, naphthalene mineralization was either unaffected or inhibited by supplements of PO_4^{3-}. Complexation and precipitation with sediment cations [e.g., Fe(II), Fe(III), and Ca^{2+}] likely limited P bioavailability. This problem highlights the necessity for characterizing the potential for sediments to complex PO_4^{3-} when designing additives for in situ bioremediation.

Results from this work are consistent with those of Swindoll et al. (1988), who also observed a significant degree of variability in the effect of nutrients on the aerobic biodegradation of organic pollutants in aquifer sediments. It is important to note that significant naphthalene mineralization can often be achieved without the addition of nutrients (Fig. 9.5). Additional evidence that nutrient addition may not necessarily enhance biodegradation of organic contaminants by aquifer bacteria was

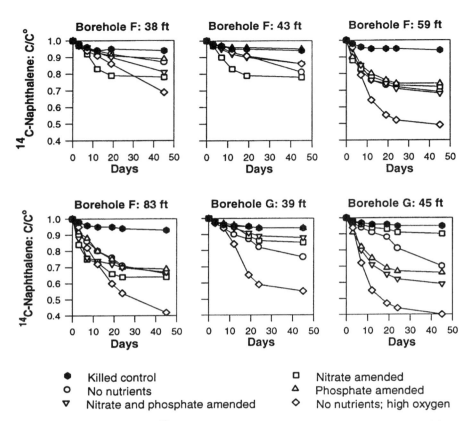

Figure 9.5 Mineralization of ^{14}C-naphthalene in sediment water microcosms prepared from several boreholes at the MGP site. The microcosms were incubated in the dark at 10°C for 6 weeks. Except for the killed control curves, each point represents the average of triplicate microcosms. Killed controls represent the average of 10 microcosms.

obtained by Balkwill and Ghiorse (1985) who found that bacteria from an uncontaminated aquifer grew better on oligotrophic media than on nutritionally rich media. In experiments on BTEX biodegradation under denitrifying conditions by bacteria from a contaminated aquifer, Hutchins et al. (1991) concluded that phosphate and ammonia addition may not have a significant effect on enhancing biodegradation. Biodegradation in the absence of added N and P is often feasible in aquifer sediments because elevated ammonia concentrations are common at most contaminated sites, and P cycling between bacteria, water, and sediment is sufficient to support limited microbial proliferation (Tiedje, 1988; Alexander, 1994). Because the vast majority of aquifer P is bound to sediments, accurate prediction of P requirements for enhanced bioremediation will be difficult if only groundwater quality data are used to estimate P requirements. Consequently, theoretical stoichiometries like those given in Table 9.2 may not always hold in the field, but can be used as a guideline in the design of field-scale bioremediation. Laboratory experimentation and pilot-scale field testing provide integration between the theory and full-scale nutrient requirements.

In addition to stoichiometry, other factors that need to be considered for selection of appropriate chemicals to supply during in situ bioremediation include pH buffering, formation of precipitates and clogging of the pores, excessive microbial growth (biofouling) at the injection well, and competition from other carbon sources (e.g., biodegradable natural organic matter). Please refer to Chapters 7 and 8 for further discussion of these subjects.

Biodegradation of organic carbon naturally present in the groundwater may result in consumption of some of the electron acceptors and nutrients intended for degradation of xenobiotics, and thus complicate estimation of stoichiometric requirements. Mihelcic and Luthy (1988) observed that biodegradation of natural organic matter adhered to surface soils reduced the nitrate available for naphthalene biodegradation in sediment slurries under denitrifying conditions. In sand and gravel aquifers, however, natural organic matter is often present only in small amounts (< 0.05%) (Curtis et al., 1986; Swindoll et al., 1988) and may not exert a significant demand on electron acceptors added to the system.

The injection of hydrogen peroxide or other oxygenated solutions into the subsurface can be complicated by abiotic processes that do not involve production of molecular oxygen. Hydrogen peroxide is a powerful oxidant, capable of oxidizing $Fe(II)$, $Mn(II)$, H_2S, as well as other inorganic and organic reductants. Water plus the oxidized form of the redox partner is the ultimate product of oxidations by hydrogen peroxide. Consequently, the usefulness of hydrogen peroxide as a source of molecular oxygen for aerobic biotransformations may be limited in aquifers by other hydrogen-peroxide-consuming reactions that compete with disproportionation. Under optimum conditions, disproportionation occurs slowly enough to provide a continual electron acceptor supply for bacterial utilization. If disproportionation occurs too quickly, evolution of oxygen gas can form bubbles that reduces its availability to bacteria and can lower aquifer permeability. Since the abundance of iron, manganese, organic matter, and other hydrogen-peroxide-reacting species varies from one soil to another, the performance of bioremediation with hydrogen peroxide varies substantially from site to site. Modifications in treatment strategy may also be necessary to minimize hydrogen peroxide toxicity. The use of hydrogen peroxide to augment the oxidant capacity of aquifers for bioremediation is reviewed by Pardieck et al. (1992). Because most contaminated sites are hypoxic or anaerobic, the groundwater may contain substantial concentrations of $Fe(II)$, a product of anaerobic iron reduction. When exposed to oxidizing conditions, $Fe(II)$ is rapidly converted to $Fe(III)$ which forms insoluble solid phases that precipitate and then may clog the injection zone. This problem can be exacerbated when orthophosphate (as a nutrient) is added with the oxygenated injectate because the phosphate can complex with the iron and other cations (e.g., calcium) to form precipitates (Robertson and Alexander, 1992; Aggarwaal et al., 1991).

The problem of phosphate precipitation and the potential for injection well plugging can be addressed in several ways. Some aquifers may naturally contain a significant amount of sediment-bound (but yet bioavailable) phosphate. A significant exchangeable phosphate content in the sediments to be treated may obviate the need for further phosphate amendments. Polyphosphates (e.g., pyrophosphate and trimetaphosphate) have been proposed as an attractive form of phosphorus addition because they decompose to orthophosphate slowly, thus minimizing the potential for orthophosphate precipitates (Aggarwaal et al., 1991; Stieof and Dott, 1995). It is recommended that geochemical equilibrium speciation models (Morel and Morgan, 1972; Jackson and Morgan, 1978; Stumm and Morgan, 1996) be used to simulate the aquifer chemistry in the presence of the proposed injectate to predict the potential for precipitates to plug the injection well. Several of the present computer codes for calculating chemical speciation are based on a program called MINEQL (Westall et al., 1976). Another approach is to avoid aerobic conditions and the associated precipitates by using nitrate as the primary electron acceptor (Hutchins et al., 1991).

Supply of Chemicals   The introduction into the subsurface environment of chemicals needed by microorganisms for growth is an important engineering challenge. The complexities of the subsurface that make it difficult to extract the contaminant by pumping (hydraulic and sorption/desorption limitations) also make it difficult to introduce growth-stimulating chemicals dissolved in water where they are needed. A common method is to inject water solutions of the chemicals into the contaminated zone. Problems with this approach include the need for large quantities of injection fluid because of the limited solubility of gases (e.g., oxygen), the difficulty to achieve good mixing in the subsurface, and biofouling (described earlier in this chapter). The rapid microbial growth near the base of the well during continuous flow of growth chemicals has two detrimental effects: (1) It inhibits the free flow of growth chemical solutions to the rest of the soil matrix, and (2) it greatly depletes the concentration of key growth chemicals in the immediate vicinity of the well. An example of the biomass production in a bioremediation process is illustrated in Figure 9.3. As a result of naphthalene biotransformation (example in Fig. 9.3), 24.7 kg of biomass would be formed with oxygen and 13 kg of biomass would be formed with nitrate per cubic meter of soil.

One method that may help to overcome the reduction in soil permeability is to alternate the injection of the chemicals to provide optimum growth conditions away from the well after the alternating pulses of chemicals have mixed. A second method is to use periodic rather than continuous injection of chemicals to promote decay of the biomass. About 80% of the biomass will be biodegradable. Decay of this portion of the biomass can be achieved if additional electron acceptor (e.g., oxygen or nitrate) is present. In a pilot test of enhanced in situ bioremediation of chlorinated compounds, Roberts et al. (1990) added pulses of electron acceptor and primary substrates (i.e., cometabolites) separately in order to avoid plugging. Computer simulations predict that pulsed input of electron acceptors and nutrients is a favorable means for avoiding plugging (Taylor and Jaffé, 1991).

Introduced Microorganisms for In Situ Bioremediation   The addition of microorganisms to the subsurface during in situ bioremediation might be beneficial when the native microbial populations lack the capability to biodegrade the contaminants or where the contaminants are present at toxic levels. Microorganisms have been added to the subsurface to "seed" and aid biodegradation, but the role of the added microorganisms has never been differentiated from that of the native microorganisms (Lee et al., 1988; Thomas and Ward, 1989). Such seed organisms may be natural or genetically engineered with additional catabolic genes to degrade contaminants. One approach for introduction of microorganisms is a closed-loop system where groundwater is extracted and treated in an above-ground bioreactor and the effluent reinjected into the subsurface. The treated groundwater that is injected contains adapted microorganisms from the bioreactor. Alternatively, water solutions amended with contaminant-degrading organisms can be injected to possibly enhance biodegradation in situ (Norris et al., 1994).

There is little documentation on the efficacy of introducing microorganisms for in situ bioremediation. In general, introduced microorganisms will not survive and persist in the subsurface without specific selection pressure (van der Meer et al., 1992). Furthermore, the use of genetically modified organisms for bioremediation raises public concerns. Another difficulty is the transport of the introduced microorganisms to the place of need. The subsurface is an efficient filter medium that generally restricts microbial transport in groundwater systems. Movement for short distances before deposition makes it difficult to disseminate introduced microorganisms widely in soil and groundwater. The groundwater chemistry, particularly ionic strength, is an important factor controlling adhesion of microbial cells to solid surfaces (Martin et al., 1992; Rijnaarts et al., 1996). Norris et al. (1994) review properties of the subsurface, microorganisms, and groundwater and environmental conditions that are other factors influencing transport and survivability of introduced microorganisms for in situ bioremediation. Stimulation of indigenous microorganisms is much preferred for implementing in situ bioremediation because they are better acclimated to conditions and selective pressures of the subsurface, such as contaminant mixtures, low nutrient levels, and low temperatures (Wilson et al., 1993). Nevertheless, introduction of microorganisms merits investigation as it will improve our knowledge of how to exploit microbial metabolism for control of contaminants.

INFLUENCE OF MASS TRANSFER ON BIOAVAILABILITY

Biodegradation rates in the field can be significantly slower than in the laboratory because of lower field temperatures and reduced bioavailability. There must be a close association between a microorganism and contaminant for biodegradation to occur. The contaminant must be available for uptake and utilization by the microorganism (Mihelcic et al., 1993). Microorganisms and hydrophobic organic pollutants are distributed among the solid, liquid, and gas phases within the subsurface. In systems where a pure organic phase (NAPL) is present, bioavailability will be controlled, in part, by the rate at which compounds dissolve from the NAPL into the aqueous phase. In locations where NAPL is absent, sorptive and desorptive processes will ultimately control bioavailability.

Many organic contaminants of interest tend to sorb onto soil such that only a small fraction of the compound may actually be in the bulk water phase. Over long contact time, the sorbing pollutants slowly diffuse into the inorganic and organic matrix and may also form bound residues. Most evidence indicates the uptake of compounds by bacteria proceeds via the liquid phase (van Loosdrecht et al., 1990). Consequently, a process such as sorption or volatilization that reduces the solution concentration tends to reduce the biotransformation rate. Furthermore, the accumulation of contaminants in fissures and cavities within subsurface solids render them inaccessible to microorganisms and their enzymes. These processes decrease the bioavailability. The important conclusion from the influence of mass transfer in terms of reduced bioavailability is that the overall reaction rate (in the absence of NAPL) is controlled by the desorption rate and not by the activity of the degrading microorganisms. The practical effect of such slow diffusion from within soil aggregates and other kinetic limitations to desorption is a decrease in the rate of removal of the contaminant, thereby increasing the time required to achieve cleanup and the amount of chemicals that must be added to sustain microbial activity.

Sorption of contaminants tends to separate the direct contact between microorganisms and contaminants, which is necessary for biodegradation to occur. For soils, sediments, and groundwater aquifers where the solid fraction and surface area are high, sorption is the primary process limiting bioavailability (Mihelcic et al., 1993). Effects of sorption on biodegradation in the subsurface can be classified into concentration effects and desorption rate limitations as discussed below.

Concentration Effects

A direct impact of sorption on biodegradation is the reduction of organic compounds in the bulk water phase. As sorbed chemicals are effectively protected from direct biodegradation, microbial growth must rely on organic substrate(s) in the bulk water phase. The rate of bulk water uptake and metabolism of the organic compound is generally given by the Monod relationship (Monod, 1949):

$$\frac{dC}{dt} = -\frac{kXC}{K_s + C} \tag{1}$$

where C is the bulk water concentration of the organic compound (mg C/L), X is the concentration of microorganisms in water (mg X/L), t is time (day), k is the maximum specific rate of biodegradation (mg C/mg X/day), and K_s is the substrate concentration when the growth rate is at half the maximum (mg C/L). Sorption reduces the bulk water concentration and as a result will prolong the time to degrade a given amount of organic pollutant as compared with a system free of sediments or soil where sorption will not occur. At very low contaminant concentrations, often in the range of μg to ng per liter, insufficient energy and carbon may be available for microbial growth and maintenance. Rittmann and McCarty (1980) defined a critical concentration (C_{min}) (mg C/L) at which microbial growth is just balanced by decay:

$$C_{min} = \frac{K_s b}{Yk - b} \tag{2}$$

where Y is the biomass yield coefficient (mg X/mg C) and b is the microbial decay coefficient (1/day). If sorption diminishes the concentration below C_{min}, biodegradation will decrease or even stop with time because there will be net decay of biomass.

On the other hand, at high concentrations, many organic compounds become toxic to microorganisms. One expression for the biodegradation rate is the Haldane equation, which is given below (Andrews, 1968):

$$\frac{dC}{dt} = -\frac{kXC}{K_s + C + \dfrac{C^2}{K_i}} \tag{3}$$

where K_i is the inhibition constant (mg C/L). Here sorption can reduce the bulk liquid concentration to lessen the toxic inhibition and increase microbial growth and biodegradation.

Desorption Rate Limitations

Due to geometrical and mass-transfer restrictions, most bacteria are present in the external surface of soil particles and in the bulk water. Only limited biological activities exist within the intraparticle pores (Jones et al., 1993). Decontamination of soils and sediments may involve three steps: (1) dissolution from the organic phase, (2) desorption of previously sorbed contaminant, and (3) biodegradation of the contaminant in the aqueous phase. In the absence of NAPL, the apparent biodegradation rate in a solid-water system can be controlled either by the desorption rate or biodegradation rate.

A mass balance for an organic compound in soil–water systems with first-order biodegradation kinetics can be written as:

$$\frac{d}{dt}[VC + mq] = -Vk_b C \tag{4}$$

where V is the volume of bulk water (L), m is the mass of soil (kg), q is the average sorbed concentration (mg C/kg), and k_b is the intrinsic first-order biodegradation rate coefficient (1/day). It is reasonable to assume that Monod kinetics can be approximated by first-order kinetics when the contaminant concentration is low. The left-hand side of Eq. (4) is the change in total mass of organic compound in the system and the right-hand side is the loss due to biodegradation. One approximation for the mass exchange between soil and water (sorption/desorption) is a first-order reaction:

$$\frac{dq}{dt} = -k_m \, (C_{\text{pore}} - C) \tag{5}$$

where k_m is the mass-transfer coefficient (L/kg day) and C_{pore} is the average organic concentration in the intraparticle pore water (mg C/L):

$$C_{\text{pore}} = \frac{q}{K_d} \tag{6}$$

Here K_d is the soil–water distribution coefficient ($K_d = q/C_{\text{pore}}$) (L/kg). Note that a high K_d indicates a tendency for the organic to remain bound to the soil, whereas a low K_d indicates a propensity to remain in, or transfer to, the aqueous phase. Coupling Eqs. (4), (5), and (6), the overall rate of biodegradation in a soil–water system can be expressed as:

$$\frac{dC}{dt} = -B_f k_b C \tag{7}$$

where B_f is dimensionless and is termed the *bioavailability factor*:

$$B_f = \frac{1}{1 + \dfrac{m}{V} K_d + \dfrac{k_b}{k_m} K_d} \tag{8}$$

Therefore, the overall rate of biodegradation in a soil–water slurry is determined by the two factors: B_f and k_b. B_f is evaluated from the extent and rate of sorption and, k_b is determined by the metabolic capability of microorganisms and environmental factors such as temperature. As applied here, the computation of k_b assumes that an excess of electron acceptor is present (i.e., biodegradation is not limited by electron acceptor availability). The time, $t_{50\%}$, to reduce the bulk water concentration in half is given by (day):

$$t_{50\%} = \frac{\ln 2}{B_f k_b} \tag{9}$$

The influence of the B_f on the extent and rate of contaminant biodegradation is illustrated in a plot of the contaminant fraction remaining (C/C_0) versus dimensionless time (product of k_b and time, t) (Fig. 9.6). A B_f value of 1.0 means there is no influence of sorption on the biodegradation, and this situation corresponds to a well-mixed liquid culture system without sorbent. As the B_f

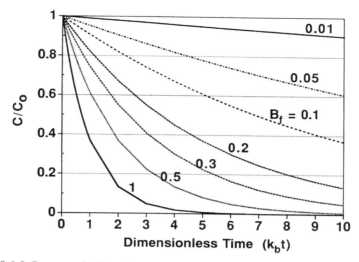

Figure 9.6 Influence of bioavailability factor on extent and rate of contaminant biotransformation.

decreases below 1.0, the removal rate declines (persistence of the chemical increases) (Fig. 9.6) due to a decreased amount of substrate in the aqueous phase (greater sorption). Measured K_d values for many organic compounds of interest in the subsurface are greater than 0.1 L/kg (Table 9.3). Using Eq. (8), K_d values >0.5 L/kg will yield B_f values that are less than 0.3 for biodegradation in the subsurface. Consequently, organic compounds that can be readily biodegraded in

Table 9.3 Measured Values of the Soil—Water Distribution Coefficient (K_d) for Some Important Organic Contaminants in the Subsurface

Compound	Conditions	K_d (L/kg)	Reference
Tetrachloroethene	Borden sand	0.35	Young and Ball (1994)
Trichloroethene	Moffett sand	2.0	Harmon et al. (1992)
Carbon tetrachloride	Moffett sand	1.0	Harmon et al. (1992)
Vinyl chloride	Moffett sand	0.9	Harmon et al. (1992)
Toluene	Glatt Valley aquifer (Switzerland)	0.37	Schwarzenbach and Westall (1981)
Chlorobenzene	Glatt Valley aquifer (Switzerland)	0.39	Schwarzenbach and Westall (1981)
1,4-Dichlorobenzene	Glatt Valley aquifer (Switzerland)	1.10	Schwarzenbach and Westall (1981)
1,2,4-Trichlorobenzene	Glatt Valley aquifer (Switzerland)	3.97	Schwarzenbach and Westall (1981)
1,2,3.4-Tetrachlorobenzene	Glatt Valley aquifer (Switzerland)	10.5	Schwarzenbach and Westall (1981)

liquid culture with half-lives ranging from a few hours to a few days (large k_b) can be very persistent in soils and groundwater aquifers (curves with $B_f < 0.3$ in Fig. 9.6). Low bioavailability (small B_f) can be a major factor responsible for slow in situ bioremediation.

Significance of Sorption During In Situ Bioremediation

A schematic of the relevant processes and contaminant concentrations occurring during in situ bioremediation is given in Figure 9.7. A typical groundwater plume is shown in Figure 7a. As chemicals in the source region move with the groundwater, concentrations of the chemicals will undergo physicochemical and biological changes. An engineered bioremediation system (Fig. 9.1) usually includes extraction and injection wells and equipment for addition and mixing of nutrients.

Two important features of bioremediation (as shown in Fig. 7b) are: (i) most bacteria are associated with solid surfaces, and (ii) organic compounds are sorbed into the solid phase. Transport of microorganisms from the bulk liquid to solid surfaces can occur by chemotaxis, advection, and diffusion. Once microorganisms contact a solid surface, they may attach permanently to the solid surfaces. As they grow, a surface film of microorganisms will accumulate. Surface growth can be removed by decay and detached by fluid shear and sloughing.

Figure 9.7c presents an idealized model of a soil particle. The solid matrix is porous and sorption sites are homogeneously distributed. Diffusion is the only major mass-transfer mechanism within the solid and occurs only in the radial direction. The pore structure within the particle is assumed to be too small for penetration of bacteria. Thus bacteria tend to be located on the outer surfaces. Distribution of biomass can be patchy colonies or a continuous surface film dependent on the net results of growth, decay, and attachment/detachment (Rittmann, 1993).

The theoretical response of organic concentrations in aqueous and solid phases as a function of time during bioremediation in the subsurface is shown in Figure 7d. Biotransformation is initially insignificant as the amount of biomass present is very small. Sorption removes a large portion of the organic compounds from the bulk water. As concentration gradients within soil aggregates diminish, organic concentrations in the bulk water increase quickly and finally level off and reach a steady state with biomass accumulated from utilization of the organic contaminant.

With engineered bioremediation, microbial activity is enhanced, which brings about a rapid decrease in the bulk water concentration (Fig. 9.7d). This lowering of the bulk water concentration will promote desorption of the contaminants by increasing the local concentration gradient for diffusion. The desorbing contaminants can be biodegraded as they pass through the attached biomass, which keeps the bulk water concentrations low. Thus, the net rate of desorption is accelerated in the presence of biotransformation. However, if the rate of desorption is much slower than the biodegradation rate (low B_f), stimulating microbial growth will only have a minimal impact on accelerating the rate of soil decontamination.

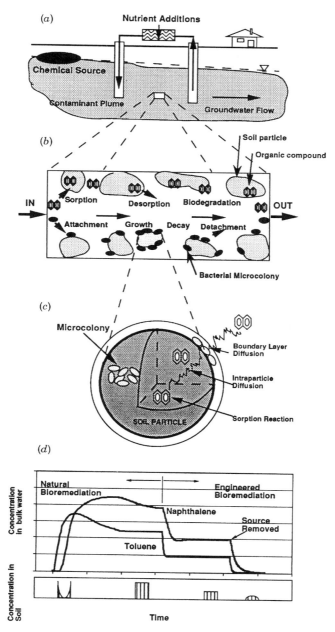

Figure 9.7 Role of sorption during in situ bioremediation. (*a*) Schematic of in situ bioremediation, (*b*) major mechanisms involved during in situ bioremediation, (*c*) idealized (model) soil particle, and (*d*) model simulations of sorption and biodegradation of toluene and naphthalene in the subsurface. The lower portion of (*d*) depicts the concentration profile along the centerline of the soil particle for different times.

Furthermore, upgradient sources of contamination must be removed (e.g., removal and containment of separate phase organic liquids) in order to achieve effective in situ bioremediation of the contaminant plume.

Enhancing Desorption-Limited Biodegradation

There is increasing interest in adding surfactants (synthetic or natural) to aqueous sediments or soils to enhance biodegradation. Surfactant molecules interact with organic contaminants to form surfactant-organic complexes by covalent bonds, charge-transfer, hydrogen bonds, or van der Waals interactions (Traina et al., 1989; Chen et al., 1992; Rouse et al., 1994). Surfactants can increase the apparent solubility and mobility of organic contaminants. The amount of chemical sorbed to soil is thus reduced. Furthermore dissociation rate of organic chemical from surfactant–organic complexes is much faster than desorption from the solid phase, so the apparent rate of biodegradation is increased (even though the surfactant–organic complex may not be directly biodegradable). The beneficial effects (by increased solubility) will be much more pronounced if NAPL is present. Please refer to Chapters 2 and 3 for further discussions on surfactants.

Another possible way to enhance desorption-limited biodegradation is to increase the rate of desorption by raising the temperature of groundwater and aquifer materials. This might be achieved by warming the extracted groundwater and recharging it upgradient of the contaminant source, thereby warming the contaminated zone to accelerate bioremediation.

Both the extent (K_d) and rate (D) (m^2/s) of sorption are strongly temperature dependent. The relationship between diffusivity and temperature in water can be expressed empirically as (Cussler, 1997):

$$D = \frac{T}{\mu} \tag{10}$$

where T is the absolute temperature (K) and μ is the viscosity (kg/ms); μ is also temperature dependent:

$$\mu \propto T^{-6.2} \tag{11}$$

So diffusivity varies with temperature by:

$$D \propto T^{+7.2} \tag{12}$$

For example, a temperature increase from 10 to 20°C (283 to 293 K) will increase the diffusivity (D) by 28%. Schwarzenbach et al. (1993) provided the following approximation for the change of K_d with temperature:

$$\frac{K_d^{T_1}}{K_d^{T_2}} \cong \exp\left[\frac{\Delta H_s}{R}\left(\frac{1}{T_1} - \frac{1}{T_2}\right)\right] \tag{13}$$

Here ΔH_s is the heat of dissolution (J/mol) in water and R is the gas constant (8.31441 J/K mol). Applying Eq. (13) for naphthalene with dissolution heat of 9900 J/mol, a 14% reduction in naphthalene sorption to soil occurs when warming a solid-water suspension by 10°C [e.g., increasing from 10 to 20°C (283 to 293 K)]. Combining the effects of temperature on both D [Eq. (12)] and K_d [Eq. (13)], increasing temperature from 10 to 20°C (283 to 293 K) can bring about a 48% increase in the apparent diffusivity for naphthalene (corresponding increase in the rate of desorption) in the soil, which will significantly increase the B_f [Eq. (8)].

Another benefit from the increased temperature is the enhancement of microbial metabolism. Rates of many microbial reactions typically double for each 10°C rise in temperature (Metcalf & Eddy, 1991). Below 10°C, many microbial reactions are slow; a temperature increase at low temperatures can significantly accelerate microbial reactions.

While addition of surfactants and increasing temperature show promise to improve bioavailability and bioremediation performance, they are in the development stage and have not yet been demonstrated.

CONCLUDING REMARKS

Subsurface microorganisms are capable of transforming many different organic contaminants. This has led to a great interest in exploiting biological processes for in situ treatment. Intrinsic bioremediation requires field and laboratory studies to characterize the occurrence and distribution of subsurface microorganisms and assessing the extent and rates at which these microorganisms degrade the contaminants of interest. Engineered in situ bioremediation often involves addition of adequate nutrients and electron acceptor(s) to stimulate microbial growth. Laboratory and field studies are recommended to obtain design parameters. The delivery of growth chemicals to the subsurface is difficult to engineer and is an area that requires more research.

Both laboratory and field studies have shown that the rates of bioremediation of hydrophobic organic contaminants, such as aliphatic hydrocarbons, polynuclear aromatic hydrocarbons (PAHs), and polychlorinated biphenyls (PCBs), are likely to be controlled by their availability to microorganisms rather than the metabolic capabilities of microorganisms. Sorption tends to reduce bioavailability of organic compounds. Slow desorption could control the overall performance of bioremediation. The practical effect of sorption is to increase the time required to achieve cleanup and amount of chemicals that must be added to sustain microbial activity. Research should be especially directed toward developing methods to improve the bioavailability of hydrophobic organic contaminants.

Acknowledgments

This work was supported in part by Cooperative Agreement ECD-8907039 between the National Science Foundation and Montana State University, in part by the

Hazardous Substance Management Research Center headquartered at the New Jersey Institute of Technology and the New Jersey Commission on Science and Technology, and in part by the Baltimore Gas and Electric Company. We are grateful to EA Engineering and Science, Inc., for assistance in gathering subsurface sediment and groundwater samples. We thank Peter C. D'Adamo and Louis DeFilippi for helpful reviews of the material.

REFERENCES

Aggarwaal, P. K., J. F. Means, and R. E. Hinchee (1991). Formulation of Nutrient Solutions for in situ Bioremediation, in *In Situ Bioremediation: Applications and Investigations for Hydrocarbon and Contaminated Site Remediation.*, R. E. Hinchee and R. F. Olfenbuttel, Eds., Butterworth-Heinemann, Stoneham, MA, pp. 51–66.

Alexander, M. (1994). *Biodegradation and Bioremediation*, Academic Press, San Diego.

Alvarez, P. J. J. and T. M. Vogel (1991). Substrate Interactions of Benzene, Toluene, and *para*-xylene During Microbial Degradation of Pure Cultures and Mixed Aquifer Slurries, *Appl. Environ. Microbiol.*, **57**:2981–2985.

Andrews, J. F. (1968). A Mathematical Model for the Continuous Culture of Microorganisms Utilizing Inhibitory Substrates, *Biotech. Bioeng.*, **10**:707–723.

Baedecker, M. J., I. M. Cozzarelli, R. P. Eganhouse, D. I. Siegel, and P. C. Bennett (1993). Crude Oil in a Shallow Sand and Gravel Aquifer—III. Biogeochemical Reactions and Mass Balance Modeling in Anoxic Groundwater, *Appl. Geochem.*, **8**:569–586.

Balkwill, D. L. and W. C. Ghiorse (1985). Characterization of Subsurface Bacteria Associated with Two Shallow Aquifers in Oklahoma, *Appl. Environ. Microbiol.*, **50**(3):580–588.

Barranco, F. T., J. L. Kocornik, I. D. MacFarlane, N. D. Durant, and L. P. Wilson (1994). Subsurface Sampling Techniques Used for a Microbiological Investigation, in *Bioremediation of Chlorinated and Polycyclic Aromatic Compounds*, R. E. Hinchee, A. Leeson, L. Semprini, and S. K. Ong, Eds., Lewis Publishers, Boca Raton, FL, pp. 474–479.

Bedient, P. B., H. S. Rifai, and C. J. Newell (1994). *Ground Water Contamination Transport and Remediation*, Prentice-Hall, Englewood Clliffs, NJ.

Bouchez, M., D. Blanchet, and J-P. Vandecasteele (1995). Degradation of Polycyclic Aromatic Hydrocarbons by Pure Strains and by Defined Strain Associations: Inhibition Phenomena and Cometabolism, *Appl. Microbiol. Biotechnol*, **43**:156–164.

Bouwer, E. J. (1992). Bioremediation of Organic Contaminants in the Subsurface, in *Environmental Microbiology*, R. Mitchell, Ed., Wiley, Chapter 11, pp. 287–318.

Bouwer, E. J. and A. J. B. Zehnder (1993). Bioremediation of Organic Compounds—Putting Microbial Metabolism to Work, *Trends Biotech*, **11**:287–318.

Bouwer, E. J., N. Durant, L. Wilson, W. Zhang, and A. Cunningham (1994). Degradation of Xenobiotic Compounds in situ: Capabilities and Limits, *FEMS Microbiol. Rev*, **15**:307–317.

Bredehoeft, J. D. (1994). Hazardous Waste Remediation: A 21st Century Problem, *Groundwater Monit. Remediation*, **14**:95–100.

Chang, M-K., T. C. Voice, and C. S. Criddle (1993). Kinetics of Competitive Inhibition and Cometabolism in the Biodegradation of Benzene, Toluene, and *p*-xylene by Two *Psuedomonas* Isolates, *Biotech. Bioeng.*, **41**:1057–1065.

Chen, S., W. P. Inskeep, S. A. Williams, and P. R. Callis (1992). Complexation of 1-Naphthol by Humic and Fulvic Acids, *Soil Sci. Soc. Am. J.*, **56**:67–73.

Cozarelli, I. M., R. P. Eganhouse, and M. J. Baedecker (1990). Transformation of Monoarmatic Hydrocarbons to Organic Acids in Anoxic Ground Water, *Environ. Geol. Wat. Sci.* **16**:135–141.

Curtis, G. P., M. Reinhard, and P. V. Roberts (1986). Sorption of Hydrophobic Organic Solutes by Sediments, in *Geochemical Processes at Mineral Surfaces*, J. A. Davis and K. F. Hayes, Eds., Am. Chem. Soc. Symp. Ser. **323**:191–216.

Cussler, E. L. (1987) *Diffusion: Mass Transfer in Fluid Systems*, 2nd Edition, Cambridge University Press, Cambridge, UK.

Davis, J. W., N. J. Klier, and C. L. Carpenter (1994). Natural Biological Attenuation of Benzene in Ground Water Beneath a Manufacturing Facility, *Ground Wat.*, **32**:215–226.

Dean, J. A. (1987). *Handbook of Organic Chemistry*, McGraw-Hill, New York.

Devlin, J. F. and J. F. Barker (1994). A Semipassive Nutrients Injection Scheme for Enhanced in situ Bioremediation, *Ground Wat.*, **32**:374–380.

Dobbins, D. C., J. R. Thorton-Manning, D. D. Jones, and T. W. Federle (1987). Mineralization Potential of Phenol in Subsurface Soil, *J. Environ. Quality*, **16**(1):54–58.

Downey, D. C., R. A. Frishmuth, S. R. Archabal, C. J. Pluhar, P. G. Blystone, and R. N. Miller (1995). Using in situ Bioventing to Minimize Soil Vapor Extraction Costs, in *In Situ Aeration: Air Sparging, Bioventing, and Related Remediation Processes*, R. E. Hinchee, R. N. Miller and P. C. Johnson, Eds., Battelle Press, Columbus, OH, pp. 247–266.

Durant, N. D., L. P. Wilson, and E. J. Bouwer (1995). Microcosm Studies of Subsurface PAH-degrading Bacteria from a Former Manufactured Gas Plant, *J. Contaminant Hydrol.*, **17**:213–237.

Durant, N. D., C. A. A. Jonkers, and E. J. Bouwer (1997). "Spatial Variability in the Naphthalene Mineralization Response to Oxygen, Nitrate, and Orthophosphate Amendments in MGP Aquifer Sediments," *Biodegradation*, in press.

Gillham, R. W. and D. R. Burris (1992). In situ Treatment Walls—Chemical Dehalogenation, Denitrification, and Bioaugmentation, *Subsurface Restoration Conference*, Dallas, Texas, June 21–24, pp. 66–68.

Gillham, R. W., R. C. Starr, and D. J. Miller (1990). A Device for in-situ Determination of Geochemical Transport Parameters, 2. Geochemical Reactions, *Ground Wat*, **82**:858–862.

Godsy, E. M., D. F. Goerlitz, and D. Grbic-Galic (1992). Methanogenic Biodegradation of Creosote Contamination in Natural and Simulated Ground Water Ecosystems, *Ground Wat.* **30**:232–242.

Hambrick, G. A. III, R. D. DeLaune, and W. H. Patrick, Jr. (1980). Effect of Estuarine Sediment pH and Oxidation-Reduction Potential on Microbial Hydrocarbon Degradation, *Appl. Environ. Microbiol.*, **40**:365–369.

Harmon, T. C., L. Semprini, and P. V. Roberts (1992). Simulating Solute Transport Using Laboratory-based Sorption Parameters, *J. Environ. Eng., ASCE*, **118**:666–689.

Hickman, G. T. and J. T. Novak (1989). Relationship between Subsurface Biodegradation Rates and Microbial Density, *Environ. Sci. Tech.*, **23**:525–532.

Holliger, C., G. Schraa, A. J. M. Stams, and A. J. B. Zehnder (1990). Reductive Dichlorination of 1,2-Dichloroethane and Chloroethane by Cell Suspensions of Methanogenic Bacteria, *Biodegradation*, **1**:253–261.

Holliger, C., G. Schraa, A. J. M. Stams, and A. J. B. Zehnder (1992). Enrichment and Properties of an Anaerobic Mixed Culture Reductively Dechlorinating 1,2,3-Trichlorobenzene to 1,3-Dichlorobenezene, *Appl. Environ. Microbiol.*, **58**:1636–1644.

Hunt, M. J., M. A. Beckman, M. A. Barlaz, and R. C. Borden (1995). Anaerobic BTEX Biodegradation in Laboratory and in situ Columns, in *Intrinsic Bioremediation*, R. E. Hinchee, J. T. Wilson, and D. C. Downey, Eds., Battelle, Columbus, OH, pp. 101–108.

Hutchins, S. R., G. W. Sewell, D. A. Kovacs, and G. A. Smith (1991). Biodegradation of Aromatic Hydrocarbons by Aquifer Microorganisms under Denitrifying Conditions, *Environ. Sci. Technol.*, **25**:68–76.

Jackson, G. A. and J. J. Morgan (1978). Trace Metal-Chelator Interactions and Phytoplankton Growth in Seawater Media: Theoretical Analysis and Comparison with Reported Observations, *Limnol. Oceanograph*, **23**(2):268–282.

Jones, W. L., J. D. Dockery, C. R. Vogel, and P. J. Sturmen (1993). Diffusion and Reaction within Porous Packing Media: A Phenomenological Model, *Biotech. Bioeng.*, **41**:947–956.

Kao, C.-M. and R. C. Borden (1994). Enhanced Aerobic Bioremediation of a Gasoline-Contaminated Aquifer by Oxygen-Releasing Barriers, in *Hydrocarbon Bioremediation*, R. E. Hinchee, B. C. Alleman, R. E. Hoeppel and R. N. Miller, Eds., Lewis Publishers, Boca Raton, FL, pp. 262–266.

King, J. H. M., P. M. DiGrazia, B. Applegate, R. Burlage, J. Sanseverino, P. Dunbar, F. Larimen, and G. S. Sayler (1990). Rapid Sensitive Bioluminescent Reporter Technology for Naphthalene Exposure and Biodegradation, *Science*, **249**:778–781.

Klecka, G. M., J. W. Davis, D. R. Gray, and S. S. Madsen (1990). Natural Bioremediation of Organic Contaminants in Groundwater: Cliffs-Dow Superfund Site, *Ground Wat.*, **28**:534–543.

Lee, M. D., J. M. Thomas, R. C. Borden, P. B. Bedient, C. H. Ward, and J. T. Wilson (1988). Biorestoration of Aquifers Contaminated with Organic Compounds, *CRC Crit. Rev. Environ. Control*, **18**:29–89.

Leeson, A., R. E. Hinchee, J. Kittel, G. Sayles, C. Vogel, and R. Miller (1993). Optimizing Bioventing in Shallow Vadose Zones in Cold Climates, *Hydrol. Sci. J.*, **38**(4):283–295.

Lovely, D. R., F. H. Chapelle, and J. C. Woodward (1994). Use of Dissolved H_2 Concentrations to Determine Distribution of Microbially Catalyzed Redox Reactions in Anoxic Groundwater, *Environ. Sci. Tech.*, **28**(7):1205–1209.

Mackay, D. M. and J. A. Cherry (1989). Groundwater Contamination Pump-and-Treat Remediation, *Environ. Sci. Tech.*, **23**:630–636.

Mackay, D., W. Y. Shui, and K. C. Ma (1992). *Illustrated Handbook of Physical-Chemical Properties and Environmental Fate for Organic Chemicals, Volume 1. Monoaromatic Hydrocarbons, Chlorobenzenes, and PCBs*, Lewis Publishers, Chelsea, MI.

Madsen, E. L. (1991). Determining in-situ Biodegradation: Facts and Challenges, *Environ. Sci. Tech.*, **25**:1663–1673.

Madsen, E. L., J. L. Sinclair, and W. C. Ghiorse (1991). In situ Biodegradation: Microbiological Patterns in a Contaminated Aquifer, *Science*, **252**:830–833.

Major, D. W., E. W. Hodgins, and B. J. Butler (1991). Field and Laboratory Evidence of in situ Biotransformation of Tetrachloroethene to Ethene and Ethane at a Chemical Transfer Facility in North Toronto, in *On-Site Bioreclamation Processes for Xenobiotic and Hydrocarbon Treatment*, R. E. Hinchee and R. F. Olfenbuttel, Eds., Butterworth-Heinemann, Boston, pp. 147–171.

Martin, R. E., E. J. Bouwer, and L. M. Hanna (1992). Application of Clean Bed Filtration Theory to Bacterial Deposition in Porous Media, *Environ. Sci. Technol.*, **26**:1053–1058.

McCarty, P. L. (1971). Energetics and Bacterial Growth, in *Organic Compounds in Aquatic Environments*, J. Faust and J. V. Hunter, Eds., Marcel Dekker, New York, pp. 495–531.

Metalf & Eddy, Inc. (1991). *Wastewater Engineering: Treatment, Disposal and Reuse*, McGraw-Hill, New York.

Milhelcic, J. R. and R. G. Luthy (1988) Degradation of Polycyclic Aromatic Hydrocarbons under Various Redox Conditions in Soil-Water Systems, *Appl. Environ. Microbiol.*, **54**:1182–1187.

Mihelcic, J. R., D. R. Lueking, R. Mitzell, and J. M. Stapleton (1993) Bioavailability of Sorbed-and Separate-Phase Chemicals, *Biodegradation.*, **4**:141–154.

Mohn, W. W. and J. M. Tiedje (1991). Evidence for Chemiosmotic Coupling of Reductive Dechlorination and ATP Synthesis in *Desulfomonile tiedjei*, *Arch. Microbiol.*, **157**:1–6.

Monod, J. (1949). The Growth of Bacterial Cultures, *Ann. Rev. of Microbiol.*, **3**:371–394

Morel, F. and J. Morgan (1972). A Numerical Method for Computing Equilibria in Aqueous Chemical Systems, *Environ. Sci. Technol.*, **6**(1):58–67.

National Research Council (1993). *In Situ Bioremediation When Does It Work?* National Academy Press, Washington, DC.

National Research Council (1994). *Alternatives for Ground Water Cleanup.*, National Academy Press, Washington, DC.

Nielsen, P. H., P. L. Bjerg, P. Nielsen, P. Smith, and T. H. Christensen (1996) In situ and Laboratory Determined First-Order Degradation Rate Constants of Specific Organic Compounds in an Aerobic Aquifer, *Environ. Sci. Tech.*, **30**:31–37.

Norris, R. D., R. Hinchee, R. Brown, P. L. McCarty, L. Semprini, J. Wilson, D. Kampbell, M. Reinhard, E. J. Bouwer, R. Borden, T. Vogel, M. Thomas, and H. Ward (1994) *Handbook of Bioremediation.*, Lewis Publishers, Boca Raton, FL.

Pankow, J. F. and J. A. Cherry (1996). *Dense Chlorinated Solvents and Other DNAPLs in Groundwater.*, Waterloo Press, Portland, OR.

Pankow, J. F., R. L. Johnson, and J. A. Cherry (1993). Air Sparging in Gate Wells in Cutoff Walls and Trenches for Control of Plumes of Volatile Organic Compounds (VOCs), *Ground Wat.*, **31**:654–663.

Pardieck, D. L., E. J. Bouwer, and A. T. Stone (1992). Hydrogen Peroxide Use to Increase Oxidant Capacity for in situ Bioremediation of Contaminated Soils and Aquifers: A Review, *J. Contaminat Hydrol.*, **9**:221–242.

Payne, W. J. (1981). *Denitrification.*, Wiley, New York.

Phelps, T. J., C. B. Fliermans, T. R. Garland, S. M. Pfiffner, and D. C. White (1989). Methods for Recovery of Deep Terrestrial Subsurface Sediments for Microbiological Studies, *J. Microbiol. Meth.*, **9**:267–279.

Plumb, R. H. (1991). The Occurrence of Appendix IX Organic Constituents in Disposal Site Groundwater, *Ground Wat. Monit. Rev.*, **11**:157–164.

Rijnaarts, H. H. M., W. Norde, E. J. Bouwer, J. Lyklema, and A. J. B. Zehnder (1996). Bacterial Deposition in Porous Media: Effects of Cell Coating, Substratum Hydrophobicity, and Electrolyte Concentration, *Environ. Sci. Technol.*, **30**(10):2877–2883.

Rittmann, B. E. (1993). The Significance of Biofilms in Porous Media, *Wat. Resour. Res.*, **29**:2195–2202.

Rittmann, B. E. and P. L. McCarty (1980). Model of Steady-State Biofilm Kinetics, *Biotech. Bioeng.*, **22**:2343–2357.

Roberts, P. V., G. D. Hopkins, D. M. Mackay, and L. Semprini (1990). A Field Evaluation of in situ Biodegradation of Chlorinated Ethenes: Part 1, Methodology and Field Site Characterization, *Ground Wat.*, **28**:591–603.

Robertson, B. K. and M. Alexander (1992). Influence of Calcium, Iron, and pH on Phosphate Availability for Microbial Mineralization of Organic Chemicals, *Appl. Environ. Microbiol.*, **58**(1):38–41.

Rouse, J. D., D. A. Sabatini, J. M. Suflita, and J. H. Harwell (1994). Influence of Surfactants on Microbial Degradation of Organic Compounds, *Crit. Rev. Environ. Sci. Tech.*, **24**:325–370.

Russell, B. F., T. J. Phelps, W. T. Griffin, and K. A. Sargent (1992). Procedures for Sampling Deep Subsurface Microbial Communities in Unconsolidated Sediments, *Ground Wat. Monitor. Remediation.*, **12**:96–104.

Schink, B. (1988). Principles and Limits of Anaerobic Degradation: Environmental and Technological Aspects, in *Biology of Anaerobic Microorganisms*, A. J. B. Zehnder, Ed., Wiley, New York, pp. 771–846.

Schwarzenbach, R. E. and J. Westall (1981). Transport of Nonpolar Organic Compounds from Surface Water to Groundwater. Laboratory Sorption Studies, *Environ. Sci. Tech.*, **11**:1360–1367.

Schwarzenbach, R. P., P. M. Gschwend, and D. M. Imboden (1993). *Environ. Organic Chem.*, Wiley, Chapter 11, p. 277.

Spain, J. C. and P. A. van Veld (1993). Adaptation of Natural Communities to Degradation of Xenobiotic Compounds: Effects of Concentration, Exposure, Time, Inoculum, and Chemical Structure, *Appl. Environ. Microbiol.*, **45**:428–435

Stieof, M. and W. Dott (1995). Application of Hexametaphosphate as a Nutrient for in situ Bioreclamation, in *Applied Bioremediation of Petroleum Hydrocarbons.*, R. E. Hinchee, J. A. Kittel and H. J. Reisinger, Eds., Battelle Press, Columbus, OH, pp. 301–310.

Stumm, W. and J. J. Morgan (1996). *Aquatic Chemistry: Chemical Equilibria and Rates in Natural Waters.*, 3rd. ed., Wiley, New York.

Sturman, P. J., P. S. Stewart, A. B. Cunningham, E. J. Bouwer, and J. H. Wolfram (1995). Engineering Scale-up of in situ Bioremediation Processes: A Review, *J. Cont. Hydrol.*, **19**, 171–203.

Swindoll, M. C., M. C. Aelion, and F. K. Pfaender Influence of Inorganic and Organic Nutrients on Aerobic Biodegradation and Adaptation Response of Subsurface Microbial Communities, *Appl. Environ. Microbiol.*, **54**(1):212–217.

Taylor, R. T., W. B. Durham, A. G. Duba, K. J. Jackson, R. B. Knapp, J. P. Knezovich A. M. Wijesinghe, D. J. Bishop, M. L. Hanna, M. C. Jovanovich, and D. R. Shannard (1992). In-situ Microbial Filters, in *In-Situ Bioremediation Symposium '92*, Niagara-on-the-Lake, Ontario, Canada, September 20–24, pp. 66–67.

Taylor, S. W. and P. R. Jaffe (1991). Enchanced in situ Biodegradation and Aquifer Permeability Reduction, *J. Envrion. Engr.*, *ASCE*, **117**,25–46.

Thierrin, J., G. B. Davis, and C. Barber (1995). A Ground-water Tracer Test with Deuterated Compounds for Monitoring in situ Biodegradation and Retardation of Aromatic Hydrocarbons, *Ground Wat.*, **33**,469–475.

Thomas, J. M. and C. H. Ward (1989). In situ Biorestoration of Organic Contaminants in the Subsurface, *Environ. Sci. Tech.*, **23**,760–766.

Tiedje, J. M. (1988). Ecology of Denitrification and Dissimilatory Nitrate Reduction to Ammonium, in *Biology of Anaerobic Microorganisms.*, A. J. B. Zehnder, Ed., Wiley, New York, pp. 179–244.

Traina, S. J., D. A. Spontac, and T. J. Logan (1989). Effects of Cations on Complexation of Naphthalene by Water-Soluble Organic Carbon, *J. Environ. Qual.*, **18**,221–227.

Travis, C. C. and C. B. Doty (1990). Can Contaminated Aquifers at Superfund Sites Be Remediated? *Environ. Sci. Tech.*, **24**,1464–1466.

van der Meer, J. R., W. M. de Vos, S. Harayama, and A. J. B. Zehnder (1992). Molecular Mechanisms of Genetic Adaptation to Xenobiotic Compounds, *Microbiol. Rev.*, **56**,677–694.

van Loosdrecht, M. C. M., J. Lyklema, W. Norde, and A. J. B. Zehnder (1990). Influence of Interfaces on Microbial Activity, *Microbiol. Rev.*, **54**(1), 75–87.

Vogel, T. M., C. S. Criddle, and P. L. McCarty (1987) Transformations of Halogenated Aliphatic Compounds, *Environ. Sci. Tech.*, **21**,722–736.

Webster, J. J., G. J. Hampton, J. T. Wilson, W. C. Ghiorse, and F. R. Leach (1985). Determination of Microbial Cell Numbers in Subsurface Samples, *Ground Wat.*, **23**, 17–25.

Weidemeier, T. H., M. A. Swanson, J. T. Wilson, D. H. Kampbell, R. N. Miller, and J. E. Hansen (1995). Patterns of Intrinsic Bioremediation in Two U.S. Air Force Bases, in *Intrinsic Bioremediation.*, R. E. Hinchee, J. T. Wilson, and D. C. Downey, Eds., Battelle, Columbus, OH, pp. 31–50.

Westall, J. C., J. Zachary, and F. Morel (1976). MINEQL: A Computer Program for the Calculation of Chemical Equilibrium Composition of Aqueous Systems, Technical Note No. 18, Ralph M. Parsons Laboratory, MIT, Cambridge, MA.

Wilson, J. T., J. F. McNabb, D. F. Balkwill, and W. C. Ghiorse (1983). Enumeration of Bacteria Indigenous to a Shallow Water-Table Aquifer, *Ground W.*, **21**,134–142.

Wilson, J. T., J. F. McNabb, J. W. Cochran, T. M. Wang, M. B. Tomson, and P. B. Bedient (1985). Influence of Microbial Adaptation on the Fate of Organic Pollutants in Ground Water, *Environ. Toxicol. Chem.*, **4**,721–726.

Wilson, J. T., G. D. Miller, W. C. Ghiorse, and F. R. Leach (1986). Relationship between the ATP Content of Subsurface Material and the Rate of Biodegradation of Alkylbenzenes and Chlorobenzene, *J. Contaminant Hydrol.*, **1**,163–170.

Wilson, J. T., G. B. Smith, J. W. Cochran, J. F. Barker, and P. V. Roberts (1987). Field Evaluation of a Simple Microcosm Simulating the Behavior of Volatile Organic Compounds in Subsurface Materials, *Wat. Resour. Res.*, **23**,1547–1553.

Wilson, L. P. and E. J. Bouwer (1997). Biodegradation of Aromatic Compounds under Mixed Oxygen/Dentrifying Conditions: A Review, *J. Ind. Microbiol. Biotech.*, **18**,116–130.

Young, D. F. and W. P. Ball (1994). A Priori Simulation of Tetrachloroethene Transport through Aquifer Material Using an Intraparticle Diffusion Model, *Environ. Prog.*, **13**,9–20.

Young, L. Y. and C. E. Cerniglia, Eds. (1995). *Microbial Transformation and Degradation of Toxic Organic Chemicals*, Wiley, New York.

Pentachlorophenol Biodegradation: Laboratory and Field Studies

Carol D. Litchfield and Madhu Rao

Department of Biology, George Mason University, Fairfax, Virginia 22030

BACKGROUND ON THE PRODUCTION AND USES OF PENTACHLOROPHENOL

Pentachlorophenol (PCP or Penta) was first produced in the 1930s. From its initial use as a biocide for the preservation of wood at more than 500 commercial sites in the United States, it has subsequently been used as a fungicide, bactericide, herbicide, molluscicide, algacide, and insecticide (Davis et al., 1994). Owing to its toxic nature, it is on the list of Priority Pollutants defined by the U.S. Environmental Protection Agency (EPA) (Moos et al., 1983). Technical-grade Penta is also contaminated with chlorinated dibenzodioxins and dibenzofurans (Bajpai and Banerji, 1992).

The world production of PCP in 1977 was around 80 million pounds (36 million kilograms) (Moos et al., 1983). The restriction of its use (along with creosote) in the United States in the early 1980s led to a significant drop in its production from 1,500,000 tons (1,360,080 megagrams) in 1981 to 50,000 tons (45,000 megagrams) in 1987. Many countries including Sweden, Finland, and Japan have banned the use of PCP as a wood preservative. Today, the primary uses of PCP are in cooling towers, paper mills, and drilling muds (Middaugh, et al., 1994a).

It is estimated that about 700 former wood-preserving sites exist in the United States. As a wood preservative PCP was applied to wood as a solution in petroleum solvents to treat poles, posts, railroad ties, and dimensional lumber. Though being practiced in the United States since the late 1800s, wood preservation processes date further back in history. For many years, treatment pits were used, which were subsequently replaced by treatment cylinders. But even these did not offer sufficient

Biological Treatment of Hazardous Wastes, Edited by Gordon A. Lewandowski and Louis J. DeFilippi
ISBN 0-471-04861-5   ©1998 John Wiley & Sons, Inc.

protection from spills. These early method used to handle, store, and treat wood resulted in a significant amount of PCP escaping into the environment. Major improvements at most plants have not prevented these historic practices from leaving their mark on the environment (Pflug and Burton, 1988).

Penta concentration levels have been reported to be high in the vicinity of wood treatment facilities. In surface and subsurface soils, the concentration of PCP has been found to be several thousand milligrams PCP per kilogram of soil at depths exceeding one meter from the surface. Around these sites, the groundwater has frequently been found to be contaminated with concentrations up to 5 mg/L (Bajpai and Banerji, 1992). The ATTIC (Alternative Treatment Technology Information Center) system, which provides technical information on innovative procedures to treat hazardous wastes, includes 132 case studies from 10 U.S. companies involved in remediation, out of which approximately 10% (13 cases) involve site contamination due to the use of PCP and creosote as wood preservatives (Devine, 1994).

Because of its widespread success as a pesticide and as a wood preservative, scientists have assumed that Penta was essentially nonbiodegradable. It was postulated that the chlorines at the 3 or 5 meta position were sterically hindered from enzymatic dehalogenation (Alexander, 1965), and Penta would therefore be a persistent chemical in the environment. This recalcitrance was a major factor in restricting the use of PCP.

PCP SOLUBILITY

One of the anomalies in the following studies is that while PCP is poorly soluble (~5 mg/L) at 20–25°C and pH 7, most investigators have performed their experiments at concentrations above this. The PCP is put into the media as an acetone, alcohol, or other solvent solution, and the resulting precipitation is generally ignored. Part of the reason for this is that PCP partitions onto the various solids (including the inoculum, flask, or test tube), thus making it difficult to state what concentration is actually in solution. Arcand et al. (1995) noted over an order of magnitude increase in solubility between pH 7.0 and 8.0, which would seem to argue for conducting the experiments at a slightly more alkaline pH. Usually, the media in the studies reported below have been buffered between pH 7.0 and 8.0, and in all cases have been above the pKa of PCP, so that the organisms were actually metabolizing the anionically charged phenate form.

TOXIC EFFECTS ON MICROORGANISMS

Adding to the expectation that PCP would not be biodegradable was the observation that Penta inhibited manganese transport in *Bacillus subtilis* (Eisenstadt et al., 1973) and amino acid transport in a different strain of *B. subtilis* (Brummett and Ordal,

1977) at 0.1 mM PCP (26.6 ppm). Further work by Nicholas and Ordal (1978) showed that chlorophenols in general acted as noncompetitive inhibitors of proline and uncompetitive inhibitors of glycine transport in *B. subtilis*. Meanwhile, Penta had been found to be an inhibitor of both oxidative phosphorylation and substrate-level phosphorylation in *Streptococcus agalactiae* at 0.01 mM (2.66 ppm) concentration (Mickelson, 1974) and later in succinate oxidation in *Mycobacterium epraemurium* (Kato et al., 1976).

That this uncoupling effect is not confined just to gram-positive organisms but also applies to gram-negative bacteria was demonstrated by Reddy and Peck (1978) for *Vibrio succinogenes*. These investigators also noted an inhibition of electron transport phosphorylation coupled to fumarate reduction by 0.3 mM PCP (79.8 ppm).

In a survey of the toxicity to several methanogens of eight aromatic compounds, Patel et al. (1991) found that PCP was the most toxic of the compounds tested with an IC_{50} (50% Inhibitory Concentration) of 8 mg/L as measured by the inhibition of methane production. Growth was also severely inhibited in *Methanospirillum hungatei* and *Methanobacterium espanolae*, but PCP was less inhibitory to the growth of *M. bryantii* M.O.H. and *Methanosaeta concilii* GP6.

Concern was also expressed that the presence of PCP in feeds, bedding materials, and the like might adversely affect the ruminant microbial population. To answer this concern, Yokoyama et al. (1988) examined 14 ruminant bacterial strains and tested the effects of increasing concentrations of PCP on their growth and production of acetate, propionate, butyrate, and total fatty acids. Total growth of a mixture of the organisms was inhibited by Penta concentrations greater than 18.8 μM (5 ppm). At concentrations greater than this, there was also inhibition of propionate formation. However, no generalized pattern of growth inhibition was observed with the individual strains.

Similarly, interest in the effects on the nitrogen cycle resulting from the use of PCP as a pesticide prompted Tam and Trevors (1981) to study the toxicity of Penta to *Azotobacter vinelandii*. The EC_{50} (50% Effective Concentration) for PCP was found to be at concentrations greater than 200 μg PCP/L, while both growth and CO_2 evolution were slightly inhibited at about 200 μg PCP/L.

The effects of Penta on eukaryotic microorganisms have also been investigated. Of several biosynthetic activities tested, RNA synthesis and free and membrane bound ribosomes were found to be the cellular systems most affected by concentrations of PCP ranging from 1 to 10 μg/mL in *Saccharomyces* strain 2200 (Ehrlich et al., 1987). Ruckdeschel and Renner (1986) examined the effects of PCP and 31 of its metabolites on 16 yeasts and mycelial fungi. From their studies they calculated minimal inhibitory concentrations. These general findings are summarized by the relative ranking depicted in Table 10.1.

Ruckdeschel and co-workers (1987) also examined the effects of PCP and 35 known or potential metabolites on 30 bacterial strains including gram positives, spore-formers, gram negatives, and anaerobes. The results were, in general, similar to those found for the fungi, with PCP being less toxic than many of its tri- and tetrachlorophenol breakdown products. As a rule of thumb, chlorines *meta* or *para* to the hydroxyl group increased the toxicity to both fungi and bacteria.

Table 10.1   Toxicity of PCP and Its Metabolic By-products in Decreasing Order[a]

COMPOUND
2,3,4,5-Tetrachlorophenol and 3,4,5-Trichlorophenol
2,3-Dichlorophenol
2,3,4,6-Tetrachlorophenol, 2,3,4-Trichlorophenol, and 2,4,5-Trichlorophenol
2,3,5-Trichlorophenol
2,3,5,6-Tetrachlorophenol & 2,5-Dichlorophenol
2,3,4,5,6-Pentachlorophenol[b]
21 Other metabolites

[a] Data adapted from Ruckdeschel and Renner (1986).
[b] **Penta** is highlighted by boldface type.

At concentrations greater than 1 mM (266 ppm) the inhibition of photosynthesis and its subsequent formation of 5-aminolevulinic acid were noted for the green alga *Scenedesmus obliquus*. Protein synthesis, however, was inhibited at concentrations greater than only 0.01 mM (2.66 ppm) (Senger and Ruhl, 1980), which is below the solubility limit.

If protein synthesis and/or photosynthesis can be inhibited by such low concentrations of the compound, then ecological effects resulting from the slow dissolution and persistence might be severe. One possible explanation for the observed toxicity was noted by Vollmuth et al. (1994) who found that polychlorinated dibenzofurans and polychlorinated dibenzo-*p*-dioxins can be formed during the photolysis of water containing about 1 mg Penta/L water. However, the incident radiation in their experiments was much higher than would be found in surface waters, and some of the observed toxicity to microorganisms could be the result of abiotic decomposition of PCP.

MICROBIAL DEGRADATION

The recalcitrance of PCP was certainly the operating paradigm until numerous investigators reported the isolation of fungi and bacteria able to degrade Penta. The remainder of this section will review the work with fungi, aerobic, and anaerobic consortia found to degrade Penta.

Fungal Degradation

As early as 1951 it had been postulated by Young and Carroll that the biodegradation of Penta might be an explanation for the loss of this preservative from soil. However, this was discounted by Leutritz (1965), who believed that all the Penta disappearance from wood was due to leaching and solubilization. By 1978, however, Kaufman was able to list 35 photolytic or metabolic by-products of PCP resulting from its decomposition in soil or by pure cultures of microorganisms.

In 1963, Lyr reported that the enzymatic detoxification of PCP by fungi was due to extracellular phenol laccases (phenol oxidases) secreted into the culture medium. Subsequently, Duncan and Deverall (1964) examined the sensitivity of *Trichoderma* sp. (P42) to 0.4% (0.015 M) PCP. Chemical analysis of the wood chips after a 12-week incubation period showed that approximately 43% of the PCP had been degraded by *Trichoderma* sp. (P42).

The mechanism for this detoxification was not reported, but Cserjesi (1967) noted that *Cephaloacus fragans* could be adapted to grow in the presence of PCP at an initial concentration of 40 ppm and at an increased concentration of 280 ppm, but there was no evidence of metabolism of the PCP. However, *T. vergatum* did metabolize PCP, initially present at 8.2 ppm, resulting in less than 1 ppm PCP remaining after only 12 days of incubation. This metabolism did not appear to be due to the phenol oxidases. Later, Cserjesi and Johnson (1972) reported that 10–20% of the PCP appeared as the methylated product penta-chloroanisole (PCA), and no free phenol or other methylated products were recovered.

Work by Lamar et al. (1990) continued this investigation into the mechanisms of Penta degradation, this time using several species from the genus *Phanerochaete*. PCP conversion to PCA occurred as a first step in the biodegradation, with *P. chrysosporium* converting 64% of the PCP to PCA, while *P. sordida* converted 71% to PCA. In aqueous media both species mineralized PCA. However, PCA volatilization was as significant as PCA mineralization (Lamar et al., 1990). Unfortunately, these authors did not investigate the recovery of tetrachlorohy-droquinone (TCHQ), which had been found to be a significant oxidation product for *P. chrysosporium* (Mileski, et al., 1988).

In addition, Mileski et al. (1988) found that under nitrogen-limited conditions, 50.5% of the ^{14}C-labeled PCP was mineralized, 24.8% was converted to a water soluble form, and 21.9% was solvent extractable. Only 2.8% was converted to mycelial biomass, indicating that under these conditions PCP was not serving as a primary carbon source for this fungus. Under non-nitrogen-limiting conditions, the major by-products were found in the hexane-extractable fraction, and 7.6% was associated with the mycelium. Concentrations greater than 0.38 mM (102 ppm) were inhibitory both to glucose respiration and the amounts of PCP mineralized.

That PCP degradation was accomplished by extracellular enzymes was conclusively demonstrated by Lin et al. (1990). With high extracellular enzyme concentrations (6.42 mg/L) and a high biomass density (2.21 gm/L), 70% of the carbon was released as ^{14}CO$_2$ and only approximately 5% was solvent extractable at the conclusion of the 22-day experiments. These authors also found a very small percentage (4.79%) of the labeled carbon associated with the fungal hyphae. A solution containing PCP, H$_2$O$_2$ and a semipurified extracellular enzyme mixture also

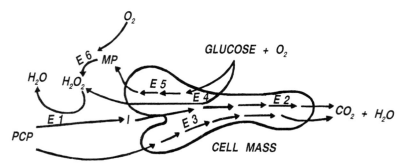

Figure 10.1 Metabolic model for the biodegradation of Penta by *P. chrysosporium*. E1: extracellular enzymes for transformation of PCP to unidentified intermediates; E2: cell-bound enzymes for metabolism of intermediates to CO_2; E3: cell-bound enzymes for the direct conversion of Penta to CO_2; E4: intracellular glucose oxidase; E5: enzymes for the generation of the metabolic products used as substrates for E6; E6: extracellular oxidases for H_2O_2 production; I: major degradation intermediate; MP: metabolic products used for H_2O_2 production. (From Lin et al., 1990, and used with the permission of John Wiley & Sons, Inc.)

resulted in the production of TCHQ, confirming that lignin peroxidases were involved in PCP degradation. However, Lin et al. (1990) also found mineralization of PCP using just *P. chrysosporium* biomass and proposed a pathway involving cell-bound enzymatic systems as well as the extracellular lignin peroxidases (Fig. 10.1).

Whether any of the numerous species of *Phanerochaete*, which produce the extracellular lignin peroxidase, are actively degrading PCP in contaminated surficial soils is unclear. Recently, Radtke et al. (1994) reported that phenazine-producing pseudomonads strongly inhibit the growth of *P. chrysosporium*. The pseudomonas isolates were from agricultural soils contaminated with trinitrotoluene (TNT) or benzo(*a*)pyrene. Despite isolating *P. chrysosporium* and antagonistic bacteria from the same soil, Radtke et al. also noted that the conditions that most favored lignin peroxidase and the Mn-peroxidase (ie., low nitrogen and sucrose as carbon source) enhanced the development of the antagonistic pseudomonades. In theory, this could prevent growth of the fungus and its production of the enzymes necessary for PCP degradation.

Seigle-Murandi and co-workers (1991, 1992, 1993) have systematically surveyed the 999 fungal strains in the Collection Mycology Pharmacy Grenoble for their ability to degrade PCP. Starting in 1991, they reported that for 45 different fungi, except for *Saccharomyces cerevisiae* X2180 1B, which had no action on PCP, 27 of the isolates degraded more that 10% of the PCP in the culture medium, and 12 strains demonstrated between 1 and 9% degradation of PCP present in the culture medium at 5000 mg PCP/L. The concentrations tested here were 10 to 200 times higher than others have used (Seigle-Murandi et al., 1991). Most of the poorer PCP degraders were in the yeast or Tubercuariales groups (Seigle-Murandi et al., 1991).

Seigle-Murandi et al. (1993) continued their studies examining 1056 fungal strains and using 100 mg PCP/L. The taxonomic groups showing the best degradation of PCP were the Dematiaceae and the Zygomycetes, with the yeasts and Basidiomycetes having the poorest percent of PCP degraders. All of the Dematiaceae degraded at least 30% of the PCP, while all of the Zygomycetes tested showed greater than 40% degradation of PCP. In additional investigations they reported either no correlation (Dematiaceae) or an inverse correlation (Zygomycetes) between the disappearance of PCP and the production of phenol oxidases. Seigle-Murandi et al. (1991) also demonstrated the importance of the abiotic (ie., light induced) destruction of PCP using *Phoma glomerata*, which degraded PCP at a lower rate in the dark than in the light.

Later Seigle-Murandi et al. (1992) described the degradation of PCP and the production of phenol oxidases by individual isolates. From this extensive study, no generalized pattern emerged as to which fungus might produce phenol oxidase and which might degrade PCP, except that activity among the yeasts was very low.

The most extensive investigation into a potential metabolic pathway for fungal metabolism of PCP is the work by Kremer et al. (1992). These authors used the basidiomycete *Mycena avenacea* TA 8480 and PCP at 100 mg/L. They isolated intermediates during growth on PCP and identified them using gas chromatography/ mass spectroscopy. Their proposed pathway for the initial destruction of Penta is shown in Figure 10.2.

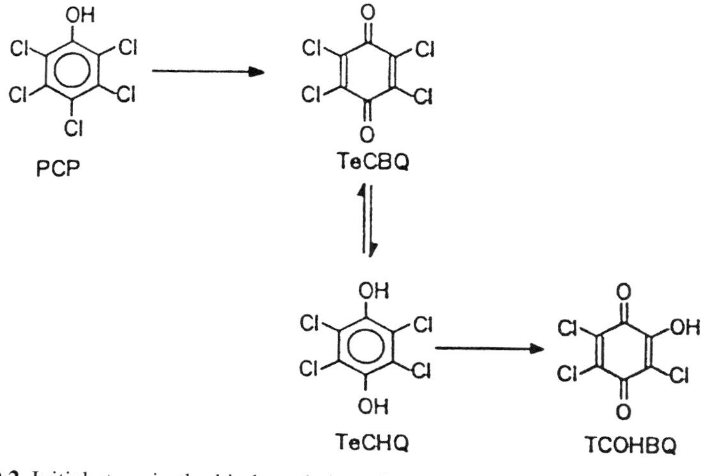

Figure 10.2 Initial steps in the biodegradation of Penta by the nonwhite rot basidiomycete *Mycena avenacea*. (From Kremer et al., 1992, and used with the permission of Verlag der Zeitschrift fur Naturforschung.)

Overview of Aerobic Bacterial Metabolism

A detailed review of the metabolism of Penta has recently been published by McAllister et al. (1996). Two reviews of the characteristics and genetics of microbial dehalogenases have also recently been published (Janssen et al., 1994; and Fetzner and Lingens, 1994), so these aspects of microbial metabolism of Penta will not be considered in this summary. In his review of microbial degradation of aromatic compounds, Evans (1963) believed that the chlorines would stay attached to the ring unless their removal was needed for hydroxylation of the ring. In 1972, however, Chu and Kirsch (1972) reported the isolation from a continuous-flow enrichment culture of an organism KC-3, which could use Penta as sole carbon and energy source. KC-3 was a gram-variable rod with the staining characteristics of the coryneform group. In a later study (1973), these same authors reported that among the many metabolites of Penta that were used by KC-3, this particular strain could also use several of the metabolites that had been shown to be more toxic to the fungi; namely, the 2,3,4,6-tetrachlorophenol, 2,3,5,6-tetrachlorophenol, and 2,3,5-trichlorophenol. This was also confirmed by Watanabe (1973) who reported that an organism closely related to *Pseudomonas* could completely dechlorinate PCP at a concentration of 40 ppm.

Besides showing that a *Pseudomonas* strain could rapidly mineralize PCP and incorporate up to 50% of the labeled carbon into cell constituents, Suzuki (1977) demonstrated that two important intermediates were formed during this metabolism: tetrachlorocatechol and TCHQ. He also demonstrated that PCA, so important in fungal metabolism of PCP, was not an intermediate in the bacterial metabolism of Penta. In 1979, Rott et al. reported the isolation and identification of 13 metabolites formed during the growth of 12 different bacterial strains from 9 different genera (both gram positive and gram negative). The most common metabolite was pentachlorophenyl acetate, which constituted up to 6.2% of the added PCP.

One interesting modification to the degradation of PCP by mixed cultures was the study reported by Ryding et al. (1994). These authors investigated the influence of toluene on PCP degradation by an enrichment culture from a municipal activated-sludge plant. They found that maximal degradation (estimated half-life of < 3 h) occurred with toluene present in the reactor. The culture also degraded 2,4,6-trichlorophenol and 2,4-dichlorophenol but not 2,6-dichlorophenol or 2,4,5-trichlorophenol. The initial attack on Penta appeared to be the formation of tetrachlorocatechol, although tetrachloroguaiacol was a minor "dead-end" product when toluene was present. This is a particularly interesting study because site contamination with PCP seldom occurs as a pure compound but more often in a mixture of oil and other constituents which may affect its metabolism. These authors reported a different end product and sequence than has been found in the pure culture studies. Thus, while the pure cultures may provide information about specific pathways, the effects of mixtures of contaminants on the metabolism of consortia is an almost unexplored research area.

The question of whether bacterial degradation of PCP had any practical importance was addressed by Moos et al. in 1983. They determined that PCP could be degraded by a microbial consortium in a continuously stirred tank reactor (CSTR) with maximum mineralization and cell incorporation occurring at about 350 μg/L. They also showed that degradation was the major process for loss of the PCP and that sorption and volatilization were insignificant factors in the mass balance of PCP disappearance. Similar low concentrations of PCP were found to be optimal for another microbial consortium consisting of protozoans and bacteria (Klecka and Maier, 1985). This consortium had been isolated from an industrial sewage treatment plant and was also obtained through continuous culture enrichment. The optimal Penta concentration for growth occurred at approximately 200–300 μg PCP/L, and greater than 99.8% of the PCP was removed. A refinement of these batch-fed or continuous culture techniques is the use of a nutristat, which continuously measures the critical nutrient concentration in the culture vessel and automatically adjusts that nutrient level. This technique was successfully used by Rutgers et al. (1993) to show PCP degradation, but the specific growth rate of 0.035/h was lower than that reported by Klecka and Maier (1985).

Studies Involving Arthrobacter Species

The complete mineralization of PCP by five strains of *Arthrobacter* was reported by Stanlake and Finn (1982). Their strains were isolated from a variety of sources: uncontaminated soil, soil contaminated with PCP, fresh water contaminated with PCP, a clean fresh-water stream, and clarified sewage sludge. The diversity of habitats of PCP degrading microorganisms indicates that the genes for chlorophenol degradation are more widespread in the microbial community than was first thought.

In addition, Stanlake and Finn (1982) demonstrated that the growth of the arthrobacters on Penta was related to the amount of undissociated PCP; at concentrations of undissociated PCP greater than approximately 200 μg PCP/L, there was no growth of the strains. They also showed that the lag phase was remarkably reduced at higher medium pH values. Therefore, pH is also important in controlling the amount of growth and hence the amount of biodegradation of Penta. Attempts to understand the metabolic pathway used by the arthrobacters showed that a pentachlorophenol dehalogenase converted PCP to TCHQ, as had been found in the fungi. However, Schenk et al. (1990) were unable to distinguish between an oxygenolytic and a hydrolytic dehalogenation because unlabeled TCHQ became labeled with $^{18}O_2$ when incubated in the presence of the enzyme and labeled water. Earlier, Schenk and co-workers (1989) had shown that the dehalogenase required NADPH and oxygen, but the clear role of oxygen has not been elucidated because of the biological reaction of TCHQ with labeled water.

Finn (1983) proceeded to test the biodegradability of PCP with *Arthrobacter* ATCC 33790. He showed that the addition of Penta at a 10% level to an operating activated-sludge unit resulted in the loss of 40 mg PCP/L in 1–2 days instead of the usual 6–7 days. He also reported that the application of 10^4–10^6 arthrobacters per gram of soil contaminated with 100 mg/L soil moisture resulted in the loss of the contaminant within 15 h compared to 2 weeks in the uninoculated control, thus demonstrating that the enrichment of a habitat with a specific degrader designed for the contaminant could have practical applications in the bioremediation of hazardous wastes.

This approach has been followed up by studies on the degradation of Penta by immobilized *Arthrobacter* cells by Wang and co-workers. Using an alginate matrix that contains *Arthrobacter* co-immobilized with activated carbon, 90% of the PCP was removed from solution within 2 h compared to 50% removal by free cells in 22 h. Mineralization by the immobilized cells, however, was slower than with the free cells at 117 μM (31.1 mg/L) PCP concentration (Lin and Wang, 1991). Further studies on the co-immobilized *Arthrobacter* system resulted in a modeling of the effects of the activated carbon and pH on bacterial degradation of PCP (Siahpush et al., 1992). They concluded that the most important aspect of the degradative process was the type of adsorbent and its affinity for the substrate.

Studies with* Pseudomonas *Species The isolation in 1982 of *Pseudomonas cepacia* strain AC1100, which could dehalogenate 2,4,5-trichlorophenoxyacetic acid (2,4,5-T) (Kilbane et al., 1982), paved the way for the study of the dehalogenases in this organism. Although oxygen uptake in the presence of PCP was only 9.3% of that compared to 2,4,5-T, there was 100% uptake (disappearance) of the substrate, 92% mineralization on the basis of CO_2 production, and 94% dehalogenation as measured by halogen released. Only 8% of the PCP carbon could be accounted for in the biomass. Maximal dechlorination occurred within 3 h at 0.2 mM (43.2 mg/L), PCP. However, *P. cepacia* strain AC1100 could not use PCP as a sole source of carbon and energy. When succinate or lactate was added to a PCP-containing medium, the dechlorination system was partially induced, but the authors concluded that this was probably because of the induction of 2,4,5-trichlorophenol dehalogenase (Karns et al., 1983a,b).

Banerji and Bajpai (1994) showed that PCP degradation by *P. cepacia* AC was indeed a cometabolic process that resulted in up to 100 mg PCP/L being degraded provided glucose was present. However, these same authors also noted that with activated sludge, PCP could be used as a sole carbon and energy source. In the presence of glucose, activated-sludge consortia produced a diauxic type of growth, again demonstrating the metabolism of PCP after glucose communation had closed (Banerji and Bajpai, 1994). However, Radehaus and Schmidt (1992) have isolated a *Pseudomonas* sp. strain RA2 that mineralizes PCP at concentrations up to 160 mg/L. To date there have been no further reports on the mechanisms for, or the control of, dehalogenation at these higher concentrations.

Studies Involving* Flavobacterium *Species In 1985, Saber and Crawford (1985) reported the isolation of *Flavobacterium* strains capable of mineralizing Penta. These strains had been isolated from five PCP-contaminated sites in Minnesota. The samples were retrieved from 30 cm below land surface. Between 73 and 83% of the labeled ^{14}C-PCP was recovered as [14]CO_2 with lag times of 30–80 h, depending on the strain. That same year, Crawford and Mohn (1985) described experiments that demonstrated that the addition of their flavobacterial strains to soils contaminated with PCP resulted in up to 80% degradation of the PCP. This led to patenting the organism, which could tolerate up to 250 mg PCP/L (Crawford, 1987).

Crawford and co-workers continued their investigations into Penta degradation by flavobacteria noting that the first step in the metabolic pathway was again TCHQ. This intermediate resulted from the hydrolytic dehalogenation of PCP with the oxygen coming from water, not molecular oxygen. This was followed by two reductive dehalogenation reactions leading first to TCHQ followed by dichlorohydroquinone (Steiert and Crawford, 1986). In response to exposure to PCP, *Flavobacterium* sp. strain ATCC 39723 synthesized a periplamic protein (PcpA), and degradative activity was lost when the membrane was disrupted (Xun and Orser, 1991a). These data indicate that the initial hydrolytic reaction requires intact membranes for activity.

Xun and Orser (1991b) later reported the isolation, purification, and properties of the pentachlorophenol hydroxylase from *Flavobacterium* sp. strain ATCC 39723. The purified hydroxylase, which was isolated by a seven-step purification scheme from whole cell paste, required both oxygen and reduced nicotinamide-adenine dinucleotide phosphate (NADPH). The enzyme also completely converted 37 μM PCP (9.8 ppm) to TCHQ in NADPH-supplemented buffer at pH 7.0 within 20 min. The involvement of molecular oxygen was also indicated by the observation that when the reaction was conducted in an anaerobic environment, there was a threefold increase in the amount of time and an approximately fourfold decrease in the amount of PCP conversion. When the samples were returned to an aerobic environment, activity resumed. The enzyme was subsequently confirmed to be a monooxygenase and not a hydrolase (Xun et al. 1992a).

The existence of a PCP specific monooxygenase as the first step in the degradation of Penta emphasizes the importance of oxygen to the rapid bioremediation of Penta in the environment. Orser and co-workers also reported the isolation of a tetrachloro-*p*-hydroquinone reductive dahalogenase (Xun et al., 1992b). Again, cell paste was the source of the enzyme, and the reactions were carried out in the presence of the reducing agent dithiotheritol. The reductive dehalogenase is a dimer with a molecular weight of 30 000 and requires reduced glutathione but not NADPH, reduced nicotinamide adenine dinucleotide (NADH), dithiothreitol, or ascorbic acid (Xun et al., 1992b). Thus, the aerobic degradation of Penta requires both an aerobic environment for the monooxygenase and the presence of reduced gluthathione for the subsequent reductive dehalogenation.

Other workers have investigated some of the factors affecting the survival and PCP-degrading activity of flavobacteria in the environment. Topp et al. (1988) noted that the presence of glutamate greatly enhanced the degradation of PCP. This study indicated that the presence of a readily metabolizable carbon source increased both the rate and amount of Penta degradation. Further studies by Topp and Hanson (1990a,b) using glucose confirmed the importance of a readily metabolizable carbon source for both cell viability in the presence of PCP and degradation of PCP. Neither nitrogen nor sulfate limitations induced the same effects as glucose limitations. In a continuous culture system, the flavobacterium was able to withstand exposures up to 60 mg PCP/L.

Several groups have further exploited the degradative abilities of the *Flavobacterium* sp. by examining bioreactor configurations for the degradation of PCP. O'Reilly and Crawford (1989) reported that immobilization of the bacteria in polyurethane foam in batch reactors allowed exposure of the cells to concentrations of PCP up to 300 mg/L. They attributed this to the polyurethane foam reversibly adsorbing the PCP and hence protecting the bacteria from these high concentrations of Penta. However, there was a gradual decrease in degradation over time in semicontinuous reactors. They also attempted to decrease the amount of readily metabolizable carbon source by reducing the amount of glutamate in continuous culture systems. This was unsuccessful.

In none of the above reports, however, did the authors address the possibility that some of the limitations to growth on PCP and/or its degradation might result from increases in the chloride concentration, or pH changes, due to Penta hydrolysis. Puhakka and Järvinen (1992) developed a continuous-flow fluidized-bed bioreactor to treat simulated wood-preservative-contaminated groundwater that contained PCP. With no additional carbon source except chlorophenols and an hydraulic retention time of about 5 h, 92.5% of the PCP was removed when feed concentrations ranged between 8 and 14 mg/L. The bioreactor was maintained with pH and chloride ion control over 130-day period.

Because PCP contamination is common at wood-treating sites that have switched to chromated-copper arsenate (CCA) mixture for wood preservation, Wall and Stratton (1994) examined the effects this CCA mixture might have on flavobacterial degradation of Penta. They found that CCA increased both the lag time and the time for a given amount of PCP to be biodegraded. This effect occurred at concentrations as low as 0.1–0.5 mg CCA/L, which is considerably below the levels used industrially.

Metabolism by* Rhodococcus *sp.   Studies similar to those performed with *Flavobacterium* sp. have been conducted by Apajalahti and Salkinoja-Salonen (1986) with the gram-positive actinomycete *Rhodococcus chlorophenolicus*. Both PCP and TCHQ were readily degraded. Concentration of the Penta was important, with greater degradation at $2 \mu M$ (0.53 ppm) than $10 \mu M$ (2.66 ppm). Chlorine substitution at the *meta* position made the polychlorophenols less degradable, with maximum inhibition caused by the dichlorophenols. However, with this rhodococcus no substitution on the 4-position was required for enzymatic attack (Apajalahti and Salkinoja-Salonen, 1987a).

In fact, with *Rhodococcus chlorophenolicus* the presence of molecular oxygen was required for primary attack on the PCP molecule, but the oxygen involved in the hydroxylation came from water (Apajalathi and Salkinoja-Salonen, 1987a). Complete reductive dechlorination by cell-free extracts was demonstrated later in 1987 with no accumulation of a trichlorophenol intermediate (Apajalathi and Salkinoja-Salonen, 1987b).

Application of *Rhodococcus chlorophenolicus* to PCP-contaminated soil was tested in the laboratory by Middeldorp et al. (1990). These authors found that *R. chlorophenolicus* mineralized 45–50% of the ^{14}C-labeled PCP to $^{14}CO_2$, while *R.*

rhodochorous, a chlorophenol *o*-methylating isolate, failed to mineralize any of the PCP present at 630 mg/Kg. However, both organisms produced a slight amount of PCA in peat (0.1–0.5 ppm) with slightly higher amounts in a sandy soil (about 0.4 ppm).

Two other factors that influenced PCP degradation were investigated. The addition of a carbon source from a distillery waste also increased the amount of Penta degradation in both inoculated and uninoculated soils. Because the distillery waste was not sterilized, it is conceivable that some of the additional degradation may have resulted from the presence of other microorganisms, which was not discussed by the authors. The other factor was the influence of inoculum size on bacterial degradation. They found that cell concentrations greater than 10^4 cells/kg dry soil were required, with optimal mineralization occurring with 10^8 cells/kg dry soil (Middeldrop et al., 1990).

Overview of Anaerobic Degradation of PCP

Because a common treatment method for publicly owned treatment works (POTWs) is anaerobic digestion, it was natural that these facilities would be the sources for the inocula used to investigate anaerobic digestion of Penta. The anticipated reductive dehalogenation was indeed found to occur. In a study involving a semicontinuously stirred tank reactor operated under methanogenic conditions, Guthrie et al. (1984) found that PCP was inhibitory to unacclimated cultures when the PCP was present at 0.3 ppm, but acclimated cultures tolerated 0.6 ppm PCP, and digester concentrations could go as high as 5 ppm. They also demonstrated that less than 2% of the PCP was sorbed or volatilized, so the major loss of PCP was through biodegradation. There was no attempt to investigate the degradative pathway involved.

Mikesell and Boyd (1985) compared three different methanogenic sludges for their ability to degrade seven chlorinated phenols. The Jackson sludge completely degraded PCP at an initial concentration of 12.5 ppm within 14 days, while 39 and 69% of the PCP remained after 70 days in the Mason and Adrian sludges, respectively. Whether these differences could be attributed to the types of sewage inputs the different plants received was not discussed by the authors. However, they did describe the primary metabolic steps in the anaerobic degradation of Penta. The first product formed was 3,4,5-trichlorophenol resulting from *ortho* dehalogenation. This was followed by *para* removal of chlorine, yielding the terminal end product 3,5-dichlorophenol with no mineralization noted.

In further studies with the Jackson and Michigan sludges, Mikesell and Boyd (1986) reported that when the anaerobic sludges were acclimated to mono-chlorophenols, the di-, tri, tetra-, and penta-chlorphenols were also metabolized with the 3-chlorophenol acclimated sludge dehalogenating at the *meta* position. PCP was 60% mineralized to CO_2 and CH_4 after the accumulation of the predicted intermediates from sequential reductive dehalogenation. Subsequently, Madsen and Aamand (1991) found that sulfate-reducing conditions inhibited the reductive dehalogenation and that this inhibition could be eliminated by the addition of hydrogen or molybdate, implying that the dehalogenation is occurring under methanogenic conditions.

As had been noted with the aerobic degradation of Penta, the addition of a supplemental carbon source was also found to increase the rate and amount of anaerobic dechlorination. Research by Hendrickson et al. (1991) demonstrated a 94% dechlorination in a glucose-amended bioreactor operated under methanogenic conditions, while the unamended control reactor showed only 20% dechlorination of PCP. The amounts of PCP removal were higher in both cases (Hendricksen et al., 1992), with tetrachlorophenol the major by-product in the unamended control reactor (Hendricksen et al., 1991).

Wu et al. (1993) developed anaerobic PCP-degrading granules for an upflow anaerobic fixed-film bioreactor. Examination of the granules showed the presence of *Methanotrix*-like rods as well as *Methanosarcina* sp. and *Methanobacterium* sp. This system also contained acetate, propionate, butyrate, and methanol, and it was able to treat up to 60 ppm PCP/L.

Finally, thermophilic anaerobic dechlorination of Penta has been successfully demonstrated in the laboratory using inocula from several sediment samples. Samples from anaerobic sludge plants, a stable laboratory anaerobic reactor, and a thermophilic anaerobic reactor treating cow manure were less successful in mineralizing the PCP (Larsen et al., 1991).

Along with developing information on the catabolic pathways of bacterial and fungal degradation of PCP, laboratory and field studies have also been conducted to show that the work described above has practical importance for the treatment of Penta-contaminated sites.

BIODEGRADATION TREATABILITY STUDIES

Prior Treatment Technologies

Improper disposal of pentachlorophenol treated woods and other activities associated with this process have led to the contamination of soils and groundwater at various locations. Due to the toxicity of Penta, its tendency to bioaccumulate and its persistance in the environment, clean up of contaminated sites has been made mandatory by regulatory agencies. A review of the literature shows that conventional treatment methods such as activated-carbon adsorption and land treatment were historically used for the treatment of pentachlorophenol-contaminated environments. The former process involved treating contaminated waters by pumping through a bed of activated charcoal. This merely transferred the PCP to the charcoal, which was handled as hazardous waste and disposed of accordingly. Landfarming (see Chapter 13) is a simple and cost-effective treatment method for contaminated soils and has been used to treat PCP with varying degrees of success (Parker and Jenkins, 1986; Harmsen, 1991; Mueller et al., 1991a). Using the traditional method of landfarming only the bioavailable fractions could be removed, and it could be years before the process is complete. For the nonbioavilable fraction, as Harmsen et al. (1994) suggested, an additional time-consuming and extensive form of landfarming would have to be employed to diffuse and desorb the pollutant before it is available to the microorganisms. Other chemical and physical methods of treatment have also been considered and have shown great potential. Middaugh et al. (1994a) in a study conducted at American Creosote Works, Pensacola, Florida, demonstrated that hyperfiltration is a viable option for treatment of creosote-and pentachlorophenol-contaminated groundwater (Table 10.2). Pignatello and co-workers (1983) reported that photolysis of PCP in surface waters accounted for a decline of 5–28% at initial PCP concentrations of 48–432 µg/L. Banerji and Bajpai (1994) also suggested photolysis as a desirable option for reducing the degree of chlorination of PCP and thereby reducing its toxicity.

Bioremediation, either in situ or in specially constructed treatment cells, is an alternative technology for contaminated soils and groundwater. It has been estimated that cleanup costs using biological treatment methods would be at least 10% less than the costs associated with physical and chemical methods (Kim and Maier, 1986). Utilization of engineered biological treatment systems seems therefore to be a promising alternative and has led to many laboratory treatability investigations and field studies (some of which are still underway).

Treatability Studies

Aerobic Treatment of PCP-Contaminated Wastewaters Chemical contaminants of surface waters often end up in wastewater treatment plants via municipal sewers. Consequently, several treatability studies on PCP using wastewater and treatment plant sludges have been reported and will be reviewed here. Biodegradation of PCP has been observed in sewage treatment plants, ranging from no degradation to greater than 75%. PCP degradation has been examined by Nyholm et al. (1992) in a laboratory-scale activated-sludge reactor fed with synthetic sewage under various operating conditions (sludge ages and sludge loading). At a sludge loading factor of 0.1–0.6, the bioreactor was spiked with 5–1000 µg PCP/L. It was concluded from this study that with long sludge ages (>10–15 days) PCP degraded best (biodegradation exceeded 80%), while degradation failed or was poor with sludge ages less than 8 days (Table 10.3). This indicates biodegradation by slow-growing specific degraders or possibly inhibition of the microbial population.

The behavior of chlorophenols in another activated-sludge plant was examined at pilot scale, and the resulting data were inconsistant with the previous results (Parker et al., 1994). PCP was adsorbed to the greatest extent, and biodegraded the least, with values for the sorption and biodegradation coefficients ranging from 0.6 to 9.6 L/g and 0.021 to 0.058 L/g h, respectively [for solids (SRTs) of 4–10 days]. The authors used mixed second-order kinetics (including biomass) using the TOX-CHEM model, coupled with the nonlinear regression routine UWHAUS. Their results suggest dependence on sludge age for removal of PCP.

The fate of pentachlorophenol in bench-scale, continuous-flow activated-sludge reactors maintained at a range of SRTs (5-, 10-, 15-, and 20-day SRTs) was studied

Table 10.2 Summary of Bench-, Pilot- and Field-Scale Treatability Studies

Reference	Treatment Mode	% Removal[a]	Comments
Pflug and Burton (1988)	BioTrol Aqueous Treatment (BAT) system	>99%	Field study: Influent PCP conc: 93 mg/L groundwater treatment using immobilized culture
Litchfield et al. (1992)	BIFAR process (upflow fluidized-bed activated carbon reactor)	85% (avg.)	Field scale: Influent PCP conc: 8–84 µg/L, under denitrifying conditions
Puhakka and Jarvinen (1992)	Continous-flow fluidized-bed reactor	92.5%	Contaminated groundwater, chlorophenol loading rate: 217 g/m3/day; HRT: 5 h, celite carrier used for immobilization
Mueller et al. (1993)	Two-stage bioreactor, continuously stirred tank reactor (1 in number), stirred tank batch reactors (3 in number)	97.4% (71.9% biodegradation)	Bench scale; Contaminated groundwater: total influent PCP conc. of 540 mg over 32 days.
Mueller et al. (1993)	Same system	77.8% (24.1% biodegradation)	Pilot scale (454 L), total influent PCP of 12000 mg over 14 days.
Middaugh et al. (1994b)	Two-stage bioreactor, continuously stirred tank reactor (1 in number), stirred tank batch reactors (3 in number)	69.4% (36.9% biodegradation, run 1); 95.4% (81% biodegradation, run 2);	Field Scale: Contaminated groundwater
Middaugh et al. (1994a)	Hyperfiltration	97.8% (predemonstration); 6.3% (field demonstration)	–

[a] Disappearance of exactable PCP in sample.

Table 10.3 Summary of Bench-, Pilot-, and Field-Scale Treatability Studies

Reference	Reactor Type	% Removal	Comments
Melcer and Bedford (1988)	Continuous-flow activated-sludge reactor	90–99%	Influent PCP conc: 0.1–12 mg/L; SRT: 10–20 days, HRT: 6 h; wastewater seeded with municipal sludge
Tokuz (1989)	Rotating biological contactors (RBCs)	52.6% (avg.)	Influent PCP conc: 1–4 mg/L, wastewater seeded with municipal sludge
Hendricksen et al. (1991)	Two anaerobic fixed-film reactors; one control and the other amended with glucose (1 g/L)	60% (control), 98% (with glucose)	Influent PCP conc: 1–2 mg/L; wastewater seeded with anaerobic digested municipal sludge; duration of operation: 6 months
Nyholm et al. (1992)	Continuously stirred tank reactor (CSTR)	>80% with sludge ages of 10–15 days, 0–20% with sludge age <8 days	Contaminants spiked into synthetic peptone sludge in activated-sludge reactors
Hendricksen et al. (1992)	Two anaerobic upflow sludge blanket reactors (USABs), one control and the other amended with glucose (0.9 g/L)	32–77% (control), 99% (with glucose)	Influent PCP conc: 4.5 mg/L; wastewater seeded with granular sludge grown on sugar containing waste; duration of operation: 10 months
Wu et al. (1993)	Anaerobic upflow sludge blanket reactor (USAB)	>99%	Loading rate: 88–97 mg/L day; HRT: 10.8–15 h; carbon source: mixture of acetate, propionate, butyrate, and methanol

by Melcer and Bedford (1988). Their work suggested that a background level of PCP was necessary to maintain a PCP degrading consortium, and an unacclimated system did not succeed in treating wastewater containing $350 \mu g$ PCP/L. In the activated-sludge systems, operated at SRTs of 10–20 days, PCP concentrations of 0.1–12 mg/L were degraded to less than $10 \mu g/L$ (Table 10.3), but at SRTs lower than 5 days only 40–60% of the PCP was removed.

The performance of a conventional activated-sludge system experiencing shock loading could be greatly enhanced by utilizing specific microorganisms with PCP biodegradation capability along with the use of acclimated biomass. Penta degradation in a continous laboratory-scale activated-sludge system with PCP concentration varying between 1 and 120 mg/L was studied in acclimated systems with and without the addition of *Arthrobacter* sp. strain ATCC 33790. The system consisted of a 6.25-L mixed liquor vessel and a 1.66 L external clarifier. Hydraulic retention times (HRTs) ranged from 8.9 to 10.4 h and a mean cell residence time (MCRT) of 6.2 days was maintained. Effects of shock loading up to 120 mg Penta/L were examined, and the *Arthrobacter*-amended system showed a brief improved response to shock loads as compared to the acclimated system without the specific strain (Rochkind-Dubinsky et al., 1987).

Treatment of PCP-Contaminated Waters   Both surface and groundwaters have been contaminated with Penta, so numerous treatability studies have been completed to demonstrate that biodegradation of the PCP can occur in such systems. A few of those reported studies will be described in this section. Puhakka and Jarvinen (1992) studied polychlorophenol degradation in two aerobic continuous-flow fluidized-bed reactors using pure oxygen for aeration and spherical silica-based microcarriers for cell immobilization. The reactors were maintained at temperatures from 24 to 29°C. A high recyle ratio provided dilution of feed and completely mixed conditions and prevented inhibition due to high concentrations of chlorophenols in the feed. Low HRTs (5 h) and high loading rates ($217 g/m^3$ day) produced removal efficiences of >99% for the chlorophenols (Table 10.2).

A further study confirmed the biotreatability of polychlorophenols, including PCP, with high removal efficiencies. At an HRT of 5 h and a chlorophenol loading rate of 445 mg/L day, 99.7% of chlorophenol removal was reported using aerobic fluidized-bed reactor systems (Mäklen et al., 1993).

BIFAR, a biological process for the destruction of hazardous wastes, was employed by Litchfield and co-workers (1994) to demonstrate its effectiveness in the biodegradation of pentachlorophenol in groundwater. BIFAR is an upflow fluidized-bed activated-carbon reactor with a recycle vessel and a fixed carbon bed for polishing before discharge and can be operated under aerobic or denitrifying conditions. The laboratory-scale system consisted of a 1-in. (2.5-cm) diameter and 5-ft (1.5-m) high glass column. It was filled with 2 ft (0.6 m) of 12-by-40 mesh activate carbon. Laboratory results under aerobic conditions showed greater than 95% reduction of influent PCP at a loading rate of 0.6 mg/g carbon/day. Results of the field pilot and full-scale applications of BIFAR are presented later in this chapter.

Bioreactor technologies offer effective treatment of pentachlorophenol-contaminated groundwater; but when the efficiency of such a system is assessed, it is important to consider factors such as partitioning to biomass and physical adsorption. Mueller et al. (1993) conducted a study on the bioremediation of PCP- and creosote-contaminated groundwater using a two-stage bioreactor. The two bioreactors were set up in series and included separate inocula for each bioreactor: one inoculum, CRE1–13, for bioreator 1, and a second inoculum, EPA505 and SR3, for bioreactor 2. These microorganisms had been selected from PCP-contaminated soils in Florida. These inocula were then applied to field tests using a continuously stirred tank reactor for bioreactor 1, while the second reactor was operated in a batch mode. The results indicated removal efficiencies of 97.4 and 77.8% in the bench- and pilot-scale studies, respectively (Table 10.2). Although volatilization was not a significant factor, loss of PCP due to adsorption and partitioning was estimated to be 25.5 and 53.7% in the bench- and pilot-scale studies, respectively.

Anaerobic Treatability Studies   A number of studies have looked at the application of advanced anaerobic treatment systems for biodegradation of highly halogenated aromatics. An upflow anaerobic sludge blanket (UASB) reactor is one such system. The performance of a laboratory-scale UASB reactor operated at 28°C was investigated by Wu et al. (1993) in treating wastewaters containing high PCP concentrations. A water-jacketed glass column reactor was used to effect temperature control. The reactor volume was 100 mL, and the settler volume was 225 mL. Anaerobic granules were developed in the presence of PCP, with volatile fatty acids and methanol as co-substrates. A continuous feed was maintained using a minipulse peristaltic pump, and the treated effluent was recycled at 1 L/h using another peristaltic pump. With a PCP loading rate of 88–97 mg/L day and HRT of 10.8–15 h, 99% removal was reported (Table 10.2). This was not as high a loading as an aerobic fixed-film reactor (99% removal with a loading rate of 360 mg/L day) but is an effective alternative for treating Penta-contaminated wastewaters with relatively high organic content [2–10 g of chemical oxygen demand (COD) per liter] since an anaerobic system would not have an oxygen demand.

Treatability Studies on PCP-Contaminated Soils   Various biotreatment processes have been employed in the past for the treatment of contaminated soils, including in situ, solid-phase bioremediation and soil slurry bioreactors. Biodegradation in reactor systems offers advantages over the other processes since they are more manageable. Mueller et al. (1991b) conducted a bench-scale study to evaluate the potential of slurry-phase bioremediation (refer to Chapter 3 for a discussion on bioslurry reactors). The study was conducted in two 1.5-L bioreactors at a temperature of 28.5°C. The bioreactors were designed to automatically maintain a pH of 7.1 ± 0.1 and 90% dissolved oxygen. The reactors were operated in a batch culture mode for 30 days. The system was successful in degrading >50% of the targeted compounds, but PCP and other high-molecular-weight compounds were not removed extensively when an indigenous microbial population was employed.

Although physicochemical processes can be controlled to a great extent in a bioreactor, it may not be the most cost-effective method of treatment for a given contaminated matrix, since the soil would have to be physically removed from the contaminated site. Otte et al. (1994) suggested an alternative technology. They cultured specially selected indigenous microorganisms in a soil–slurry reactor before introducing them into static soil microcosms. PCP loading rate was increased in steps from 0 to 500 mg/L day^{-1} to culture the biomass in a 15-L fed-batch soil–slurry bioreactor. This increased the activity and upper tolerance limit to PCP of the consortium. The inoculum was then introduced into static soil microcosms (125 mL). The results showed a 10-fold increase in the rate of PCP mineralization compared to the uninoculated control microcosms.

Other laboratory treatability studies to evaluate feasibility for field pilot and full-scale studies have also been conducted for the treatment of PCP-contaminated soils. Trudell et al. (1994) conducted a feasibility study of in situ bioremediation at an abandoned PCP site for 35 weeks, at temperatures of 5, 15, and 25°C. Initial PCP concentrations in the range of 300–900 mg/kg were reduced to 10–200 mg/kg. Based on the laboratory treatability results, a field pilot study was designed for a 2500-m^2 treatment plot. The design included a plastic covering over the test field to control infiltration and evaporation, drip irrigation system to maintain optimum moisture content, and a leachate collection system at the base of the excavation. Results of the field pilot study are not yet available.

Another in situ bioremediation field pilot study was conducted by Piontek and Simkin (1994) to test and evaluate the practicability of several treatment applications, including bioventing, use of nitrate as an electron acceptor, bioremediation in combination with chemically enhanced soil flushing, and water flooding. This last mentioned method involves the installation of dual horizontal recovery drain lines and parallel delivery drain lines. This improves water recovery in the upper drain lines, and DNAPL (see Advanced Fixed Film Reactor, Chapter 2) recovery occurs in the lower drain lines. Based on the test results, water flooding combined with soil flushing was capable of reducing the subsurface oxygen demand. However, the use of nitrate as an alternative electron acceptor did not reduce the oxygen demand significantly. Bioventing, was found to be a feasible approach but under limited conditions.

Factors Influencing PCP Degradation in Model Systems

Acclimated Microbial Populations   In many cases the presence of acclimated microbial cultures is responsible for improving the performance of biological reactors used for wastewater treatment. PCP spiked into synthetic peptone sewage sludge was studied in a laboratory-scale activated-sludge reactor under a range of operating conditions. It was observed that biodegradation rates in the case of an acclimated microbial consortium were one order of magnitude higher than initially unacclimated cultures (Nyholm et al., 1992). In another case of anaerobic sewage sludge digestors, PCP was found to be inhibitory to unacclimated methanogenic bacteria at a concentration of around 200 μg/L. The threshold value rose to

600 μg/L after acclimation. Bajpai and Banerji (1992) while exploring biodegradation of soil slurries using a mixed culture from activated sludge also concluded that adaptation significantly reduced the lag periods for cell growth.

Immobilization of the Microorganisms   Use of immobilized biomass is an effective way to improve performance by protecting the organisms from the toxicity of the contaminant and increasing the hydraulic retention time in bioreactors (see Chapter 1 and 2). O'Reilly and Crawford have conducted a number of studies to examine the degradation of pentachlorophenol by using polymeric immobilization matrices, namely calcium alginate and polyurethane (O'Reilly and Crawford, 1988, 1989; O'Reilly et al., 1988; Frick et al., 1988; and Crawford et al., 1990). In one study a comparison was made between the degrading capability of flavobacteria isolated from environments contaminated with PCP. The cells were either immobilized in calcium alginate or tested as free cell suspensions. After 42 days the immobilized cells degraded 71% of the Penta present at 10 ppm, whereas the cell suspension degraded less than 0.1% of the PCP over the same period of time. Complete degradation of pentachlorophenol was also reported using polyurethane-immobilized flavobacteria. PCP concentrations of 100, 200, and 300 ppm were degraded in 48 h, 4 days, and 5 days, respectively (Crawford et al., 1990). It was also suggested that polyurethane was a better immobilization matrix compared to calcium alginate because the polyurethane has greater mechanical strength and improved oxygen transfer characteristics (O'Reilly and Crawford, 1989). In another study Hu et al. (1994) reported degradation of 200–250 mg PCP/L using alginate-immobilized cells and up to 700 mg PCP/L using polyurethane-immobilized cells.

Inhibition of Microbial Activity in Bioreactors   The biodegradation of PCP to a certain extent depends on the matrix in which it is contained. In a study to determine the fate of priority pollutants in biological wastewater treatment systems it was concluded that PCP degradation followed an inhibition-type function. The maximum rate of biodegradation occurred at 350 μg PCP/L and was a function of biomass concentration (Moos et al., 1983). In another study, synthetic wastewater was treated in a laboratory-scale continuous, aerobic, fiber-wall reactor, using reagent grade (98% pure), commercial grade (75–85% pure), and improved commercial grade PCP (75–85% pure, with chlorodioxins substantially reduced). The results indicated relative inhibition of commercial grade PCP degradation. This was attributed to the presence of impurities such as chlorodioxins in the commercial-grade Penta (Rochkind-Dubinsky et al., 1987).

Effects of Supplemental Carbon Sources on Bioreactor Degradation Rates   Hendricksen et al. (1991) assessed the anaerobic degradation of PCP and phenol in two fixed-film reactors; one served as the control while the other was amended with a dose of 1 g glucose/L and 1–2 mg PCP/L. The reactors were operated for 6 months, at the end of which it was calculated that the removal of PCP in the reactor with added glucose was 98%, while the removal in the control reactor was approximately 60% (Table 10.3). Likewise in a study conducted in two upflow anaerobic sewage sludge blanket reactors, removal of PCP reached 99% in the glucose-amended reactor, while the control showed efficiencies between 32 and 77% (Table 10.3). It was further observed that the glucose-amended reactor had a higher concentration of biomass. As a result of that and the stimulatory effect of glucose (as described earlier), 94% of the PCP was completely dechlorinated in reactors with added glucose as compared to 20% (maximum) in the control reactor (Hendricksen et al., 1992).

In their study of biodegradation of soil slurries by mixed cultures from activated sludge under aerobic conditions, Bajpai and Banerji (1992), observed that the highest rate of PCP degradation occurred in the absence of glucose, suggesting preferential uptake. This may seem contradictory to the studies of Hendrickson et al. (described above) but it emphasizes the importance of laboratory treatability studies to optimize conditions before pilot or full-scale implementation of a biotreatment system.

Treatment Train Approaches Biotreatment in combination with one or more physical/chemical processes may be an effective way for the removal and degradation of pentachlorophenol from the environment. Khodadoust et al. (1994) evaluated the coupling of soil solvent washing with anaerobic biodegradation of the washed extract for the treatment of PCP-contaminated soils. Both in situ solvent washing, by continuously flushing solvent through a packed bed of soil, and ex situ solvent washing procedures were conducted. The wash fluid was fed into an expanded-bed anaerobic granular activated-carbon reactor, with successful bio-degradation of the extract. Mahaffey and Sanford (1991) employed a similar technology for in situ treatment of contaminated soils. The process involved soil washing coupled with screening to obtain a target clean up level 0.5 mg/kg of soil. About 80% of the soil did not need further processing and could be directly discharged into the disposal pit. The remaining 20% of the contaminated soil, combined with the wash extract, could then be biotreated in a slurry reactor.

FIELD STUDIES

As mentioned, pentachlorophenol has been successfully biodegraded in various environments in laboratory-scale studies, some of them with recommendations for field treatment.

In 1985, BioTrol Corp. acquired a patent for a process that could be effectively used for treating pentachlorophenol-contaminated environments. On further development, they engineered a family of treatment systems and used them at different wood-preserving sites for treating PCP- and creosote–contaminated groundwater, lagoon water, and soils. The technology used for treating contaminated groundwater was an immobilized bacterial system (see Chapter 1 and 2 for discussion of fixed film reactor). The groundwater at the site had on an average 80–120 ppm of PCP. After one pass through their system, the contaminated water was treated to levels permitted for discharge into a sewer system (Table 10.2). It was also reported that the treatment system did not result in the formation of intermediates and could be used for meeting different discharge standards. A more complex system was used for treating contaminated lagoon water. In this process, COD and biological oxygen demand (BOD) were removed prior to PCP treatment. Treatability study data showed that the system had reduced the PCP concentrations from 10–36 to 0–0.05 ppm (Table 10.2), Pflug and Burton, 1988). A third type of treatment system was the BioTrol Soil Treatment System used for the decontamination of soils. This process utilized water for extracting the contaminants from the soils, which was later treated as above. In pilot studies at wood treatment sites, 3% total organic contamination was reduced to 20–25 ppm. At the time of the report further information on the use of this system was not provided (Pflug and Burton, 1988).

A comprehensive biological treatment plan was initiated for an abandoned wood treatment site located in Libby, Montana, in the spring and summer of 1989. The bioremediation project consisted of a soil treatment program including: in situ land treatment of de-rocked, contaminated soils stored in waste pits, ex situ land treatment of de-rocked contaminated soils in land treatment units (LTUs), and irrigation treatment of contaminated rock piles. It also included a groundwater extraction and ex situ treatment system, and two oxygen injection systems for in situ treatment of the contaminated aquifer. Piotrowski et al. (1994) in their status report suggested reasonable success with the operation of these systems. Approximately 5686 yd ($4347 \, m^3$) of contaminated soil have been effectively treated in the LTUs over a 4-year period. To date, results of the bioreactor performance indicate greater than 88% removal of the pentachlorophenol in the ex situ groundwater treatment program. The authors mentioned that at the time of the report, they observed reduced concentrations of contaminants in the in situ groundwater treatment program as well.

In another field study PCP has been reported (Litchfield et al., 1994) to be biodegraded at a site in North Central United States, by indigenous microorganisms in soil and groundwater operating with nitrate as a major electron acceptor. The site was used to treat wooden window frames. Soils in the unsaturated zone had been contaminated with up to 1% PCP, and the groundwater had PCP concentrations ranging between 13 and 90 mg/L. The remedial system used for the ex situ treatment of groundwater was an upflow fluidized-bed granular activated-carbon unit. Pretreatment with polymer addition, to remove oil, grease, and suspended particles, was carried out as required. The secondary and tertiary treatment units consisted of the BIFAR biological reactor and activated-carbon reactors, respectively, with nutrient additions made prior to the treatment of contaminated water in the BIFAR unit. Treatment of the soil around and under the former dip treatment room was accomplished by constructing a shallow subsurface seepage bed system. After addition of nutrients, a portion of the groundwater from the BIFAR unit was passed through the seepage beds for treatment. At the time of the report, Litchfield and co-workers estimated biodegradation of greater than 2800 kg PCP in the above-ground treatment unit and 23–51 kg of PCP in the soils under the seepage beds over 4 years of operation (Table 10.2).

EMERGING TECHNOLOGIES

Even though biological treatment systems are now used to treat PCP-contaminated waters, wastewaters, and soils, it is imperative that new methods and technologies be combined with the biotreatment technology to increase the removal efficiencies and decrease the time required to reach target levels. Several improvements in bioremediation are developing that should accomplish these goals. One of these improvements involves the addition of solid-phase, biodegradable organic amendments to hydrophobic soils to increase their moisture holding capacity and oxygen diffusion to support a larger microbial population. Seech et al. (1994) in their study to demonstrate the efficiency of this technology reported 99% PCP removal in solids with an initial PCP concentration of approximately 680 mg/kg of soil.

Dutta and Singh (1991) introduced a membrane aeration system for optimizing the aeration efficiency of a biological activated-carbon (BAC) system. Membrane aeration provides fast, bubble-free aeration without flooding or loading. It is achieved by placing a thin film of synthetic polymer membrane between the gas and liquid phase (see Chapter 4 for a discussion of membrane reactor). To date, no reports have been published on the application of this technology to Penta biodegradation, but it certainly has potential to improve aerobic bioremediation.

Finally, chemical oxidation of chlorinated compounds in soils using hydrogen peroxide and ferrous salts (Fenton's reagent) was studied by Martens and Frankenberger (1994). They reported the partial chemical oxidation of certain chlorinated compounds, with the possibility of further degradation by soil microflora. They also concluded that Fenton's Reagent did not significantly reduce the indigenous microbial population in the amended soils. Hence, bioremediation in combination with chemical oxidation may be a promising technology.

CONCLUSIONS

During the last 20 years, a great deal of basic information on the biodegradation of PCP has been accumulated. Once Penta was shown to be biodegradable, applied research began to demonstrate the possibility of using a less costly and more natural process for destroying PCP, namely bioremediation. Thus basic research coupled with field pilot studies have made public and governmental acceptance of biotreatment of Penta-contaminated soils and water possible. The story of PCP bioremediation is a classic example of both basic and applied research combining to solve a serious environmental problem.

REFERENCES

Alexander, M. (1965). Persistence and Biological Reactions of Pesticides in Soils, *Soil Sci. Soc. Am. Proceed.*, **29**(1):1–7.

Apajalahti, J. H. A. and M. S. Salkinoja-Salonen (1986). Degradation of Polychlorinated Phenols by *Rhodococcus chlorophenolicus*, *Appl. Microbiol. Biotech.*, **25**:62–67.

Apajalahti, J. H. A. and M. S. Salkinoja-Salonen (1987a). Dechlorination and *para*-Hydroxylation of Polychlorinated Phenols by *Rhodococcus chlorophenolicus*, *J. Bacteriol.*, **169**(2):675–681.

Apajalahti, J. H. A. and M. S. Salkinoja-Salonen (1987b). Complete Dechlorination of Tetrachloroquinone by Cell Extracts of Pentachlorophenol-Induced *Rhodococcus chlorophenolicus*, *J. Bacteriol.*, **169**(11):5125–5130.

Arcand, Y., J. Hawari, and S. R. Guiot (1995). Solubility of Pentachlorophenol in Aqueous Solutions: The pH Effect, *Wat. Res.*, **29**(1):131–136.

Bajpai, R. and S. Banerji (1992). Bioremediation of Soils Contaminated with Pentachlorophenol, *Ann. New York Acad. Sci.*, **665**:423–434.

Banerji, S. K. and R. K. Bajpai (1994). Cometabolism of Penachlorophenol by Microbial Species, *J. Hazard. Mat.*, **39**:19–31.

Brummett, T. B. and G. W. Ordal (1977). Inhibition of Amino Acid Transport in *Bacillus subtilis* by Uncouplers of Oxidative Phosphorylation, *Arch. Biochem. Biophys.*, **178**:368–372.

Chu, J. P. and E. J. Kirsch (1972). Metabolism of Pentachlorophenol by an Axenic Bacterial Culture, *Appl. Microbiol.*, **23**(5):1033–1035.

Chu, J. P. and E. J. Kirsch (1973). Utilization of Halophenols by a Pentachlorophenol Metabolizing Bacterium, *Develop. Ind. Microbiol.*, **14**:264–273.

Crawford, R. L. (1987). Biodegradation of Pentachlorophenol, U. S. Pat. 4,713,340.

Crawford, R. L. and W. W. Mohn (1985). Microbiological Removal of Pentachlorophenol from Soil Using a *Flavobacterium, Enzy. Microbiol. Tech.*, **7**:617–620.

Crawford, R. L., K. T. O'Reilly, and H. L. Tao (1990). Microorganism Stabilization for In Situ Degradation of Toxic Chemicals, in *Advances in Applied Biotechnology Series*, Vol. 4, D. Kamely, A. Chakrabarty, and G. S. Omenn, Eds., Portfolio Publishing, Woodlands, Texas, pp. 203–211.

Cserjesi, A. J. (1967). The Adaptation of Fungi to Pentachlorophenol and its Biodegradation, Can. J. Microbiol., **13**:1243–1249.

Cserjesi, A. J. and E. L. Johnson (1972). Methylation of Pentachlorophenol by *Trichoderma virgatum, Can. J. Microbiol.*, **18**:45–49.

Davis, A., J. Campbell, C. Gilbert, M. V. Ruby, M. Bennett, and S. Tobin (1994). Attenuation and Biodegradation of Chlorophenols in Ground Water at a former Wood Treating Facility, *Groundwater*, **32**(2):248–257.

Devine, K. (1994). Bioremediation: The State of Usage, in *Applied Biotechnology for Site Remediation*, R. E. Hinchee, D. B. Anderson, F. Metting, Jr., and G. D. Sayles, Eds., Lewis Publishers, FL, Boca Raton, pp. 435–438.

Duncan, C. G. and F. J. Deverall (1964). Degradation of Wood Preservatives by Fungi *Appl. Microbiol.*, **12**(1):57–62.

Dutta, C. and R. Singh (1991). Recent Developments in Hydrocyclones and Biotreatment of Wastes Incorporating Polymer Membrane and Activated Carbon, in, *Gas, Oil, Coal, and Environmental Biotechnology*, Vol. 3, C. Akin and J. Smith, Eds., Institute of Gas Technology, Chicago, pp. 239–254.

Eisenstadt, E., S. Fisher, C. L. Der, and S. Silver (1973). Manganese Transport in *Bacillus subtilis* W23 During Growth and Sporulation, *J. Bacteriol.*, **113**(3):1363–1372.

Ehrlich, W., M. Mangir, and E. R. Lochmann (1987). The Effect of Pentachlorophenol and Its Metabolite Tetrachloroquinone on RNA, Protein, and Ribosome Synthesis in Saccharomyces Cells, *Ecotoxicol. Environ. Safety*, **13**:7–12.

Evans, W. C. (1963). The Microbiological Degradation of Aromatic Compounds, *J. Gen. Microbiol.*, **32**:177–184.

Fetzner, S. and F. Lingens (1994). Bacterial Dehalogenases: Biochemistry, Genetics, and Biotechnological Applications, *Microbiol. Rev.*, **58**(4):641–685.

Finn, R. K. (1983). Use of Specialized Microbial Strains in the Treatment of Industrial Waste and in Soil Decontamination, *Experentia*, **39**:1231–1236.

Frick, T. D., R. L. Crawford, M. Martinson, T. Chresand, and G. Bateson (1988). Microbiological Cleanup of Groundwater Contaminated by Pentachlorophenol, in *Environmental Biotechnology: Reducing Risks from Environmental Chemicals through Biotechnology*, G. S. Omenn, Ed., Plenum Press, New York, pp. 173–190.

Guthrie, M. A., E. J. Kirsch, R. F. Wukasch, and C. P. L. Grady, Jr. (1984). Pentachlorophenol Biodegradation—II Anaerobic, *Wat. Res.*, **18**(4):451–461.

Harmsen, J. (1991). Possibilities and Limitations of Landfarming for Cleaning Contaminated Soils, in, *On-Site Bioreclamation: Processes for Xenobiotic and Hydrocarbon Treatment*, R. E. Hinchee and R. F. Olfenbuttel, Eds., Butterworth-Heinemann, Boston, MA, pp. 255–272.

Harmsen, J., H. J. Velthorst, and I. P. A. M. Bennehey (1994). Cleaning of Residual Concentrations with an Extensive Form of Landfarming in *Applied Biotechnology for Site Remediation*, R. E. Hinchee, D. B. Anderson, F. B. Metting, Jr., and G. D. Sayles, Eds., Lewis Publisher, Boca Raton FL, pp. 84–91.

Hendriksen, H. V., S. Larsen, and B. K. Ahring (1991). Anaerobic Degradation of PCP and Phenol in Fixed-Film Reactors: The Influence of An Additional Substrate. *Wat. Sci. Tech.*, **24**(3/4):431–436.

Hendriksen, H. V., S. Larsen, and B. K. Ahring (1992). Influence of a Supplemental Carbon Source on Anaerobic Dechlorination of Pentachlorophenol in Granular Sludge, *Appl. Environ. Microbiol.*, **58**(1):365–370.

Hu, Z. C., R. A. Korus, W. E. Levinson, and R. L. Crawford (1994). Adsorption and Biodegradation of Pentachlorophenol by Polyurethane-Immoblized *Flavobacterium*, *Environ. Sci. Tech.*, **28**:491–496.

Janssen, D. B., F. Pries, and J. R. van der Ploeg (1994). Genetics and Biochemistry of Dehalogenating Enzymes, *Ann. Rev. Microbiol.*, **48**:163–191.

Karns, J. S., J. J. Kilbane, S. Duttagupta, and A. M. Chakrabarty (1983a). Metabolism of Halophenols by 2,4,5-Trichlorophenoxyacetic Acid-Degrading *Pseudomonas cepacia*, *Appl. Environm. Microbiol.*, **46**(5):1176–1181.

Karns, J. S., S. Duttagupta, and A. M. Chakrabarty (1983b). Regulation of 2,4,5-Trichlorophenoxyacetic Acid and Chlorophenol Metabolism in *Pseudomonas cepacia* AC1100, *Appl. Environ. Microbiol.*, **46**(5):1182–1186.

Kato, L., C. Adapoe, and M. Ishaque (1976). The Respiratory Metabolism of *Mycobacterium lepraemurium*, *Can. J. Microbiol.*, **22**:1293–1299.

Kaufman., D. D. (1978). Degradation of Pentachlorophenol in Soil and by Soil Microorganisms, in *Pentachlorophenol: Chemistry, Pharmacology, and Environmental Toxicology*, K. R. Rao, Ed., Plenum Press, New York, pp. 27–39.

Khodadoust, A. P., J. A. Wagner, M. T. Suidan, and S. I. Safferman (1994). Solvent Washing of PCP Contaminated Soils with Anaerobic Treatment of Wash Fluids, *Wat. Environ. Res.*, **66**(5):692–697.

Kilbane, J. J., D. K. Chatterjee, J. S. Karns, S. T. Kellogg, and A. M. Chakrabarty (1982). Biodegradation of 2,4,5-Trichlorophenoxyacetic Acid by a pure Culture of *Pseudomonas cepacia. Appl. Environ. Microb.*, **44**:72–78.

Kim, C. J. and W. J. Maier (1986). Biodegradation of Pentachlorophenol in Soil Environments, *41st. Purdue University Industrial Waste Conference Proceedings*, pp. 303–312.

Klecka, G. M. and W. J. Maier (1985). Kinetics of Microbial Growth on Pentachlorophenol, *Appl. Environ. Microbiol.*, **49**(1):46–53.

Kremer, S., O. Sterner, and H. Anke (1992). Degradation of Pentachlorophenol by *Mycena avenacea* TA 8480-Identification of Initial Dechlorinated Metabolites, *Zeitschrift Naturforschung*, **47c**, 561–566.

Lamar, R. T., M. J. Larsen, and T. K. Kirk (1990). Sensitivity to and Degradation of Pentachlorophenol by *Phanaerochaete* spp. *Appl., Environ. Microbiol.*, **56**(11):3519–3526.

Larsen, S., H. V. Hendriksen, and B. K. Ahring (1991). Potential for Thermophilic (50 C) Anaerobic Dechlorination of Pentachlorophenol in Different Ecosystems, *Appl. Environ. Microbiol.*, **57**(7):2085–2090.

Leutritz, J., Jr. (1965). Biodegradability of Pentachlorophenol, *Forest Products J.*, **15**:269–272.

Lin, J. E. and H. Y. Wang (1991). Degradation of Pentachlorophenol by Non-Immobilized, Immobilized and Co-Immobilized *Arthrobacter* Cells, *J. Ferment. Bioeng.*, **72**(4):311–314.

Lin, J. E., R. F. Hickey, and H. Y. Wang (1990). Degradation Kinetics of Pentachlorophenol by *Phanaerochaete chrysosporium. Biotech. Bioeng.* **35**:1125–1134.

Litchfield, C. D., S. Huang, G. J. Gromicko, and C. Steffy (1992). BIFAR: A Biological Process for the Destruction of Hazardous Wastes, *Ground Wat. Manage., Book 13*, Proceedings of the Focus Conference on Eastern Regional Ground Water Issues. pp. 71–78.

Litchfield, C. D., G. O. Chieruzzi, D. R. Foster, and D. L. Middleton (1994). A Biotreatment Train Approach to a PCP-Contaminated Site: In Situ Bioremediation Coupled with an Above Ground BIFAR System Using Nitrate as the Electron Acceptor, in *Bioremediation of Chlorinated and Polycyclic Aromatic Hydrocarbon Compounds*, R. E. Hinchee, A. Lesson, L. Semprini, and S.K. Ong, Eds., Lewis Publishers, Boca Raton, FL, pp. 155–163. 155–163.

Lyr, H. (1963). Enzymatische Detoxifikation Chlorierter Phenole Phytopathal, *Zeitung*, **47**:73–83.

Madsen, T. and J. Aamand (1991). Effects of Sulfuroxy Anions on Degradation of Pentachlorophenol by a Methanogenic Enrichment Culture, *Appl. Environ. Microbiol.*, **57**(9):2453–2458.

Mahaffey, W. R. and R. A. Sanford (1991). Bioremediation of PCP-Contaminated Soil: Bench to Full-Scale Implementation, *Remediation*, Summer, 305–323.

Mäklnen, P. M., T. J. Theno, J. F. Ferguson, J. E. Ongerth, and J. A. Puhakka (1993). Chlorophenol Toxicity Removal and Monitoring in Aerobic Treatment: Recovery from Process Upsets, *Environ. Sci. Tech.*, **27**:1434–1439.

Martens, D. A. and W. T. Frankenberger, Jr. (1994). Feasibility of In Situ Chemical Oxidation of Refractile Chlorinated Organics by Hydrogen Peroxide Generated Oxidative Radicals in Soil, in *Emerging Technology for Bioremediation of Metals*, J. L. Means and R. E. Hinchee, Eds., Lewis Publishers, Boca Raton, FL, pp. 74–83.

McAllister, K. A., H. Lee, and J. T. Trevors (1996). Microbial Degradation of Pentachlorophenol, *Biodegradation*, **7**(1):1–40.

Melcer, H. and W. K. Bedford (1988). Removal of Pentachlorophenol in Municipal Activated Sludge Systems, *J. Wat. Pollution Control Fed.*, **60**(5):622–626.

Mickelson, M. N. (1974). Effect of Uncoupling Agents and Respiratory Inhibitors on the Growth of *Streptococcus agalactiae*, *J. Bacteriol.*, **120**(2):733–740.

Middaugh, D. P., R. L. Thomas, S. E. Lantz, C. S. Heard, and J. G. Mueller (1994a). Field Scale Testing of a Hyperfiltration Unit for Removal of Creosote and Pentachlorophenol from Ground Water: Chemical and Biological Assessment, *Arch. Environ. Contamination Toxicol.*, **26**:309–319.

Middaugh, D. P., S. E. Lantz, C. S. Heard, and J. G. Mueller (1994b). Field-Scale Testing of a Two-Stage Bioreactor for Removal of Creosote and Pentachlorophenol from Ground Water: Chemical and Biological Assessment, *Arch. Environ. Contamination Toxicol.*, **26**:320–328.

Middeldorp, P. J. M., M. Briglia, and M. S. Salkinoja-Salonen (1990). Biodegradation of Pentachlorophenol in Natural Soil by Inoculated *Rhodococcus chlorophenolicus*, *Microbial Ecol.*, **20**:123–139.

Mikesell, M. D. and S. A. Boyd (1985). Reductive Dechlorination of the Pesticides 2,4-D, 2,4,5-T, and Pentachlorophenol in Anaerobic Sludges, *J. Environ. Qual.*, **14**(3):337–340.

Mikesell, M. D. and S. A. Boyd (1986). Complete Reductive Dechlorination and Mineralization of Pentachlorophenol by Anaerobic Microorganisms, *Appl. Environ. Microbiol.*, **52**(4):861–865.

Mileski, G. J., J. A. Bumpus, M. A. Jurek, and S. D. Aust (1988). Biodegradation of Pentachlorophenol by the White Rot Fungus *Phanaerochaete chrysosporium*, *Appl. Environ. Microbiol.*, **54**(12):2885–2889.

Moos, L. P., E. J. Kirsch, R. F. Wukasch, and C. P. L. Grady, Jr. (1983). Pentachlorophenol Biodegradation—I, *Wat. Res.*, **17**(11):1575–1584.

Mueller, J. G., S. E. Lantz, B. O. Blattmann, and P. J. Chapman (1991a). Bench-Scale Evaluation of Alternative Biological Treatment Processes for the Remediation of Pentachlorophenol- and Creosote-Contaminated Materials: Solid-Phase Bioremediation, Environ. Sci. Tech., **25**(6):1045–1054.

Mueller, J. G., S. E. Lantz, B. O. Blattmann, and P. J. Chapman (1991b). Bench-Scale Evaluation of Alternative Biological Treatment Processes for the Remediation of Pentachlorophenol- and Creosote-Contaminated Materials: Slurry-Phase Bioremediation, Environ. Sci. Tech., **25**(6):1055–1061.

Mueller, J. G., S. E. Lantz, D. Ross, R. J. Colvin, D. P. Middaugh, and P. H. Pritchard (1993). Strategy Using Bioreactors and Specially Selected Microorganisms for Bioremediation of Groundwater Contaminated with Creosote and Pentachlorophenol, *Environ. Sci. Tech.*, **27**(4):691–698.

Nicholas, R. A. and G. W. Ordal (1978). Inhibition of Bacterial Transport by Uncouplers of Oxidative Phosphorylation, *Biochem. J.*, **176**:639–647.

Nyholm, N., B. N. Jacobsen, B. M. Pedersen, O. Poulsen, A. Damborg, and B. Schultz (1992). Removal of Organic Micropollutants at PPB Levels in Laboratory Activated Sludge Reactors Under Various Operating Conditions, *Wat. Res.*, **26**(3):339–353.

O'Reilly, K. T. and R. L. Crawford (1988). Degradation of Substituted Phenols by Immobilized Bacteria, in *Environmental Biotechnology: Reducing Risks from Environmental Chemicals through Biotechnology*, G. S. Omenn, Ed., Plenum Press, New York, pp. 457–458.

O'Reilly, K. T. and R. L. Crawford (1989). Degradation of Pentachlorophenol by Polyurethane-Immobilized *Flavobacterium* Cells, *Appl. Environ. Microbiol.*, **55**(9):2113–2118.

O'Reilly, K. T., R. Kadakia, R. A. Korus, and R. L. Crawford (1988). Utilization of Immobilized Bacteria to Degrade Aromatic Compounds Common to Wood Treatment Wastewaters, *Wat. Sci. Tech.*, **20**(11/12):95–100.

Otte, M. P., J. Gagnon, Y. Comeau, N. Matte, C. W. Greer, and R. Samson (1994). Activation of an Indigenous Microbial Consortium for Bioaugmentation of Pentachlorophenol/Creosote Contaminated Soils, *Appl. Microbiol. Biotech.*, **40**:926–932.

Parker, L. V. and T. F. Jenkins (1986). Removal of Trace-Level Organics by Slow-Rate Land Treatment, *Wat. Res.*, **20**(11):1417–1426.

Parker, W. J., J. P. Bell, and H. Melcer (1994). Modelling the Fate of Chlorinated Phenols in Wastewater Treatment Plants, *Environ. Prog.*, **13**(2):98–104.

Patel, G. B., B. J. Agnew, and C. J. Dicaire (1991). Inhibition of Pure Cultures of Methanogens by Benzene Ring Compounds, *Appl. Environ. Microbiol.*, **57**(10):2969–2974.

Pflug, A. D. and M. B. Burton (1988). Remediation of Mutimedia Contamination from the Wood Preserving Industry, in *Environmental Biotechnology: Reducing Risks from Environmental Chemicals through Biotechnology*, G. S. Omenn, Ed., Plenum Press, New York, pp. 193–201.

Pignatello, J. J., M. M. Martinson, J. G. Steiert, R. E. Carlson, and R. L. Crawford (1983). Biodegradation and Photolysis of Pentachlorophenol in Artificial Freshwater Streams, *Appl. Environ. Microbiol.*, **46**(5):1024–1031.

Piontek, K. R. and T. J. Simpkin (1994). Practicability of in situ Bioremediation at a Wood-Preserving Site, in *Bioremediation of Chlorinated and Polycyclic Aromatic Hydrocarbon Compounds*, R. E. Hinchee, A. Lesson, L. Semprini, and S. K. Ong, Eds, Lewis Publishers, Boca Raton, FL, pp. 117–128.

Piotrowski, M. R., J. R. Doyle, D. Cosgriff, and M. C. Parsons (1994). Bioremedial Progress at the Libby, Montana, Superfund Site, in *Applied Biotechnology for Site Remediation*, R. E. Hinchee, D. B. Anderson, F. B. Metting, Jr., and G. D. Sayles, Eds., Lewis Publishers, New York, pp. 240–255.

Puhakka, J. A. and K. Järvinen (1992). Aerobic Fluidized-Bed Treatment of Polychlorinated Phenolic Wood Preservative Constitutents, *Wat. Res.*, **26**(6):765–770.

Radehaus, P. M. and S. K. Schmidt (1992). Characterization of a Novel *Pseudomonas* sp. That Mineralizes High Concentrations of Pentachlorophenol, *Appl. Environ. Microbiol.*, **58**(9):2879–2885.

Radtke, C., W. S. Cook, and A. Anderson (1994). Factors Affecting Antagonism of the Growth of *Phanaerochaete chrysosporium* by Bacteria Isolated from Soils *Appl. Microbiol. Biotech.*, **41**:274–280.

Reddy, C. A. and H. D. Peck, Jr. (1978). Electron Transport Phosphorylation Coupled to Fumarate Reduction by H_2- and Mg^{2+}-Dependent Adenosine Triphosphatase Activity in Extracts of the Rumen Anaerobe *Vibrio succinogenes*, *J. Bacteriol.*, **134**(3):982–991.

Rochkind-Dubinsky, M. L., G. S. Sayler, and J. W. Blackburn (1987). Pentachlorophenol, in *Microbiological Decomposition of Chlorinated Aromatic Compounds*, Marcel Dekker, New York, pp. 102–107.

Rott, B., S. Nitz, and F. Korte (1979). Microbial Decomposition of Sodium Pentachlorophenolate, *J. Agricult. Food Chem.*, **27**(2):306–310.

Ruckdeschel, G., and G. Renner (1986). Effects of Pentachlorophenol and Some of its Known and Possible Metabolites on Fungi, *Appl. Environ. Microbiol.*, **51**(6):1370–1372.

Ruckdeschel, G., G. Renner, and K. Schwarz (1987). Effects of Pentachlorophenol and Some of Its Known and Possible Metabolites on Different Species of Bacteria, *Appl. Environ. Microbiol.*, **53**(11):2689–2692.

Rutgers, M., J. J. Bogte, A. M. Breure, and J. G. van Andel (1993). Growth and Enrichment of Pentachlorophenol-Degrading Microorganisms in the Nutristat, Substrate Concentration-Controlled Continuous Culture, *Appl. Environ. Microbiol.*, **59**(10):3373–3377.

Ryding, J. M., J. A. Puhakka, S. E. Strand, and J. F. Ferguson (1994). Degradation of Chlorinated Phenols by a Toluene Enriched Microbial Culture, *Wat. Res.*, **28**(9):1987–1906.

Saber, D. L. and R. L. Crawford (1985). Isolation and Characterization of *Flavobacterium* Strains That Degrade Pentachlorophenol, *Appl. Environ. Microbiol.*, **50**(6):1512–1518.

Schenk, T., R. Müller, F. Mörsberger, M. K. Otto, and F. Lingens (1989). Enzymatic Dehalogenation of Pentachlorophenol by Extracts from *Arthrobacter* sp. Strain ATCC 33790, *J. Bacteriol.*, **171**(10):5487–5491.

Schenk, T., R. Müller, and F. Lingens (1990). Mechanism of Enzymatic Dehalogenation of Pentachlorophenol by *Arthrobacter sp.* Strain ATCC 33790, *J. Bacteriol*, **172**(12):7272–7274.

Seech, T., I. J. Marvan, and J. T. Trevors (1994). On Site/Ex Situ Bioremediation of Industrial Soils Containing Chlorinated Phenols and Polycyclic Aromatic Hydrocarbons, in *Bioremediation of Chlorinated and Polycyclic Aromatic Hydrocarbon Compounds*, R. E. Hinchee, A. Leeson, L. Semprini, and S. K. Ong, Eds., Lewis Publishers, Boca Raton, FL, pp. 451–455.

Seigle-Murandi, F., R. Steiman, and J. L. Benoit-Guyod (1991). Biodegradation Potential of Some Micromycetes for Pentachlorophenol, *Ecotoxicol. Environ. Safety*, **21**:290–300.

Seigle-Murandi, F., R. Steiman, J. L. Benoit-Guyod, B. Muntalif, and L. Sage (1992). Relationship Between the Biodegradative Capability of Soil Micromycetes for Pentachlorophenol and for Pentachloronitrobenzene, *Sci. Total Environ.*, **123/124**:291–298.

Seigle-Murandi, F., R. Steiman, J. L. Benoit-Guyod, and P. Guiraud (1993). Fungal Degradation of Pentachlorophenol by Micromycetes, *J. Biotechnol.*, **30**:27–35.

Senger, H., and D. Rühl (1980). The Influence of Pentachlorophenol on the Biosynthesis of 5-Aminolevulinic Acid and Chlorophyll, *J. Biochem.*, **12**:1045–1048.

Siahpush, A. R., J. E. Lin, and H. Y. Wang (1992). Effect of Adsorbents of Degradation of Toxic Organic Compounds by Coimmobilized Systems, *Biotech. Bioeng.*, **39**:619–628.

Stanlake, G. J. and R. K. Finn (1982). Isolation and Characterization of a Pentachlorophenol-Degrading Bacterium, *Appl. Environ. Microbiol.*, **44**(6):1421–1427.

Steiert, G. J. and R. L. Crawford (1986). Catabolism of Pentachlorophenol by a *Flavobacterium* sp., *Biochem. Biophys. Res. Commun.*, **141**(2):825–830.

Suzuki, T. (1977). Metabolism of Pentachlorophenol by a Soil Microbe, *J. Environ. Sci. Hlth.*, **B12**(2):113–127.

Tam, T. Y. and J. T. Trevors (1981). Toxicity of Pentachlorophenol to *Azotobacter vinelandii*, *Bull. Environ. Contamin. Toxicol.*, **27**:230–234.

Tokuz, R. Y. (1989). Biological Systems-B. Aerobic: Biological Treatment of Chlorinated Phenols using a Rotating Biological Contactor, *43rd Purdue Industrial Waste Conference Proceedings*, Lafayette IN 283–288.

Topp, E. and R. S. Hanson (1990a). Factors Influencing the Survival and Activity of a Pentachlorophenol-Degrading *Flavobacterium* sp. in Soil Slurries, *Can. J. Soil Sci.*, **70**:83–91.

Topp, E. and R. S. Hanson (1990b). Degradation of Pentachlorophenol by a *Flavobacterium* Species Grown in Continuous Culture under Various Nutrient Limitations, *Appl. Environ. Microbiol.*, **56**(2):541–544.

Topp, E., R. L. Crawford, and R. S. Hanson (1988). Influence of Readily Metabolizable Carbon on Pentachlorophenol Metabolism by a Pentachlorophenol-Degrading *Flavobacterium* sp., *Appl. Environ. Microbiol.*, **54**(10):2452–2459.

Trudell, M. R., J. M. Marowitch, D. G. Thompson, C. W. Fulton, and R. E. Hoffmann (1994). In Situ Bioremediation at a Wood-Preserving Site in a Cold, Semi-Arid Climate: Feasibility and Field Pilot Design, in *Bioremediation of Chlorinated and Polycyclic Aromatic Hydrocarbon Compounds*, R. E. Hinchee, A. Leeson, I. Semprini, and S. K. Ong, Eds, Lewis Publishers, Boca Raton, FL, pp. 99–116.

Vollmuth, S., A. Zajc, and R. Niessner (1994). Formation of Polychlorinated Dibenzo-*p*-dioxins and Polychlorinated Dibenzofurans during the Photolysis of Pentachlorophenol-Containing Water, *Environ. Sci. Tech.*, **28**:1145–1149.

Wall, A. J. and G. W. Stratton (1994). Effects of a Chromated-Copper-Arsenate Wood Preservative on the Bacterial Degradation of Pentachlorophenol, *Can. J. Microbiol.*, **40**:388–392.

Watanabe, I. (1973). Isolation of Pentachlorophenol Decomposing Bacteria from Soil, *Soil Sci. Plant Nutri.*, **19**(2):109–116.

Windholz, M., S. Budavari, L. Y. Stroumtsos, and M. N. Fertig (1976). *The Merck Index*, 9th ed., Rahway, Merck, NJ, pg. 6904.

Wu, W. M., L. Bhatnagar, and J. G. Zeikus (1993). Performance of Anaerobic Granules for Degradation of Pentachlorophenol, *Appl. Environ. Microbiol.*, **59**(2):389–397.

Xun, L. and C. S. Orser (1991a). Purification of a *Flavobacterium* Pentachlorophenol-Induced Periplasmic Protein (PcpA) and Nucleotide Sequence of the Corresponding Gene (pcpA), *J. Bacteriol.*, **173**(9):2920–2926.

Xun, L. and C. S. Orser (1991b). Purification and Properties of a Pentachlorophenol Hydroxylase, a Flavoprotein from *Flavobacterium* sp. Strain ATCC 39723, *J. Bacteriol.*, **173**(14):4447–4453.

Xun, L., E. Topp, and C. S. Orser (1992a). Confirmation of Oxidative Dehalogenation of Pentachlorophenol by a *Flavobacterium* Pentachlorophenol Hydroxylase, *J. Bacteriol.*, **174**(17):5745–5747.

Xun, L., E. Topp, and C. S. Orser (1992b). Purification and Characterization of a Tetrachloro-*p*-Hydroquinone Reductive Dehalogenase from a *Flavobacterium* sp., *J. Bacteriol.*, **174**(24):8003–8007.

Yokoyama, M. T., K. A. Johnson, and J. Gierzak (1988). Sensitivity of Ruminal Microorganisms to Pentachlorophenol, *Appl. Environ. Microbiol.*, **54**(11):2619–2624.

Young, H. C. and J. C. Carroll (1951). The Decomposition of Pentachlorophenol When Applied as a Residual Pre-emergence Herbicide, *J. Am. Soc. Agronomy*, **43**:504–507.

CHAPTER 11

Natural Restoration of PCB-Contaminated Hudson River Sediments

Frank J. Mondello, Daniel A. Abramowicz, and James R. Rhea

General Electric Corporate Research and Development, Schenectady, New York 12301 (F.J.M.), (D.A.A.); HydroQual Inc., 4914 West Genesee Street, Suite 119, Camillus, New York 13031 (J.R.R.)

INTRODUCTION

PCB Nomenclature and Applications

Polychlorinated biphenyls (PCBs) are a family of 209 related chemical compounds synthesized by the direct chlorination of biphenyl. These compounds were manufactured and sold as complex mixtures (under the trade names Aroclor, Phenoclor, Clophen, and Kanechlor) with formulations produced for varied applications that differed in their average chlorine level. The individual PCB isomers (congeners) in these mixtures are referred to according to the position of the chlorine substitution, for example, 2,3,4,3′,4′-pentachlorobiphenyl (the shorthand 234–34-CB will be used in this chapter).

The physical and chemical properties of PCBs (excellent dielectric and flame-resistant properties, chemical and thermal stability) led to their extensive industrial use as heat transfer fluids, hydraulic fluids, solvent extenders, plasticizers, flame retardants, organic diluents, and dielectric fluids (Hutzinger et al., 1974). Extensive application of these chemically and thermally stable compounds has resulted in widespread contamination (Buckley, 1982; Tanabe et al., 1983). It is estimated that several hundred million pounds have been released to the environment (Hutzinger and Veerkamp, 1981). The more highly chlorinated PCB congeners are lipophilic, resulting in their accumulation in fatty tissues and bioconcentration in the food chain (Safe, 1980, 1992).

Biological Treatment of Hazardous Wastes, Edited by Gordon A. Lewandowski and Louis J. DeFilippi
ISBN 0-471-04861-5 ©1998 John Wiley & Sons, Inc.

The regulatory history concerning PCBs dates to the 1970s when national PCB phase outs occurred in several industrial nations, although some countries continued to permit PCB production into the 1980s. At that time, little was known concerning the potential health effects of the individual PCB congeners, and all 209 separate chemicals were treated identically by initial legislation and regulations. Moreover, the misperception that PCBs were nonbiodegradable contributed to the strong regulatory stance taken to eliminate the production and use of these chemicals. Much has been learned over the last decade concerning the potential health effects (reviewed in Kimbrough, 1995, Safe, 1992; James et al., 1993) and the relative biotransformability of different PCB congeners (reviewed in Abramowicz, 1990; Bedard, 1990; Furukawa, 1982, 1986). This chapter will focus on the natural restoration processes occurring in the Hudson River, principally on the benefits realized upon anaerobic PCB dechlorination through conversion to less chlorinated PCB congeners that exhibit reduced potentials for dioxin-like toxicity, carcinogenicity, and exposure.

PCB Analysis

A variety of different Aroclor mixtures were produced for different commercial applications. These mixtures differ significantly in average chlorine content (e.g., Aroclor 1242, 42% chlorine by weight; Aroclor 1254, 54% chlorine by weight; Aroclor 1260, 60% chlorine by weight). PCB mixtures can be chemically characterized to determine the congener distribution by gas chromatography with an electron capture detector. This detector displays enhanced sensitivity to more highly chlorinated chemical species, and relative peak areas must be adjusted for differences in detector response to determine individual congener concentrations. High-resolution capillary analysis (e.g., using a fused silica capillary coated with a bonded liquid phase of polydimethylsiloxane, or DB-1) resolves these complex mixtures into 118 peaks (Brown et al., 1987b). Figure 11.1 displays the resultant chromatogram obtained upon analysis of Aroclor 1242 on a DB-1 column, revealing major peaks identified in the 118-peak numbering system. Each analysis of Aroclor 1242 results in a nearly identical tracing, with reproducible retention time, area, and congener composition for each peak. Specific peak assignments are included in Table 11.1, although complete congener assignments have been given elsewhere (Brown et al., 1987b).

Figure 11.1 shows chromatograms of Aroclor 1242 (trace a) and altered Aroclor 1242 (traces b–f) after transformation by known physical, chemical, and biological processes. These chromatograms contain detailed information that can be used to "fingerprint" the PCB mixture and therefore any alteration process it has undergone. For example, evaporation (trace b) preferentially removes lightly chlorinated biphenyls (the most volatile components), resulting in a general increase in the relative concentrations of more highly chlorinated congeners. This loss of lightly chlorinated PCBs can be easily distinguished from other processes that also result in the preferential removal of less chlorinated congeners (e.g., aerobic biodegradation). This is demonstrated in Figure 11.1 by analysis of peaks 46, 47, and 48.

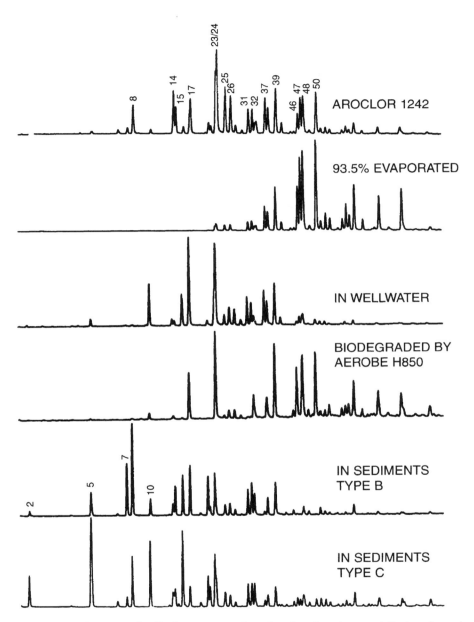

Figure 11.1 Congener distribution patterns of unaltered and environmentally transformed Aroclor 1242. Patterns characteristic of various physical and biological alterations processes are shown. Type B and C dechlorination results from anaerobic bacteria in Hudson River sediment.

Table 11.1 Selected Capillary PCB Congener Assignments via High- Resolution DB-1 Column (118-peak resolvable peaks)

Peak Number	Congener Assignment (major component)
2	2-CB
5	2-2-, 26-CB
7	2-3-CB
8	2-4-, 23-CB
10	26-2-CB
14/15	25-2-/24-2-CB
23/24	25-4-/24-4-CB
31	25-25-, 35-26-CB
33	24-24-CB
46	245-4-, 235-26-CB
47	25-34-, 345-2-CB
48	24-34-, 236-25-, 245-26-CB
50	23-34-, 234-4-CB

Evaporation (trace b) increases the level of these three tetrachlorinated congeners in the same relative proportion as in the original Aroclor 1242, since these congeners all have similar boiling points (trace b). In contrast, dramatic differences are observed in the relative levels of peaks 46, 47, and 48 after aerobic biodegradation (trace d), when compared to the original mixture (note the absence of peak 47 after biodegradation). Although aerobic biodegradation and evaporation both result in enhanced levels of highly chlorinated congeners, the congener-specific activity displayed by biodegradation results in unique changes or signatures (e.g., see peaks 31, 37, and 47).

Chromatographic analysis can also be used to distinguish processes that increase the relative levels of the lightly chlorinated congeners (e.g., dissolution into groundwater or alteration by microbial anaerobic dechlorination, labeled type B and type C; see traces e and f). The biological dechlorination of PCBs results in the production of *ortho*-enriched congeners that are present at low levels or absent in the original Aroclor 1242 (see peaks 2, 5, 7, 8 and (10). Unique patterns result that can be used to distinguish these biological processes from each other or from nonbiological processes.

History of PCBs in the Upper Hudson River

Two manufacturing plants located in the upper Hudson River (Fig. 11.2, river mile 195 and 197) produced PCB-containing capacitors for approximately 30 years, ending in 1976. These plants primarily utilized Aroclor 1242 before 1970, and exclusively used Aroclor 1016 from 1970–1976. For many years, PCB discharges

UPPER HUDSON RIVER

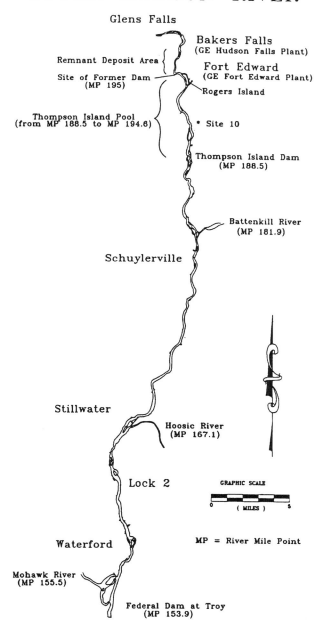

Glens Falls

Bakers Falls
(GE Hudson Falls Plant)

Remnant Deposit Area

Fort Edward
(GE Fort Edward Plant)

Site of Former Dam
(MP 195)

Rogers Island

Thompson Island Pool
(from MP 188.5 to MP 194.6)

• Site 10

Thompson Island Dam
(MP 188.5)

Battenkill River
(MP 181.9)

Schuylerville

Stillwater

Hoosic River
(MP 167.1)

Lock 2

GRAPHIC SCALE

(MILES)
0   5

Waterford

MP = River Mile Point

Mohawk River
(MP 155.5)

Federal Dam at Troy
(MP 153.9)

Figure 11.2 Map of upper Hudson River. Downstream is toward the bottom of the figure. Locations are indicated as mile points (MP), which represent the miles from the mouth of the river, which terminates at the Battery (0 MP), near New York City.

from these plants were localized immediately downstream at the Ft. Edward dam (river mile 194.9). In 1973, this dam was removed due to structural problems, thus destabilizing the sediments, which underwent extensive scouring and redeposition in the Thompson Island Pool (river mile 188.5–194). An extensive survey of the PCB distribution in the Thompson Island Pool (reach 8) was conducted by the NYS DEC in 1984 (~2000 cores). Additional sampling and analysis using high-resolution chromatography was performed by GE in 1991–1992 (~1000 cores) throughout the 40-mile stretch of the upper Hudson River.

In 1984, the upper Hudson River was placed on the National Priorities List (Superfund). The Environmental Protection Agency (EPA) investigated this site, and a Record of Decision in 1984 determined that no action should be taken to remediate the contaminated sediments, concluding that the environmental damage caused by dredging would outweigh the benefits obtained upon their removal. In addition, the EPA determined that the remnant deposits should be capped in place (this action has been completed). The remnant deposits represent an estimated 1.5 million cubic yards of sediment deposits that became exposed above the waterline when the Ft. Edward dam was removed in 1973 (NUS, 1984). Historical trends for PCBs in fish and water have been obtained by the U.S. Geological Surgery (USGS) (Fig. 11.3)

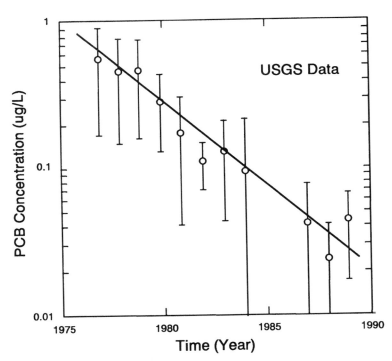

Figure 11.3 Change in average water column PCB concentration at the Stillwater Sampling Site in the Hudson River (USGS data).

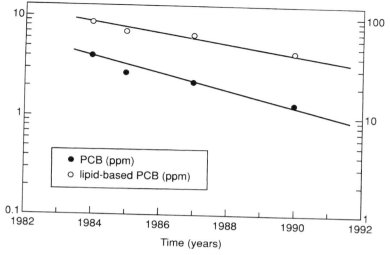

Figure 11.4 Change in average PCB concentration in striped bass from the lower Hudson River between 1984 and 1990.

and the New York State Department of Environmental Conservation (NYS DEC) (Fig. 11.4). In general, strong declines occurred from the late 1970s to the early 1990s in all media tested.

PCB Biodegradation (Laboratory and Environmental Results)

In the laboratory and in the environment, PCBs are known to undergo biotransformation under a variety of conditions (reviewed in Abramowicz, 1990; Bedard, 1990; Bedard and Quensen, 1995; Furukawa, 1982, 1986). At least two distinct biological processes are involved in the natural restoration of PCB-contaminated sites, aerobic degradation and anaerobic reductive dechlorination.

Aerobic PCB Biodegradation The aerobic bacterial biodegradation of PCBs is widely recognized and has been well studied (reviewed in Abramowicz, 1990; Bedard, 1990; Furukawa, 1982, 1986). Numerous microorganisms have been isolated that can aerobically degrade a wide variety of PCBs, although the more lightly chlorinated congeners are preferentially degraded. These organisms attack PCBs via the well-known 2.3-dioxygenase pathway, converting PCB congeners to their corresponding chlorobenzoic acids (Fig. 11.5). These chlorobenzoic acids can then be readily mineralized by many indigenous bacteria, resulting in the production of carbon dioxide, water, chloride, and biomass. Moreover, laboratory analyses and isolations have demonstrated that upper Hudson River sediments contain a broad spectrum of indigenous microorganisms capable of degrading a wide range of PCBs and chlorobenzoic acids Harkness et al., 1993).

There are two lines of evidence that strongly indicate the occurrence of naturally occurring aerobic PCB biodegradation in the Upper Hudson River. First, in a

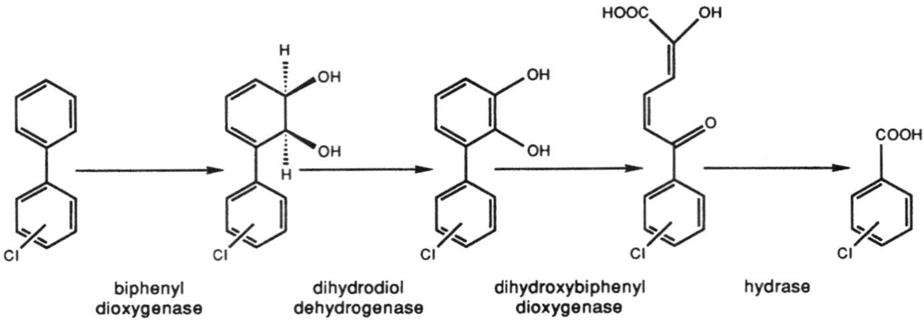

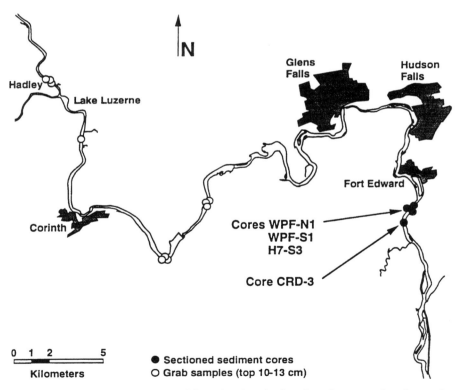

Figure 11.5 Biphenyl and PCB biodegradation pathway. Aerobic PCB biodegradation by a 2,3-dioxygenase enzyme system.

Figure 11.6 Map of upper Hudson River showing the location of cores and grab samples analyzed for PCB metabolites. Sites upstream of Hudson Falls are essentially free of PCBs. Samples from these sites served as negative controls.

recently completed field test in the upper Hudson River (Harkness et al., 1993), significant aerobic PCB biodegradation was observed without the addition of microorganisms, nutrients, or supplemental oxygen (although mixing was performed). This result suggested that these sediments contained all the necessary elements for in situ aerobic activity. To prove this hypothesis, a sensitive analytical method was developed to detect chlorobenzoic acids (the intermediate products of aerobic PCB biodegradation), in undisturbed cores taken from the upper Hudson River. Sampling locations are shown in Figure 11.6. Note that PCB metabolites (the chlorobenzoic acids) were found in all PCB-contaminated samples, but not in the

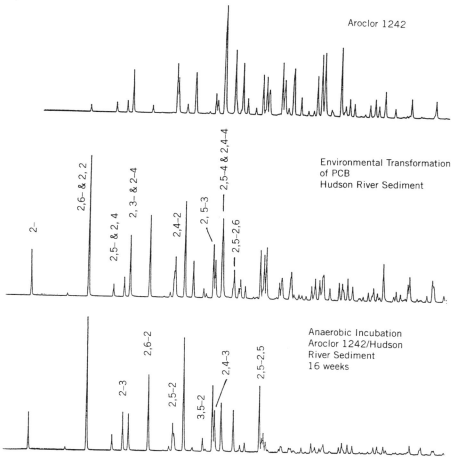

Figure 11.7 Chromatograms showing similar patterns of PCB dechlorination in the laboratory and the environment. *Top*, Aroclor 1242: *middle*, PCBs from Hudson River sediment; *bottom*, Aroclor 1242 incubated under anaerobic conditions for 16 weeks with Hudson River sediment in the laboratory.

uncontaminated sediments from further upstream (Flanagan and May, 1993). Moreover, the concentrations and congener distributions of the observed chlorobenzoic acids closely matched the expected degradation products from the PCBs mixture in the samples. Aerobic biodegradation is an important natural restoration process that, in the Hudson River, is facilitated by the widespread presence of lightly chlorinated PCB congeners resulting from anaerobic dechlorination. These congeners are much more susceptable to aerobic degradation than highly chlorinated forms.

Anaerobic PCB Dechlorination   Anaerobic bacteria attack more highly chlorinated PCB congeners through reductive dechlorination. In general, this

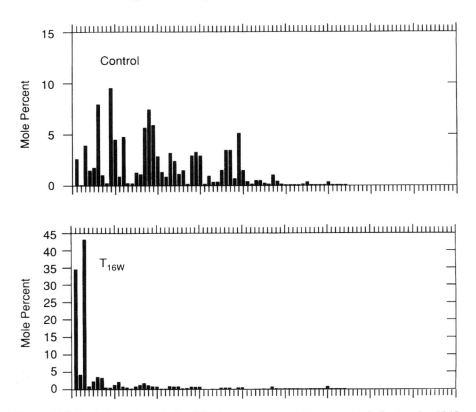

Figure 11.8 Peak-by-peak analysis of PCB content on a mole percent basis for Aroclor 1242 incubated for 16 weeks with Hudson River sediment. *X* axis indicates gas chromatography peak number (begining with peak 2). *Y* axis indicates the percentage of the total PCB represented by that peak on a mole basis. *Top chart* is an autoclaved control showing no significant alteration. *Bottom chart* (live sediment) shows extensive alteration of congener distribution after the 16-week incubation period. The congeners in peaks 2 and 4 (2-CB and 2,2'-, 2,6-CB, respectively) are shown to contain greater than 75% of the PCBs in the sample on a mole basis.

microbial process preferentially removes *meta* and *para* chlorines, resulting in a conversion of highly chlorinated PCB congeners to lower chlorinated, *ortho*-substituted congeners. The altered congener distribution of residual PCB contamination observed in several aquatic sediments was the earliest evidence of the anaerobic dechlorination of PCBs (Brown et al., 1984; Brown et al., 1987a,b). This same activity has been observed in the laboratory (Quensen et al., 1988; 1990; Abramowicz et al., 1989), where the selective removal of *meta* and *para* chlorines was also noted. The dechlorination activity observed in laboratory incubations with Aroclor 1242 are very similar to that found naturally in Hudson River sediments, as noted by the strong similarity in the observed dechlorination products (Fig. 11.7).

Many different dechlorination activities have been observed in sediments, and Hudson River sediments have been shown to contain a mixture of activities capable of extensive PCB dechlorination (Quensen et al., 1990). Figure 11.8 is a histogram quantitating the PCB content of each peak in the chromatograms for untransformed Aroclor 1242 and that incubated anaerobically with Hudson River sediment for 16 weeks (from Fig. 11.7). Note that the broad dechlorination activity has reduced the concentration of nearly all of the original PCB congeners, resulting in the formation of two major dechlorination products (2-CB and 2–2'-CB, peaks 2 and 4, respectively). This demonstrates that Hudson River sediment contains microorganisms capable of dechlorinating a wide range of PCB congeners. The extensive dechlorination displayed in Figures 11.7 and 11.8 are similar to Pattern C dechlorination observed in undisturbed Hudson River sediments (Brown et al., 1987a,b).

The naturally occurring PCB dechlorination in the Upper Hudson River is both extensive and widespread, as shown in a survey of ∼ 1000 sampling locations in a 6-mile stretch of the river (Fig. 11.9, mile point 194.5–188.5). Extensive changes had occurred in sediments exhibiting a broad range of PCB concentrations, even as low as 5 ppm (Abramowicz et al., 1991). Moreover, additional sampling by General Electric (GE) in the early 1990s confirmed with congener-specific high-resolution analyses that natural dechlorination had occurred throughout the entire 40-mile stretch of the upper Hudson River (see below). Widespread natural dechlorination of PCBs in aquatic sediments has also been documented for several other river systems Abramowicz et al., 1991; Brown and Wagner, 1990; Lake et al., 1992; Risatti, 1992). These additional surveys demonstrate that PCB dechlorination is prevalent in many aquatic sediments (both fresh and salt-water systems).

A listing of sites where PCB-dechlorinating activity has been found to date is shown in Table 11.2, including a number of PCB-contaminated and uncontaminated locations. Activity in uncontaminated sediments was detected after the addition of PCBs to clean sediments in the laboratory to assay for the presence of microorganisms capable of PCB dechlorination (Abramowicz et al., 1990). This suggests that PCB-dechlorinating activity may be the result of a common reductive pathway present in many different anaerobic microorganisms located throughout the environment. Support for this hypothesis comes from recent efforts demonstrating that several iron and cobalt heme cofactor systems are capable of reductively dechlorinating a wide variety of chlorinated organic compounds (Krone et al., 1989;

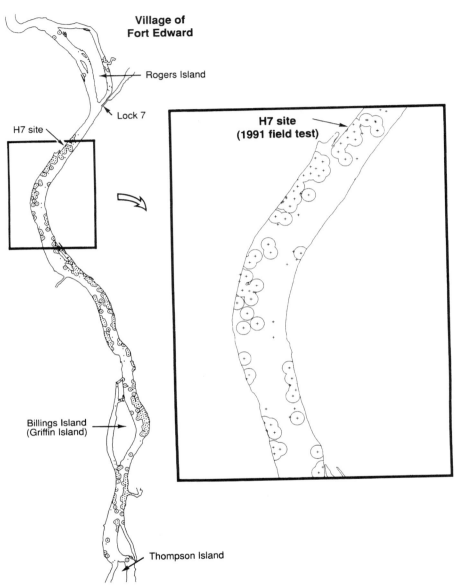

Figure 11.9 Locations of sediment PCB accumulations in the upper Hudson River, based on reanalysis of the 1984 New York State survey of the Thompson Island Pool: (+) indicates samples containing $\geq$ 10 ppm PCB, (O) indicate samples displaying extensive dechlorination.

Table 11.2   Sites Showing Evidence of Anaerobic PCB Dechlorination Activity

Sediments	PCB Contamination	Location
Adirondack Marsh	No	New York
Center Pond	No	Massachusetts
Escambia Bay	Yes	Florida
Hoosic River	Yes	Massachusetts
Hudson River	Yes	New York
Hudson River	No	New York
Kalamazoo	Yes	Michigan
Lake Hartwell	Yes	Hartwell, South Carolina
Moreau Drag Strip	Yes	New York
New Bedford Harbor	Yes	Massachusetts
Red Cedar River	No	Michigan
Rhine River	Yes	Germany
Rhine River	Yes	the Netherlands
Saint Lawrence River	Yes	Massena, New York
Saint Lawrence River	No	Massena, New York
Saline River	No	Michigan
Sheboygan River	Yes	Wisconsin
Silver Lake	Yes	Massachusetts
Waukegan Harbor	Yes	Illinois
Woods Pond	Yes	Massachusetts

Gantzer and Wackett, 1991), including PCBs under anaerobic conditions (Assaf-Anid et al., 1992). In general, environmental dechlorination is more extensive at higher PCB concentrations, consistent with the faster dechlorination rates observed at higher PCB concentrations in the laboratory (Abramowicz et al., 1993; Fish, 1996; Kim and Rhee, 1997).

PCBs IN SEDIMENTS

GE's 1991 Hudson River Sediment Sampling and Analysis Program

The principal objective of GE's 1991 sediment sampling and analysis program was to gather data to assess the nature and extent to which indigenous microorganisms had altered the chemical composition of PCBs in Hudson River sediments and to confirm widespread PCB dechlorination (Brown et al., 1987a, b; Quensen et al., 1990, Abramowicz et al., 1989). The objectives, sampling, and analysis methods are more completely described in O'Brien and Gere (1993a). A brief summary of the program methodology is presented below.

The 1991 sediment program included the sampling of approximately 1000 sediment cores from the 40-mile section of the upper Hudson River extending from Fort Edward to Troy, New York (Table 11.3). Over 500 of these cores were collected

Table 11.3 Summary of 1991 Upper Hudson River Sediment Sampling and Analysis[a]

Reach Number[b]	Approx. River Mile	Sediment (mg/kg)						Porewater (µg/L)					
		Total No. of Samples	Total No. of Analyses	Geometric Mean	Std. Dev.	Max.	Min.	Total No. of Samples	Total No. of Analyses	Geometric Mean	Std. Dev.	Max.	Min.
8	193.5–188.5	507	208	18.6	53.7	285.9	0.1	250	51	8.7	13.6	57.7	0.7
7	188.5–186.3	27	9	68.8	33.9	132.1	35.4	—	0	—	—	—	—
6	186.3–183.4	47	20	15.8	25.3	97.6	3.3	8	3	22.7	2.5	25.7	19.6
5	183.4–168.0	216	69	5.7	17.1	96.9	0.3	57	18	5.9	5.2	18.9	1.3
4	168.0–166.0	25	6	4.7	30	84.9	0.7	6	3	5.8	7.7	19.5	2.5
3	166.0–163.6	31	11	3.3	3	9.3	0.3	9	3	4.7	1.9	7.4	2.6
2	163.6–159.7	55	11	2.9	7.5	22.7	0.4	—	0	—	—	—	—
1	159.7–153.6	72	13	2.2	7.4	24	0.4	14	8	2.3	3.4	11.3	0.2
Total	193.5–153.6	980	347	7.6	22.2	285.9	0.1	344	86	6.5	5.7	57.7	0.2

[a] Porewater samples not collected for reaches 2 and 7.

[b] A "reach" is a section of the river between dams or other barriers.

from the Thompson Island Pool (TIP), the section of the river located approximately 2 miles downstream of the plant sites located in Hudson Falls and Fort Edward, New York (Fig. 11.2). This section has been the principal focus of sediment dredging proposals by the State of New York (Brown et al., 1988). On the average, 87 samples were collected per mile in the TIP, and 30 samples per mile in downstream reaches (sections of the river between dams or other barriers).

Sediment cores were collected from a boat using a vibracoring technique. A grab sample was collected using a Ponar Dredge where insufficient sediment was present to obtain a core. Cores were sectioned into 0–5, 5–10, and 10–25 cm sections. Sediment pore water was extracted by centrifugation and filtration on approximately 25% of the samples collected. Bulk sediment and sediment pore water were separately composited based on sediment depth, river location, and sediment texture at an approximately 10:1 ratio resulting in 347 bulk sediment and 86 pore water PCB analyses (Table 11.3).

Bulk sediment and sediment pore water samples were analyzed by high-resolution capillary analysis using a DB-1 column and electron capture detector (Brown et al., 1987b). The resultant 118-peak chromatograph enabled the identification of PCB congeners, PCB homologs, as well as total PCB. Deviations in congener distributions from the presumed source Aroclor distributions were used to identify processes affecting PCB fate in the sediment based on trends observed in previous field investigations and laboratory studies of PCB dechlorination (Brown et al., 1987a,b; Quensen et al., 1988, 1990; Abramowicz et al., 1993; Nies and Vogel, 1990; Alder et al., 1993).

Spatial Distribution of PCB in Upper Hudson River Sediments

The broad area coverage of the 1991 sediment survey permitted an assessment of the horizontal and vertical distribution of PCBs in upper Hudson River sediments. Although the sample compositing scheme employed during the survey limited the precision of such an analysis, strong vertical and horizontal gradients were observed in the data.

Bulk density weighted mean PCB concentrations of composite samples collected from the top 25 cm of sediment ranged from approximately 2.2 mg/kg in reach 1 to 68.8 mg/kg in reach 7. The distribution of individual composite sample PCB concentrations was highly skewed with more than 70% containing less than 25 mg/kg (Fig. 11.10). The 347 composite sediment samples collected over the 40-mile sampling area possessed a geometric mean and standard deviation for PCB concentration of 7.6 and 22.2 mg/kg, respectively (Table 11.3). Although the data exhibited marked variability, PCB concentrations generally decreased with increased distance downstream. This trend is consistent with the data collected in the late 1970s (Tofflemire et al., 1979). PCB concentrations were highest near the source in reaches 7 and 8 and lowest in reaches 1 and 2 (Table 11.3).

Sediment burial through the deposition of water-borne particulate appears to be an important fate-determining process sequestering PCB in the upper Hudson River. This is evident in PCB profiles that consistently show PCB concentrations

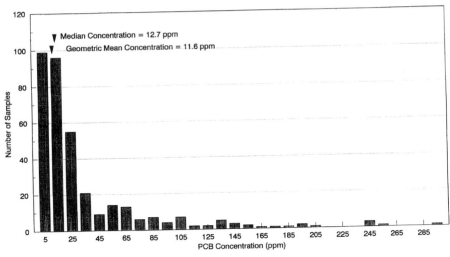

Figure 11.10 Frequency distribution of total PCB in upper Hudson River sediments collected during 1991.

increasing with sediment depth with maximum concentrations at the deepest interval sampled (10–25 cm; Fig. 11.11). Differences in total PCB concentrations between the different depths were statistically significant at the 95% confidence level ($\alpha = 0.05$). These data indicate that sediments within the upper Hudson are a sink for PCB. This sink, however, does not represent a direct PCB source to fish or

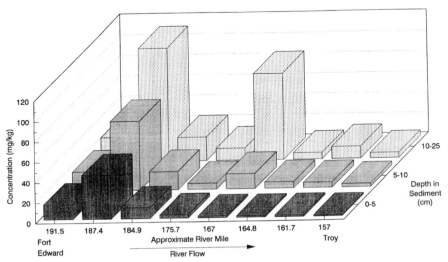

Figure 11.11 Geometric mean total PCB distribution in surficial 25 cm of sediment in upper Hudson River (Fort Edward to Troy Dam) during 1991.

other aquatic organisms. This is due to both the physical separation of these PCBs and the benthic invertebrate members of the aquatic food chain, which do not generally burrow below approximately 5 cm of the surface, as well as the physiochemical separation (due to reduced bioaccumulation), resulting from in situ dechlorination.

Temporal Trends

Surveys similar to the 1991 program were conducted by the NYS DEC in a 40-mile segment of the upper Hudson River in 1976–1978 (Tofflemire et al., 1979) and in the Thompson Island Pool only in 1984 (Brown et al., 1988). Differences in PCB analytical technique, sampling design, and sample collection procedures limit comparisons between the three surveys to a qualitative assessment of total PCB concentration changes in the intervening years.

Mean PCB concentrations within the surficial 25 cm of sediment calculated for the three surveys suggest that PCB concentrations within Thompson Island Pool sediment have declined between 1976 and 1991. However, no clear temporal trends can be established because of the marked variability of these data. Nonetheless, a reduction in mean PCB concentrations in surficial sediments is consistent with established PCB fate and transport mechanisms that remove PCBs from surficial sediments including burial, biodegradation, reductive dechlorination, and volatilization.

Impact of Reductive Dechlorination on PCB Distributions

PCB reductive dechlorination involves the microbially mediated transformation of higher chlorinated congeners to those with lower chlorination. Several different dechlorination patterns have been identified in Hudson River sediments and are presumed to be the result of the activities of different microbial consortia (Brown et al., 1987a,b). Although subtle differences exist between the patterns, each produces a PCB mixture lower in tri-, tetra-, and pentachlorobiphenyls and higher in mono- and dichlorinated biphenyls relative to the starting Aroclor mixture. Additionally, dechlorination patterns are congener specific, demonstrating selective removal of *meta* and *para*-substituted PCBs resulting in the accumulation of *ortho*-substituted congeners.

The 1991 sediment survey confirmed on a river scale that microbially mediated reductive dechlorination processes have substantially altered the PCBs in Hudson River sediments. Figure 11.12 presents the homolog distributions for PCBs in the top 25 cm of sediment between Fort Edward and the TIP. These data indicate that as much as half of the total PCBs presently in the TIP are mono- and dichlorinated congeners. By contrast Aroclor 1242 contains less than 18% mono- and dichlorinated biphenyls. Laboratory and field experiments conducted by numerous investigators have demonstrated that the principal products of reductive dechlorination of Aroclor 1242 are 2-CB, 2,2'-CB, and 2,6-CB (Brown et al., 1987a,b; Quensen et al., 1988, 1990; Abramowicz et al., 1993). That these congeners also predominate in upper Hudson River sediments is demonstrated in Figure 11.13,

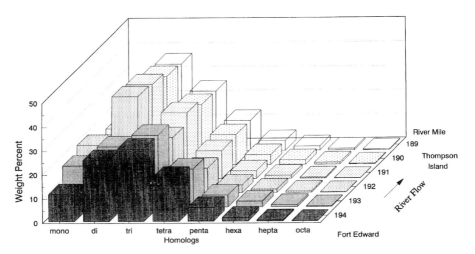

Figure 11.12 Mean PCB homolog distribution in surficial 25 cm of sediment within reach 8 of the upper Hudson River.

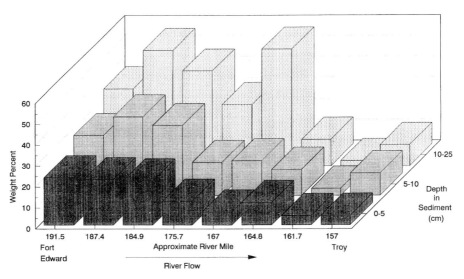

Figure 11.13 Geometric mean sum weight percent of 2-, 2,2′-, and 2,6-chlorobiphenyls (DB-1 column peaks 2, 5, and 10), in surficial 25 cm of sediment in the upper Hudson River during 1991.

which shows the total weight percent of these congeners at various river locations. The figure also shows that the amount of these congeners tends to increase with depth. This is expected since the PCBs in deeper sediment have been undergoing dechlorination for a longer period of time.

Reductive dechlorination preferentially removes chlorines occupying the *meta* and *para* positions on the biphenyl rings. This action has dramatically altered the *ortho* to *meta* and *para* ratio (O:M&P) of PCB within Hudson River sediments. Data from the 1000-core survey indicate changes in chlorine substitution patterns with sediment depth. At each depth sampled, the average O:M&P increased from 0.77 (the ratio of the presumed starting Aroclor 1242) to approximately 1.0, 1.3, and 1.5. Differences in the ratio between the sediment depths and Aroclor 1242 were significant at the 95% confidence level ($\alpha < 0.05$).

PCBs found within sediment pore water consisted of the lightly chlorinated and more soluble products of reductive dechlorination. PCBs within pore water extracted from sediments from the upper Hudson River had a geometric mean and median concentration of 7.1 and 7.2 µg/L, respectively (Fig. 11.14). Mono- and dichlorinated PCB produced by dechlorination accounted for greater than 80% of the total PCB within sediment pore water (Fig. 11.15). The several orders of magnitude difference between pore water and water column PCB concentrations provided the driving force for the diffusive flux of mono- and dichlorinated PCB from the sediments into Hudson River water. This is significant because these forms of PCB do not readily bioaccumulate in fish in the Hudson River (Jones et al., 1989; O'Brien and Gere, 1993b), thereby reducing the potential for human exposure.

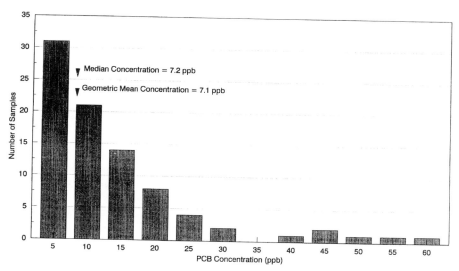

Figure 11.14 Total PCB Porewater frequency distribution in the upper Hudson River during 1991.

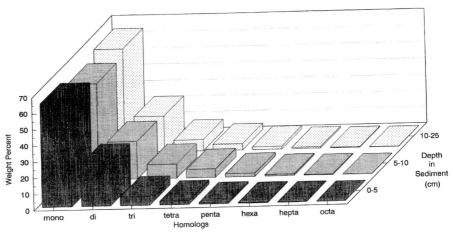

Figure 11.15 Mean PCB homolog distribution of porewater extracted from Thompson Island Pool sediments in 1991.

Benefits of Anaerobic PCB Dechlorination

Sediment surveys have demonstrated that anaerobic PCB dechlorination in the upper Hudson River is widespread and extensive. The benefits of this natural restoration process includes reducing the potential risk and potential exposure to PCBs. Among the potential risk reductions are decreased dioxin-like toxicity and reduced carcinogenicity. Structure–function relationships have shown that two *para*, two or more *meta*, and no *ortho* substituents are required for maximum dioxin-like activities (Safe, 1992). Coplanar PCB congeners like 345–34-, 34–34-, and 345–345-CB exhibit strong binding to the dioxin receptor (Safe, 1992).

The preferential loss of *meta* and *para* chlorines catalyzed by anaerobic dechlorination results in dramatic reductions in the levels of coplanar, dioxin-like congeners in the PCB mixtures (Quensen, et al., 1992a,b). These reductions in concentrations correlate with reductions in ethoxyresorufin-O-deethylase (EROD) induction potency and toxic equivalency factors for these mixtures. The reduced carcinogenicity as a result of dechlorination is supported by the recent re-analysis of the original rat cancer studies with various PCB mixtures (Moore, 1991, 1994). In these studies, only the most highly chlorinated PCB mixture containing 60% chlorine (e.g., Aroclor 1260) resulted in observable cancer potencies. Exposure to PCB mixtures containing 54 and 42% chlorine did not result in statistically increased tumor levels (Moore, 1991). Therefore carcinogenic potential correlates with total chlorine levels. Microbial anaerobic PCB dechlorination both decreases *meta* and *para* chlorine levels and total chlorine levels, reducing potential dioxin-like toxicity and carcinogenicity in PCB mixtures. It is now known that these same extensive reductions are occurring naturally in the environment (Quensen, Boyd et al., 1992a,b).

Additional decreases in risk associated with dechlorinated PCBs result from reduced PCB exposure. This is manifested in two different ways. First, the lightly chlorinated PCB congeners produced by dechlorination can be readily degraded by indigenous aerobic bacteria (Harkness et al., 1993). Moreover, new evidence indicates that this process is occurring naturally in undisturbed Hudson River sediments (Flanagan and May, 1993). Second, dechlorination significantly reduces the bioconcentration potential of the PCB mixture through conversion to congeners that do not significantly bioconcentrate in the food chain. The lightly chlorinated PCB congeners resulting from dechlorination in Hudson River sediments (e.g., 2-CB and 2–2-CB) display an ~450-fold reduction in their tendency to bioconcentrate in fish, as compared to the more highly chlorinated tri- and tetrachlorinated PCBs present in the original Aroclor 1242 mixture. Thus, the naturally occurring anaerobic PCB dechlorination reduces the potential risk associated with PCBs via direct reductions in carcinogenic potency, dioxin-like toxicity, and exposure.

An additional reduction in PCB exposure results from long-term contact of PCBs with sediment particles, and consequent reductions in bioavailability. Direct desorption experiments performed on dechlorinated sediment samples taken from the upper Hudson River demonstrate that ~50% of the PCBs in these environmental samples are present in a recalcitrant fraction that is markedly less available than freshly spiked PCBs (Abramowicz et al., 1992). Therefore the old PCB deposits represent an additional reduced risk when compared to freshly deposited PCBs since ~50% of the buried material would be unavailable even if it were in direct contact with the river biota.

In summary, a variety of natural restoration processes have significantly reduced the risks from PCBs in the Hudson River relative to that of freshly deposited Aroclor 1242. In addition, both biological and physical mechanisms contribute to reduced PCB exposure through:

Greater spatial isolation (burial)

Increased aerobic biodegradation

Reduced bioconcentration and bioaccumulation.

REFERENCES

Abramowicz, D. A. (1990). Aerobic and Anaerobic Biodegradation of PCBs: A Review, in *CRC Critical Reviews in Biotechnology*, Vol. 10, G. G. Steward and I. Russell, Eds., CRC Press, Boca Raton, FL p. 241.

Abramowicz, D. A., M. J. Brennan, and H. M. Van Dort (1989). Anaerobic Biodegradation of Polychlorinated Biphenyls, in *Extended Abstracts: 198th American Chem. Soc. National Meeting*, Div. Environ. Chem., Vol. 29(2), pp. 377–379.

Abramowicz, D. A., M. J. Brennan, and H. M. Van Dort (1990). Anaerobic and Aerobic Biodegradation of Endogenous PCBs, in *General Electric Company Research and Development Program for the Destruction of PCBs, Ninth Progress Report*, General Electric Corporate Research and Development, Schenectady, NY, Chapter 6.

Abramowicz, D. A., J. F. Brown, Jr., and M. K. O'Donnell (1991). Anaerobic PCB Dechlorination in Hudson River Sediments, in *General Electric Company Research and Development Program for the Destruction of PCBs, Tenth Progress Report*, General Electric Corporate Research and Development, Schenectady, NY, Chapter 17.

Abramowicz, D. A., M. R. Harkness, J. B. McDermott, and J. J. Salvo (1992). 1991 in situ Hudson River Research Study: A Field Study on Biodegradation of PCBs in Hudson River Sediments, in *General Electric Company Research and Development Program for the Destruction of PCBs, Eleventh Progress Report*. General Electric Corporate Research and Development, Schenectady, NY, Chapter 1.

Abramowicz, D. A., M. J. Brennan, H. M. Van Dort, and E. L. Gallagher (1993). Factors Influencing the Rate of PCB Dechlorination in Hudson River Sediments, *Environ. Sci. Tech.*, **27**:1125–1131.

Alder, A. C., M. M. Haggblom, S. R. Oppenheimer, and L. Y. Young (1993). Reductive Dechlorination of Polychlorinated Biphenyls in Anaerobic Sediments, *Environ. Sci. Tech.*, **27**:530–538.

Assaf-Anid, N., L. Nies, and T. M. Vogel (1992). Reductive Dechlorination of a Polychlorinated Biphenyl Congener and Hexachlorobenzene by Vitamin B_{12}, *Appl. Environ. Microbiol.* **58**:1057.

Bedard, D. L. (1990). Bacterial Transformations of Polychlorinated Biphenyls, in *Biotechnology and Biodegradation*, D. Kamely, A. Chakrabarty, and G. S. Omenn, Eds., Adv. Appl. Biotechnol. Series, Vol. 4, Portfolio, Woodlands, TX, p. 369.

Bedard, J. L. and J. F. Quensen III (1995). "Microbial Reductive Dechlorination of Polychlorinated Biphenyls." In *Microbial Transformation and Degradation of Toxic Organic Chemicals*. Young, L. Y.; Cerniglia, C., Eds., John Wiley & Sons, Inc., New York, pages 127–216.

Brown, J. F., R. E. Wagner, D. L. Bedard, M. J. Brennan, J. C. Carnahan, and R. J. May (1984). PCB Transformations in Upper Hudson Sediments, *Northeast. Environ. Sci.*, **3**:167–179.

Brown, J. F., D. L. Bedard, M. J. Brennan, J. C. Carnahan, H. Feng, and R. E. Wagner (1987a). PCB Dechlorination in Aquatic Sediments, *Science*, **236**:709–712.

Brown, J. F., Jr., R. E. Wagner, H. Feng, D. L. Bedard, M. J. Brennan, J. C. Carnahan, and R. J. May (1987b). Environmental Dechlorination of PCBs, *Environ. Toxicol. Chem.*, **6**:579–593.

Brown, M. P., M. B. Werner, C. R. Carusone, and M. Klein (1988). *Distribution of PCBs in the Thompson Island Pool of the Hudson River*, Final Report of the Hudson River PCB Reclamation Demonstration Project Sediment Survey, Bureau of Technical Services and Research, Division of Water, New York State Department of Environmental Conservation, May.

Brown, J. F., Jr., and R. E. Wagner (1990). PCB Movement, Dechlorination, and Detoxication in the Acushnet Estuary, *Environ. Toxicol. Chem.*, **9**:1215–1233.

Buckley, E. H. (1982). Accumulation of Airborne Polychlorinated Biphenyls in Foliage, *Science*, **216**:520.

Fish, K. M. (1996). "Influence of Aroclor 1242 Concentration on Polychlorinated Biphenyl Biotransformations in Hudson River Test Tube Microcosms." *App. Env. Microbiol.*, **62**, 3014–3016.

Flanagan, W. P. and R. J. May (1993). Metabolite Formation as Evidence for *in situ* Aerobic Biodegradation of Polychlorinated Biphenyls, *Environ. Sci. Technol.*, **27**:2207–2212.

Furukawa, K. (1982). Microbial Degradation of Polychlorinated Biphenyls (PCBs), in *Biodegradation and Detoxification of Environmental Pollutants*, A. M. Chakrabarty, Ed., CRC Press, Boca Raton, FL, p. 33.

Furukawa, K. (1986). Modifications of PCBs by Bacteria and Other Microorganisms, in *PCBs and the Environment*, Vol. 2, J. S. Waid, Ed., CRC Press, Boca Raton, FL, p. 89.

Gantzer, C. J. and L. P. Wackett (1991). Reductive Dechlorination Catalyzed by Bacterial Transition-Metal Coenzymes, *Environ. Sci. Technol.*, **25**:715.

Harkness, M. R., J. B. McDermott, D. A. Abramowicz, J. J. Salvo, W. P. Flanagan, M. Stephens, F. J. Mondello, R. J. May, J. H. Lobos, K. M. Carroll, M. J. Brennan, A. A. Bracco, K. M. Fish, G. L. Warner, P. R. Wilson, D. K. Dietrich, D. T. Lin, C. B. Morgan, and W. L. Gately (1993). *In situ* Stimulation of Aerobic PCB Biodegradation in Hudson River Sediments, *Science*, **259**:503–507.

Hutzinger, O., S. Safe, and V. Zitko (1974). *The Chemistry of PCBs*, CRC Press, Cleveland, OH, p. 7.

Hutzinger, O. and W. Veerkamp (1981). Xenobiotic Chemicals with Pollution Potential, in *Microbial Degradation of Xenobiotics and Recalcitrant Compounds*, T. Leisinger, R. Hutter, A. M. Cook, and J. Nuesch, Eds., Academic Press, New York, pp. 3–45.

Jones, P. A., R. J. Sloan, and M. P. Brown (1989). PCB Congeners to Monitor with Caged Juvenile Fish in the Upper Hudson River, *Environ. Toxicol. Chem.*, **8**:793–803.

Krone, U. E., R. K. Thauer, and H. P. C. Hogenkamp (1989). Reductive Dehalogenation of Chlorinated C_1-Hydrocarbons Mediated by Corrinoids, *Biochemistry*, **28**:4908.

Lake, J. L., R. J. Pruell, and F. A. Osterman (1992). An Examination of Dechlorination Processes and Pathways in New Bedford Harbor Sediments, *Marine Environ. Res.*, **33**:1.

Moore, J. A. (1991). *Reassessment of Liver Findings in PCB Studies for Rats*, Institute for Evaluating Health Risks, Waslington D. C., July 1.

Moore, J. A., J. F. Hardesty, D. A. Banas, and M. A. Smith (1994). "A Comparison of Liver Tumor Diagnoses from Seven PCB Studies in Rats." *Reg. Toxicol. Pharmacol.*, **20**:362–370.

Nies, L. and T. M. Vogel (1990). Effects of Organic Substrate on Dechlorination of Aroclor 1242 in Anaerobic Sediments, *Appl. Environ. Microbiol.*, **56**:2616–2617.

O'Brien & Gere Engineers, Inc. (1993a). Hudson River Project, 1991 Sediment Sampling and Analysis Program, Data Summary Report, General Electric Company, Corporate Environmental Programs, Albany, NY, May.

O'Brien & Gere Engineers, Inc. (1993b). Hudson River Project, 1992 Food Chain Study, Data Summary Report, General Electric Company, Corporate Environmental Programs, Albany, NY, May.

NUS (1984). *Feasibility Study-Hudson River PCB Site*, Vol. I, U.S. Environmental Protection Agency, Region II Office, New York.

Quensen, J. F. III, J. M. Tiedje, and S. A. Boyd (1988). Reductive Dechlorination of PCBs by Anaerobic Microorganisms from Sediments, *Science*, **242**:752–754.

Quensen, J. F. III, S. A. Boyd, and J. M. Tiedje (1990). Dechlorination of Four Commercial Polychlorinated Biphenyl Mixtures (Aroclors) by Anaerobic Microorganisms from Sediments, *Appl. Environ. Microbiol*, **56**:2360–2369.

Quensen, J., S. A. Boyd, J. M. Tiedje, R. F. Lopshire, and C. G. Enke (1992a). Expected Dioxin-like Toxicity Reduction as a Result of the Dechlorination of Aroclors, in *General Electric Company Research and Development Program for the Destruction of PCBs,*

Eleventh Progress Report, General Electric Corporate Research and Development, Schenectady, NY, Chapter 17, p. 189.

Quensen, J. F. III, J. M. Tiedje, S. A. Boyd, C. Enke, R. Lopshire, J. Giesyl, M. Mora, R. Crawford, and D. Tillitt (1992b). Evaluation of the Suitability of Reductive Dechlorination for the Bioremediation of PCB-Contaminated Soils and Sediments, in *Preprints: International Symposium on Soil Decontamination Using Biological Processes*, Karlsruhe/D, 6–9 December 1992, pp. 91–100.

Risatti, J. B. (1992). *Rates of Microbial Dechlorination of PCBs in Anaerobic Sediments from Waukegan Harbor*, Illinois Hazardous Waste Research and Information Center, Champaign, IL, HWRIC Project HWR 86–010, Report RR-061.

Safe, S. (1980). in *Halogenated Biphenyls, Naphthalenes, Dibenzodioxins, and Related Products*, R. Kimbrough, (ed.), Elsevier/North Holland, Amsterdam, p. 77.

Safe, S. (1992). Toxicology, Structure-Function Relationship, and Human and Environmental Health Impacts of Polychlorinated Biphenyls: Progress and Problems. *Environ. Hlth. Perspect*, **100**: 259–268

Tanabe, S., H. Hidaka, and R. Tatsukawa (1983). PCBs and Chlorinated Hydrocarbon Pesticides in Antarctic Atmosphere and Hydrosphere, *Chemosphere*, **12**(2):277–288.

Tofflemire, T. J., L. J. Hetling, and S. O. Quinn (1979). *PCB in the Upper Hudson River: Sediment Distribution, Water Interactions and Dredging*, Technical Paper number 55, New York State Department of Environmental Conservation, Bureau of Water Research.

■■■■■■ CHAPTER 12

Microbes in the Muck: A Look into the Anaerobic World

David Kafkewitz and Monica T. Togna

Department of Biological Sciences, Rutgers University, Newark, New Jersey 07102 (D.K.); Center for Agricultural Molecular Biology, Cook College, Rutgers University, New Brunswick, New Jersey 08903 (M.T.T.)

The treatment of wastes by microbial processes is a subject that is beyond coverage in a single chapter or even in a single volume. A narrowly defined topic is all that can be meaningfully covered in a limited space. The usual approach is to focus on the degradation of a specific class of compounds or on a single genus of microorganisms or upon a particular method of cultivation, such as a specific type of bioreactor. In this chapter we take a different approach: We explore the world of anaerobic microorganisms, a world that is unknown to many microbiologists and that contains metabolic capabilities that may very well prove useful in engineered processes. Our focus is on the organisms and their diversity and versatility; where appropriate, we cite examples of applications or potential applications.

We are taking this approach because the anaerobic world today looks very different than it looked just a few years ago. As a result of the application of appropriate anaerobic technique to environmental inocula, there has been a striking change in our understanding of the metabolic capabilities of the microorganisms that live in oxygen-free environments. This new knowledge has only recently begun to move from the specialists in anaerobic microbiology to the engineers and environmental scientists who may usefully apply it to real world problem solving. The goals of this chapter are to let the nonspecialist know what has been happening in the field of anaerobic microbiology and to provide some basis for the development of new processes involving anaerobes.

Biological Treatment of Hazardous Wastes, Edited by Gordon A. Lewandowski and Louis J. DeFilippi
ISBN 0-471-04861-5 ©1998 John Wiley & Sons, Inc.

WHY CONSIDER ANAEROBES?

Before plunging into the anoxic world it is worth asking the question why one might want to consider anaerobic organisms for a biodegradation process. To assist in the determination of when anaerobes may be the population of choice, it would be useful to compare aerobic and anaerobic biodegradation.

Aerobic biodegradation, if feasible, is almost always the method of choice. Aerobic microorganisms grow more quickly than anaerobes and can sustain higher degradation rates. Aerobic catabolism of organic compounds usually leads to mineralization, as this term is used in its traditional sense. That is, the original compound is converted to carbon dioxide and water. Other elements that may have been part of the compound, for example, halogen atoms, are usually converted to an inorganic form. As a rule mineralization of an organic compound occurs within a single cell. This means that every cell that has the ability to attack the starting compound also has the ability to carry the process of mineralization to completion. Studies of aerobic biodegradation have shown that even complex xenobiotic compounds often require only a few unique enzyme-catalyzed reactions to be converted to familiar metabolic intermediates. No matter how complex the starting compound may be, if enzymatic attack converts it to intermediates of glycolysis or the citric acid cycle, metabolism proceeds to completion just as if glucose had been the original substrate. Mass-balance determinations are relatively simple for aerobic cultures; oxygen consumption and carbon dioxide production are easily measured. Cometabolism of xenobiotic compounds, that is, the adventitious attack on a compound while another, usually simpler, compound is the organism's true target, is well described and reasonably common among aerobes. Multiorganism interactions among aerobes are also well described (McInerney, 1986).

Another reason for favoring aerobic treatment is that genetic manipulation of aerobic and facultative bacteria is a fairly routine undertaking. The genetic information that confers the ability to degrade toxic or unusual compounds is often located on small, mobile chromosomes called plasmids, and these are easily manipulated by standard genetic techniques. A potential advantage of aerobic biodegradation is that if the genes for the desired catabolic enzymes belong to an organism that may be unsuitable for use in the process being developed, say for reasons of temperature or pH, the relevant genes can be transferred, in principle at least, to a more suitable organism.

There are certain disadvantages to aerobic biodegradation that must also be considered in evaluating treatment strategy. Most significant is the need to supply oxygen. The solubility of oxygen in water is low and decreases with increasing temperature. Dense aerobic reactor cultures consume oxygen rapidly and evolve heat in amounts that can prove troublesome. Adequate aeration of reactors can be a costly process and can lead to the volatilization of toxic compounds. In situ aeration is difficult to accomplish and can also lead to volatilization. This process is often referred to as *air stripping* and can give the illusion that biodegradation has occurred when all that has been accomplished is the conversion of soil or water pollution to air pollution. The efficiency of aerobic growth also means that there will be a significant accumulation of biomass that will have to be disposed of in an acceptable manner.

Anaerobic biodegradation, by contrast, often requires mixed populations that act sequentially with intermediates produced from the catabolism of the starting compound being passed from cell to cell among organisms of different metabolic types acting as a "consortium." An organism may perform one or two chemical steps and excrete the resulting product. This compound may accumulate or may serve as the starting compound for another anaerobe. This metabolic pattern has several important practical consequences. Analytical procedures that measure the loss of the starting compound cannot be relied upon to evaluate biodegradation. Disappearance of substrate is not a sufficient criterion to conclude that degradation has occurred; it is also necessary to determine the products produced. Closing the mass balance for substrate and products, whether by chemical analysis or radio-labeled determinations, is often essential. The products excreted may be simpler than the original compound, but this is no guarantee that they are less toxic or objectionable.

The genetics of anaerobic bacteria are at a very early stage of development. This is not surprising since many recent publications describing anaerobes with hitherto unrecognized catabolic abilities are also describing species and frequently genera that are new to science.

WHEN IS OXYGEN ABSOLUTELY REQUIRED?

Beyond the factors just discussed, there is obviously no point considering the use of anaerobes for biodegradation if molecular oxygen is required for the attack on the target compounds. It is reasonable, therefore, to ask whether there are compounds that can only be enzymatically attacked if molecular oxygen is served as a co-substrate? A few years ago the answer to this question was an almost definite yes; today, it is an almost certain no.

Schink has discussed the limits of anaerobic degradation of organic compounds from the perspective of the relationship of molecular structure to degradability (Schink, 1988). Alkane degradation has long been assumed to require molecular oxygen for the addition of an atom of oxygen to a terminal methyl group. From an ecological point of view this requirement is open to question since the methanogenic degradation of alkanes is exergonic and natural selection in microbial ecosystems is driven by the exploitation of available energy sources. Consequently, it would seem unlikely for nature to ignore an energetically favorable reaction. Schink has pointed out that the hypothetical methanogenic degradation of ethane: $8C_2H_6 + 6H_2O \rightarrow 14CH_4 + 2HCO_3^- + 2H^+$ has a free energy change of $-34.4\,\text{kJ/mol}$ ethane and that the reaction becomes more exergonic as the hydrocarbons' chain length increases. Nevertheless, such reactions have not been reproducibly demonstrated in anaerobic systems. Nor has the degradation of methane coupled to the reduction of sulfate to sulfide been reproducibly demonstrated in spite of environmental evidence that it occurs (Oremland, 1988). The sulfidogenic degradation of hexadecane has, however, been unequivocally demonstrated and a previously unknown sulfidogenic bacterium isolated in pure culture (Aeckersberg et al., 1991). This isolate proves that anoxic biodegradation of alkanes is possible and supports the notion that energetically favorable reactions are not ignored by nature.

Until recently it was generally believed that molecular oxygen was required for the metabolic activation of aromatic rings. This is now known not to be true. It has been shown by a number of workers that toluene and xylenes (Zeyer et al., 1986; Kuhn et al., 1988; Dolfing et al., 1990; Evans et al., 1991; Schocher et al., 1991; Coschigano and Young, 1997), almost certainly benzene (Edwards and Grbic-Galic, 1992; Lovley et al., 1994a), and many other aromatic compounds are degradable by denitrification or under methanogenic conditions in the complete absence of oxygen. Benzene, toluene, and a number of other aromatic compounds can apparently be rapidly degraded by ferric-iron-reducing bacteria under anoxic conditions (Lovley et al., 1994b, 1993c). In those cases that have been characterized, ring activation occurs by the addition of an oxygen atom that originates in a hydroxide ion (Vogel and Grbic-Galic, 1986).

The metabolic pathways involved in anaerobic benzene biodegradation have not yet been characterized. Anaerobic degradation of polyaromatic hydrocarbons (PAHs) has been demonstrated under denitrifying conditions (Mihelcic and Luthy, 1988a,b), and thermodynamic analysis has been performed on the energetics of PAH oxidation coupled to a variety of terminal electron acceptors under anoxic conditions (McFarland and Sims, 1991). The low bioavailability of PAHs caused by their extreme hydrophobicity may very well prove to be more significant than the metabolic incompetence of microorganisms.

Schink's discussion (Schink, 1988), as well as the review by Thauer et al. (1977), of the thermodynamics of a wide range of anaerobic metabolic reactions should be consulted for more detailed information on the limitations of anaerobic metabolism.

DENITRIFICATION—A MIDDLE GROUND

Denitrification may be viewed as a sort of middle ground between aerobic and anaerobic biodegradation. The process, according to conventional wisdom, occurs in the absence of oxygen but is performed by aerobic microorganisms. Many aerobic and facultative bacteria have the ability to utilize various oxides of nitrogen as their respiratory electron acceptor. Nitrate (NO_3^-), nitrite (NO_2^-), or nitrous oxide, (N_2O) can be used as acceptors. Some organisms can catalyze only one or two reductions; classical denitrifying organisms carry the process as far as N_2; many obligate anaerobes carry the reduction to NH_4^+. Aerobic respiration of bacteria is similar to that of mammals in that it occurs via a cytochrome system and terminates with the enzyme cytochrome oxidase. In the absence of oxygen other enzymes may function to terminate respiration by reducing an oxide of nitrogen. Nitrate ion reduction has a redox value of +0.42 V; oxygen reduction occurs at +0.82 V. It therefore makes energetic sense for an organism to ignore nitrate if oxygen is present. And this is usually the case; oxygen, when present, represses denitrification

(Tiedje, 1988; Ferguson, 1994). The appeal of denitrification lies in the fact that many of the aerobes that are noted for their biodegradative abilities are denitrifyers. The genus *Pseudomonas*, as an example, combines these two abilities. This means, in principle at least, that one can use the biodegradative pathways and robust physiology of a *Pseudomonad* in the absence of oxygen. Of course, a suitable oxide of nitrogen must be present and the products of denitrification must be dealt with in some acceptable manner.

A potential limitation to the use of denitrification is the requirement for molecular oxygen as a reactant in a degradation pathway. If the catabolic pathway of interest requires oxygen because of the participation of an oxygenase, oxidase, or mixed function oxidase enzyme, then biodegradation will not occur in the absence of oxygen. When oxygen is required as an enzyme substrate in a catabolic pathway, the requirement is quantitatively small compared to the amount of oxygen consumed as an aerobic electron sink. Nevertheless, the presence of oxygen has the potential to repress denitrification, and low levels of oxygen may be used as a respiratory electron sink even in the presence of excess oxides of nitrogen. This would produce diauxic growth (sequential utilization) with respect to terminal electron sink, rather than the use of the oxygen for enzymatic attack on a recalcitrant substrate. As discussed above it no longer seems likely that there are many instances of an obligatory requirement for molecular oxygen. Furthermore, the strength of oxygen repression has been called into question, and this is a point of some practical significance.

There is no obvious reason why denitrification as a respiratory electron sink should be confined to aerobic organisms. In fact one obligately anaerobic denitrifying strain of bacteria has already been isolated and characterized (Gorny et al., 1992). This organism, which could not be assigned to any known genus, degrades phenol, resorcinol, and a number of other compounds. Lithotrophic growth of anaerobic denitrifyers has been demonstrated with enrichments and pure cultures that couple the oxidation of ferrous iron to denitrification (Straub et al., 1996).

Denitrifying bacteria, typically facultative denitrifyers of the genus *Pseudomonas*, have long been exploited to remove nitrogen from wastewaters. Wastewater containing ammonium ions or other forms of reduced nitrogen are conventionally treated by aerobic nitrification systems that utilize aerobes of the genera *Nitrobacter, Nitrosobacter*, and other similar bacteria to oxidize the nitrogen to nitrate. The nitrate is then reduced under anaerobic conditions to elemental dinitrogen by denitrifying bacteria. There are a number of physical arrangements that can be used to achieve this two-step process; but, however it is accomplished, oxic and anoxic zones or reactors are required (Zitomer and Speece, 1993). An interesting recent development that has great potential significance for wastewater treatment is the recognition of anaerobic, autotrophic, and lithotrophic ammonium oxidation coupled to denitrification. Recent work has shown that under anaerobic conditions the reaction: $5NH_4^+ + 3NO_3^- \rightarrow 4N_2 + 9H_2O + 2H^+$, which effectively combines aerobic nitrification and anaerobic denitrification, occurs and is biologically mediated (Van de Graaf et al., 1995, 1996; Mulder et al., 1995). This reaction is exergonic and has been among many reactions known to be able to, in theory at least, support microbial growth. The microorganism(s) responsible for this activity have not yet been isolated, and it is not clear from the published data whether the responsible organisms are obligate anaerobes or facultative organisms. The process has been patented in Europe (Mulder et al., 1995) as the Anammox process.

Beyond the potential applications of the Anammox process, the descriptions of an obligately anaerobic denitrifyer and the lithotrophic oxidation of ammonium ion coupled to denitrification are evidence that the conventional view of denitrification simply as an alternative respiratory pathway for aerobes, may ultimately prove to be an artifact of inadequate anaerobic methodology (Bregnard et al., 1997).

ACTIVITIES OF ANAEROBES

Reports of unusual or previously undetected metabolic activities attributable to anaerobic bacteria have been appearing in the literature at an increasing rate. The degradation of hydrocarbons and oxidation of ammonium ions under anoxic conditions have already been discussed. Reductive dehalogenation of a wide variety of chlorinated compounds such as chlorophenols, chlorobenzoates, chlorobenzenes, polychlorinated biphenyls (PCBs), perchloroethylene, trichloroethylene, and many other halogenated compounds has been intensively investigated (Mohn and Tiedje, 1992; Häggblom, 1992; Fantroussi et al., 1997; Kim and Rhee, 1997). In addition, the role of anaerobic bacteria in large-scale geochemical processes also become clearer in recent years. The work of Lovley and his colleagues has demonstrated the role of hitherto unknown groups of anaerobic bacteria in the reduction of iron and manganese on a geologic scale (Lovley et al., 1993a,b, 1994a).

In this section we provide an overview of the recent work on anaerobes and to describe the nature of anaerobic ecosystems. We will also provide a brief description of the culture methodology that underlies the recent advances in this area of microbiology. For the purposes of this chapter we define anaerobes as organisms that grow in the absence of molecular oxygen and may be irreversibly inhibited by exposure to oxygen or elevated redox potential.

METABOLIC STRATEGIES OF ANAEROBES

Biodegradation by anaerobes often differs from aerobic degradation (Zehnder and Svensson, 1986). Most of the differences result from the fact that anaerobic energy metabolism usually yields much less energy than its aerobic equivalent. When bacteria attack an organic compound, they usually do so to obtain energy. Because biologically useful energy is harvested by means of oxidations, the activities of an ecosystem can be analyzed by characterizing the electron donors, electron acceptors, and the processes that move electrons from donors to acceptors. Oxidizable substrates may be organic or inorganic compounds or ions. Hydrogen and elemental sulfur are obvious examples of inorganic electron donors.

Anaerobic degradation is often an ecosystem phenomenon involving the sequential actions of a number of microorganisms. Mineralization under anaerobic conditions, whether accomplished by a single organism or a consortium, adds to the environment the reduced form of the electron acceptor, which unlike water, the product of aerobic mineralization, can have its own environmental impact. Food chains, wherein the products of one organism are the substrates of a second organism, can be found in anaerobic ecosystems. If the first organism derives benefit from participating in a food chain, it is usually due to its metabolic waste products being removed from its environment (McInerney, 1986).

Consortia generally differ from simple food chains in that the second, third, or subsequent organisms alter the activities of the first organism or actually make possible activities that that would not be possible in isolation. The term "food web" has been applied to complex interrelationships among anaerobic bacteria.

An important fact about anaerobic ecosystems is that more than one type of mineralization is possible. The identity of the sink organisms determines the mineralization products. Methanogenic consortia dispose of electrons and carbon in the form of methane, which rises up and out of the anaerobic environment. Sulfidogenic consortia discard their electrons as sulfide and their carbon as carbon dioxide or other low-molecular-weight compound.

The diversity of organic compounds is so great that there is little point in attempting to analyze a microbial population in terms of substrate oxidized. Instead, anaerobic ecosystems are characterized by the terminal electron acceptor, or electron sink, that functions at the end of the microbial food chain. From the perspective of biodegradation complex organic compounds are usually electron sources. However, sites of unsaturation and oxidized substituents of organic molecules may serve as electron sinks.

Electron Sources and Sinks

Energy metabolism is an oxidative process; consequently there must always be a final electron acceptor (sink) into which the electrons are dumped after they have given up their useful energy. There are two possible sources of electron sinks: the environment or the cell's metabolic pathways. Electron acceptors provided by the environment are usually, but not necessarily, inorganic molecules or ions. The phrase "from the environment" implies that the acceptor is available in amounts that are in excess of the cells' needs. That is, growth is not usually limited by a shortage of terminal electron acceptors. The type of energy metabolism that uses environmentally supplied inorganic terminal electron acceptors is usually referred to as a respiratory energy metabolism. A characteristic of this type of energy metabolism is that energy harvesting [e.g., adenosine triphosphate (ATP) synthesis] occurs during electron transport in the cytoplasmic membrane. According to this scheme, oxygen is just one example of a respiratory terminal electron acceptor. Other sinks available to bacteria are sulfur oxides, elemental sulfur, nitrogen oxides and ions, carbon dioxide, ions of iron and manganese, organic compounds with reducible sites, and a variety of ions, particularly metal oxide ions, and protons. It seems reasonable to assume that any oxidized substance, whether molecular or ionic, can in theory serve as a terminal electron sink for some type of bacteria provided only that the substance's redox potential is appropriate and that neither the sink nor its reduced product is overtly poisonous or violently reactive in aqueous systems.

Electron sinks that are generated inside the cell are almost always organic compounds that are derived from the organic compounds the organism is degrading. This type of energy metabolism is called fermentative, and the reduced forms of the sink compounds are referred to as fermentation products. Energy is generally harvested by substrate-level phosphorylations. That is, and occurs in the cytoplasm. Because the sink compound is derived from the starting compound, there can never be an excess of electron sink available during a fermentation. It should be noted that in the following discussion the word "fermentation" is used as defined above; that is, as a type of energy metabolism; whereas industrial microbiologists use the word "fermentation" to describe any process that involves microorganisms.

Fermentations

Fermentations are closed systems that operate according to fixed stoichiometries, and the rules of chemical bookkeeping can be used to analyze a fermentation (Barker, 1981; Sokatch, 1969; Gottschalk, 1985; Caldwell, 1995; White, 1995). The limited availability of terminal electron sinks determines the energetic yields of fermentations. Fermentations are among the lowest energy-yielding catabolic pathways in biology. The simple lactic acid fermentation (*Lactobacillus, Streptococcus*, etc.) operates according to a fixed stoichiometry: 1 glucose 2 lactate. While there is no net oxidation or reduction involved, energy is harvested by oxidizing the intermediate compound, 3-phosphoglyceraldehyde. Pyruvate, the last intermediate in glycolysis, is reduced to lactate by the two electrons generated by the oxidation, and the fermentation is complete. The energy yield is 2 ATP/glucose. By contrast, the oxidation of glucose to carbon dioxide and water when oxygen serves as an environmentally supplied electron sink yields about 38 ATP/glucose. This simple and familiar comparison illustrates the energetic advantage of dumping electrons into an externally supplied electron sink. The significance of this metabolic strategy in the analysis of anaerobic ecosystems is becoming increasingly apparent.

Fermentations provide limited opportunities to increase energy harvest. Most fermentations are more complex than the lactic acid fermentation and typically yield a mixture of reduced and oxidized organic compounds. Short-chain fatty acids and alcohols (C_3-C_6) are usually produced because their synthesis, via complex pathways, consumes more electrons than does the direct reduction of pyruvate to lactate, thus sparing some of the remaining pyruvate (or other α-oxo acid) for further oxidation and energy harvest. The details of these pathways are available elsewhere (Zehnder et al., 1986; Barker, 1981; Sokatch, 1969; Gottschalk, 1985; Caldwell, 1995; White, 1995).

One of the most significant developments in our understanding of bacterial metabolism is the recognition of the fact that many fermentative anaerobes have some, and often considerable, ability to shift to a respiratory type of metabolism if conditions permit. This method of increasing energy harvest is of extreme importance in the field of biodegradation. The ability of fermentative organisms to shift their metabolic systems means that they are potentially much more versatile than had been previously believed. It also means that biodegradation can occur via routes that are quite different than had been previously assumed.

Hydrogen

Many fermentative bacteria produce hydrogen. A molecule of hydrogen is a reduced fermentation product that carries away two electrons. Hydrogen is important. In fact in the context of anaerobic biodegradation hydrogen is a key intermediate and is perhaps the most important metabolic product that must be considered. A molecule of hydrogen is produced by reducing two protons with two electrons. This reaction has a redox potential of $E_0' = -420$ mV. This is a very negative value for a biological reaction and means that it takes a strong reducing agent with an even more negative redox value to reduce protons to hydrogen. There are not many such reducing agents in metabolism. The most common type is exemplified by pyruvate: an α-oxo acid (often called an α-ketoacid). The oxidative decarboxylation of a compound containing this grouping generates low redox potential electrons ($E_0' = -680$ mV) that can, and usually do, reduce protons to hydrogen. "Low redox potential" is a synonym for "high potential energy" when describing electrons participating in metabolism.

Protons are an obvious sink for electrons generated by reactions that have a redox potential below -420 mV. But aside from α-oxo acids (and photosynthesis) there are not many sources of these energetic electrons. Most of the electrons that move through metabolism are carried by nicotinamide adenine dinucleotide (phosphate), or $NAD(P)^+$, or carriers of higher potential. The redox level of the reduction of $NAD(P)^+$ is -320 mV. Consequently, assuming the standard conditions used to calculate redox potentials, these electrons cannot reduce protons to hydrogen; to the contrary, the predicted direction of the reaction would be the movement of electrons from H_2 to NAD^+, the opposite of what is desired by anaerobes attempting to increase their energy yields.

Because most of the biologically important oxidations occur above the redox level of hydrogen, it would seem that protons would be a quantitatively minor electron sink for anaerobic organisms. This was in fact the prevailing view among microbiologists until the late 1960s, when R. E. Hungate attempted to account for the fact that pure cultures of rumen anaerobes produce ethanol whereas the rumen population does not (Hungate, 1966). Furthermore, the mixed population produces more acidic products and greater total biomass than would be predicted from pure culture data. Hungate suggested that hydrogen removal via methane production accounted for these changes and increased the energy-harvesting efficiency of the population (Hungate, 1950). Hungate recognized that methane-producing bacteria use hydrogen and therefore remove it from the environment of the fermentative organisms. Removal of the hydrogen means that standard conditions do not prevail in nature, and it follows that equilibrium constants calculated for chemical reactions occurring in isolation, that is, using standard redox voltages for coupled half reactions, are not likely to be valid in nature. In fact, electrons carried by NADH+H$^+$, the reduced form of NAD$^+$, can be used to reduce protons to hydrogen provided the fermentative organisms possess the appropriate enzyme (many do, some do not), and the hydrogen is removed as quickly as it is formed. The required enzyme is an NAD$^+$ linked hydrogenase and the phenomenon is called *interspecies hydrogen transfer*. A better name is *interspecies electron transfer* since hydrogen serves simply as a chemically stable form of low potential electrons (Bélaich and Bruschi, 1990; Stams, 1994).

Bacteria that oxidize hydrogen and dump the electrons into an acceptor of higher redox potential have a major effect on the chemical transformations that occur in an anaerobic ecosystem. The utilization of hydrogen is extremely important in anaerobic environments and is critical to understanding the bio-degradation potential of anaerobes. Hydrogen is an energy-rich molecule because it is carrying energy-rich, low potential electrons. The energy content of hydrogen is so great that evolution has developed many metabolic strategies to use hydrogen. All of these strategies involve oxidizing hydrogen and moving the electrons from their starting redox level of – 420 mV to a sink of higher potential.

Hydrogen is the link between interacting organisms in nature. Hydrogen produced by fermentative organisms is consumed by what are termed sink organisms. *Sink organisms* use hydrogen as fast as it is produced and maintain an environmental concentration close to zero. Thermodynamic values calculated for standard conditions no longer apply, and the production of hydrogen from NADH+H$^+$ becomes energetically favorable (Wolin, 1976). If a fermentative organism has the NADH+H$^+$ linked hydrogenase, electron disposal no longer requires organic acceptors, reduced fermentation products no longer have to be synthesized, and there is a great increase in organism's energetic efficiency. Thus an anaerobe that produces products such as ethanol or butanol when growing in pure culture can potentially stop making these products when mixed with a sink organism. Oxidized products, notably acetate and carbon dioxide, become the predominant fermentation products. So important are sink organisms that anaerobic ecosystems are classified according to the predominant sink population, for example, methanogenic, sulfidogenic, acetogenic, and the like.

This type of multiorganism interaction is called a consortium. Participation in consortia appears to be the rule in natural anaerobic environments. It is possible that the fermentation pathways leading to reduced products may be a survival mechanism for anaerobes unable to form a consortium with sink organisms. The amount of energy extracted from the substrate molecule is very much greater when a consortium is operating than when individual organisms are degrading an organic compound.

The environmental significance of hydrogen consumption by sink organisms is so profound that there exist types of bacteria that can only survive as part of consortia (Mclnerney, 1986). These bacteria are known as syntrophic bacteria or sometimes as obligate symbionts. The existence of these organisms was not demonstrated until the late 1960s, and even today only a relative handful have been isolated and characterized. Some syntrophic bacteria cannot grow in pure culture, while others have alternative energy-yielding pathways that permit pure culture isolation and growth. Syntrophic growth depends on energy-yielding pathways that involve the production of H_2 from NADH+H^+. As an example of such a pathway, consider the fate of fat or phospholipid in an anaerobic environment. Hydrolysis quickly separates the fatty acids from the glycerol, which is readily fermented. The fatty acids are another matter. There is no way fatty acids can be fermented. They are too reduced to be rearranged into fermentation products. Yet fats do not accumulate in anaerobic environments. Syntrophic bacteria attack fatty acids by a β-oxidation-type pathway with protons serving as the electron acceptors for the reoxidation of $NADH_2$. This reaction can only occur if a sink organism removes hydrogen as fast as it is released by the syntrophic bacteria. Consequently the steady-state concentration of hydrogen is critical to the metabolic functioning of microbial consortia (Mormile et al., 1996)

Syntrophic bacteria have an obligate requirement to participate in interspecies hydrogen transfer. Hydrogen transfer by means of syntrophic formate production has also been demonstrated (Dong et al., 1994). There are only a few syntrophic bacteria that have been isolated, and the details of how they extract energy from their degradative pathways are still largely unknown (Dolfing, 1988; Lowe et al., 1993e; Angelidaki and Ahring, 1995; Auburger and Winter, 1996; Schnürer et al., 1996).

It is possible to construct a hierarchy of anaerobic degraders as a means of summarizing what has been discussed thus far. At the bottom are the conventional fermentative bacteria. They eke out a modest living by rearranging the atoms in organic molecules. They are limited to substrate molecules that have enough oxygen atoms to make energy-yielding rearrangements possible. One level up are fermentative bacteria that have the ability to produce hydrogen from low redox potential sources such as α-oxo acids. They gain an energetic advantage from this ability. The next level up are those fermentative bacteria that can produce H_2 from NADH+H^+. These organisms obtain a great increase in their energetic efficiency. They also gain versatility. If consortia are possible, they benefit. If consortia are not possible, they grow as conventional fermentative anaerobes. The next level are organisms that not only can participate in consortia but must. These organisms do not have conventional fermentative pathways and specialize in compounds not readily attacked by other organisms. To accomplish this they must have sink organisms as collaborators.

RESPIRATIONS: ELECTRON SINKS AND SINK ORGANISMS

The most abundant electron acceptor found in anaerobic environments is often carbon dioxide; the product of its reduction is methane. Sulfate ions are another very important sink, particularly in marine and estuarine environments; reduction produces hydrogen sulfide. Sulfate can also be looked at as the model of a vast array of possible electron sinks. Compounds or anions with the general formula of X_aO_b are utilized by an enormous array of anaerobes. Sites of unsaturation within organic compounds are very commonly reduced. Oxidized substituents can be reduced and cleaved off the parent molecule. Carbon–halogen bonds have been recognized as electron sinks. These sinks will be described later. To complete of discussion of consortia it is necessary to follow the flow of electrons into hydrogen utilizing sink organisms.

Methanogens

The methanogens are common sink organism in terrestrial microbial ecosystems. This is so because their sink compound, carbon dioxide, is ubiquitous. Decarboxylation reactions are widespread among fermentative bacteria, and the energetically favorable attack on pyruvate that leads to hydrogen production produces equimolar amounts of CO_2. Methanogens are members of the archeae. Once thought to be a small group of prokaryotes, the methanogens are known to be a large and varied group of organisms adapted to a wide variety of environmental conditions. Linking this diverse group of organisms is the possession of a unique and limited energy metabolism: The production of methane is their only significant source of energy (Balch et al., 1976, 1979; Vogels et al., 1988; Reeve, 1992). Methane is produced from only a limited number of substrates with hydrogen and carbon dioxide being the most important. Another important substrate is formate, which is widely produced in anaerobic environments. Formate is the equivalent of an equimolar mix of hydrogen and CO_2 because many anaerobes possess an enzyme called formate hydrogen lyase, which catalyzes the freely reversible interconversion. Acetate is converted to an equimolar mix of methane and CO_2 by many methanogens. Methanol, which can arise from demethoxylation of pectin-type polysaccharides, and choline, derived from the anaerobic degradation of phospholipids, are usually minor sources of methane. Carbon monoxide is also a minor substrate for methanogenesis.

In the absence of a geochemical source of hydrogen and carbon dioxide, methanogens depend on the activity of other anaerobes for both their electron donor and sink compounds. Therefore, it is not surprising that they participate in consortia. Consortia described as methanogenic involve one or more methanogens and have the ability to mineralize organic compounds to methane and carbon dioxide. The rumen, swamps, landfills, and anaerobic sewage digesters are examples of well-studied methanogenic consortia.

Sulfidogens

The sulfidogens use oxides of sulfur, most commonly sulfate ions, as their terminal electron acceptors. The sulfidogens are eubacteria, rather than archaea. *Desulfovibrio* and *Desulfotomaculum* are the genera that have been longest known and are generally cited when sulfidogens are described (Odom and Singleton, 1993) However, the number of sulfidogens known to science has been increasing at a dramatic rate. Sulfidogens, also referred to as dissimilatory sulfate reducers (DSRs), were long thought like the methanogens to posses a unique energy metabolism. It is now apparent that the ability to use oxides of sulfur or elemental sulfur as respiratory electron sinks is very widely distributed among anaerobes. Consequently our understanding of sulfidogens and their role in the environment is undergoing major change at present. The traditional view of the sulfidogens has been described in detail by Pfenning and Widdel (1982), Postgate, 1984) and Pfenning (1989). A comprehansive view of the sulfate-reducing bacteria and many of their metaolic capabilities was provided by Widdell (1988) more recent reviews have been provided by Hansen (1994) and by Colleran et al. (1995). Briefly summarized, the traditional view that sulfate-reducing bacteria are restricted to using hydrogen and simple organic acids as their energy sources has given way to the view that there is enormous metabolic diversity among these bacteria. It would not be surprising if the metabolic capabilities of sulfidogens ultimately is shown to rival that of aerobic genera such as *Pseudomonas* (Rabus et al., 1993; Dalsgaard and Bak, 1994). Sulfate ions are present in seawater; consequently,in anoxic environments containing seawater microbial communities usually function as sulfidogenic consortia. Methanogens are also present but in head-to-head competition the sulfidogens usually win. This is so, in part, because the uptake hydrogenases, the enzyme that captures H_2 of sulfidogens generally have a higher affinity (lower K_m) than the hydrogenases of methanogens. Sulfidogens produce sulfide ions as their reduced product. Depending on circumstances the sulfide may evolve as H_2S, may stay in solution or, as is common, precipitate out as extremely insoluble pyritic compounds of divalent and trivalent metals.

The sulfate-reducing bacteria are an extremely diverse group of organisms that defy simple description. Some sulfate-reducing strains can grow as syntrophs in the absence of sulfate but the presence of a methanogen. The electrons generated during the oxidation of organic compounds that would normally be used to reduce sulfer oxide ions are instead ussed to reduce protons to hydrogen. This is energetically favorable only if there is a hydrogen-utilizing organism such as a methanogen growing with the sulfidogen. Thus sulfidogens may participate in consortia as sink organism or as syntrophs. There are also sulfidogens that obtain their energy by disproportionating inorganic sulfur ions. Typically thiosufate or sulfite are disproportionated to a mixture of sulfate and sulfide. The sulfate can then be used as the sink organic compound oxidation (Widdel, 1988). To further complicate the picture there are sulfidogens that can utilize nitrate as their electron sink when sulfate is not available. Unlike denitrifyers that produce N_2 the sulfidogens produce NH_4^+ ions (Dalsgaard et al 1994).

Sulfidogenic bacteria that do not use sulfate or most other oxide ions of sulfur have also been described. Typically these bacteria utilize elemental sulfur as their terminal electron sink and yield sulfide ions as their waste product (Widdel, 1988; Schauder and Kroger, 1993).

Hyperthermophilic archaea, some of which grow at temperatures above 100°C, are obligate anaerobes that use sulfur as their electron sink (Kelly and Adams, 1994). These are marine micoorganism isolated from thermal vents at the ocean bottom. In principle these organisms could function as sink organism in consortia; in practice their minimum growth temperatures are well above the mesophilic and even the "normal" thermophilic temperature range of other organisms (Jannasch et al., 1988) Nevertheless the metabolic potential of these organisms is worth exploring. The chemistry of sulfur at temperatures near the boiling point of water is complex, and the exact mechanism of sulfur or polysulfide reduction is not yet clear (Belkin et al., 1985).

The diversity of sulfidogens is such that they are likely to prove to be useful in many areas. The anoxic reduction of sulfate is the most obvious activity of these organisms. Postgate (1984, 1992) described the deleterious effects of that long ago when he brought these organisms to the attention of civil engineers. The usefulness of these organisms has been demostrated in artificial lagoons used to treat acid mine drainage as well as other types of toxic runoffs. The reduction of sulfate to sulfide raises the pH and renders insoluble toxic heavy metals both by a pH effect on solubility and chemical reduction of certain metals by the microbially generated H_2S (Kafkewitz et at., 1994; DeFilippi, 1994; White and Gadd, 1996).

Looking beyond sulfate reduction, it has become clear that potentially useful metabolic versatility exists among the sulfate-reducing bacteria. The anoxic oxidation of alkanes and polyaromatic hydrocarbons under sulfate-reducing conditions has been demostrated (Rueter 1994; Coates 1996). Whether these activities can be developed into applications remains to be seen. Finally, as will be described below, many of the anaerobic dehalogenating bacteria thus far taxonomically evaluated have proven to be sulfidogens. The genetic relationship, if any, of the dehalogenating enzymes to the sulfate-reducing enzymes remains to be determined.

Acetogens

Acetate is the most common oxidized fermentation product, and because it is readily utilized, it is the organic compound that links fermentative organisms and syntrophs with sink organisms. Acetate production by fermentative bacteria is accompanied by ATP synthesis; consequently fermentative organisms strive to maximize acetate production. Syntrophic bacteria, degrading fatty acids, cleave off acetate units as their organic end products. They utilize a pathway similar to the β-oxidation pathway but utilize protons rather than NAD^+ as their electron sink.

In addition to these sources of acetate there is another group of organisms known collectively as homoacetogens (Drake, 1994; Harriott and Frazer, 1997). Some are fermentative bacteria that produce acetate as their sole fermentation product. Certain of the *Clostridia*, for example, ferment hexoses to two molecules of pyruvate, which are then oxidatively decarboxylated to two acetates in the usual fashion. The carbon dioxide and protons produced by the decarboxylations are used for the de novo synthesis of a third molecule of acetate, rather than being released as CO_2 and H_2 or as formate.

Among the anaerobes grouped together as homoacetogens, there are also organisms that use hydrogen to reduce carbon dioxide but synthesize acetate instead of methane as their end products. These are not yet a well-studied group of organisms and are unfamiliar to many microbiologists. It is premature to talk about the possible significance of these organisms in consortia. But the existence of this group of organisms complicates the analysis of a complex consortium. The acetogens compete with methanogens and other sink organisms for substrates, but produce acetate which is degraded by some methanogens, sulfidogens, and likely many other sink organisms (Widdel, 1988; Dolfing, 1988).

Inorganic Ion Reduction

The ability of bacteria to reduce metal and metal oxide ions has been known since at least the nineteenth century (Ghiorse, 1988; Lovley, 1991). The reductive abilities are widely distributed among bacteria of all physiological types and has been usefully exploited. There are bacteria of clinical significance that can, for example, reduce tellurite to metallic tellurium; others can reduce selenite to elemental selenium. Diagnostic media that permit the detection and rapid identification of these bacteria have long been in widespread use. This aspect of metal ion reduction is outside the scope of this review. The following discussion will focus on the role of inorganic ions in anaerobic environments and will include examples of both obligate and facultative anaerobes.

Sulfate and nitrate ions are not the only inorganic ions that can serve as electron sinks for anaerobic microorganisms. Anaerobic environments tend to be electron-donor rich and electron-acceptor poor. Consequently organisms are frequently limited with respect to the compounds they can oxidize but promiscuous with respect to their electron sinks. This was demonstrated in the 1960s when it was shown that extracts of cells of *Micrococcus lactilyticus (Veillonella alcalescens)* couple the oxidation of hydrogen to the reduction of selenium, selenite, molybdate, tellurite, tellurate, vanadate, arsenate, bismuthate, lead dioxide, manganese dioxide, osmium tetroxide, osmium dioxide, cupric hydroxide, and a number of other inorganic substances (Woolfolk and Whitely, 1962).

In recent years a very wide range of inorganic substances have been shown to serve as electron sinks for many obligate anaerobes as well as for aerobic and facultative organisms that can exploit alternate electron sinks for their respiratory pathways. There are two general patterns discernible: fermentative anaerobes that dump their electrons into external acceptors to increase the energy yield of their fermentations and respiratory organisms in which the transfer of electrons to an external acceptor is coupled to energy harvesting (Ghiorse, 1988; Lovley, 1991).

The factors that affect the energy yield of fermentations were described earlier. The synthesis of oxidized fermentation products is coupled to energy harvesting, and any method of dumping electrons into an external sink reduces the cell's need to use fermentation pathway intermediates as electron acceptors. Interspecies hydrogen transfer is one strategy fermentative anaerobes use to dump electrons. More generally, obligate anaerobes and many facultative anaerobes growing fermentatively have the ability to exploit environmental electron acceptors that may be available to them. The transfer of electrons from reduced ferredoxin, an electron carrier at the redox level of hydrogen ($E_0' = -420$ mV), or from NADH+H$^+$ to an inorganic acceptor is usually not directly coupled to energy harvest; rather the organism gains an energetic advantage by reducing the requirement to use fermentation intermediates as electron sinks. Although reduction of external sinks may be a quantitatively minor activity, the energy yields of most fermentations are so low that even small increases can be advantageous in a competitive environment.

In no area of environmental microbiology are the effects of proper anaerobic technique more evident than in the work of Lovley and his co-workers on inorganic respirations (Lovley et al., 1994a,b, 1993a,b; Lovley, 1991, 1993a,b). The use of stringent anaerobic technique in sampling and cultivation of environmental inocula has led to findings that have greatly altered our understanding of the role of anaerobic respirations in natural ecosystems and provided strong evidence for a major geochemical role for these processes. In addition, new organisms with degradative abilities long thought to be restricted to aerobic environments have been described, and in some cases, isolated in pure culture. Relationships among seemingly unrelated organisms, and metabolic versatility in anaerobes has led to significant reconsideration of the function of anaerobic ecosystems.

The recent advances in our understanding of the role of metal ion respiration in anaerobic environments highlight the significance of anaerobic technique in obtaining meaningful data. It is interesting to compare the recent findings with earlier results to see how important adequate anaerobic technique is in evaluating environmental samples. Most of the early isolates were obtained with some level of anaerobic technique, though less rigorous than the prereduced techniques that will be described below. The isolates were usually nitrate-reducing organisms that could reduce Fe(III) when nitrate levels were low or depleted (Ottow and Glathe, 1971; Ottow, 1971; Sørensen, 1982; Jones et al., 1983, 1984). The organisms appear to have gained little or no energy from the Fe(III) reduction. These isolates were either slow growing or difficult to culture and therefore were not well characterized. In most of these studies, the culture inocula were exposed to air for a significant amount of time (from a few hours to overnight) before they were subjected to anaerobic handling and storage. This is likely to have had a significant impact on the available pool of organisms in the samples. Strictly anaerobic organisms, unable to compete with the facultative or aerobic organisms, or sensitive to oxygen in the samples, may have been completely lost or had their numbers reduced to insignificant levels. This could have eliminated organisms that were most representative of the in situ capabilities of the environmental flora. In addition, the exposure to air may have permitted Fe(II) present in the environmental sample to oxidize to Fe(III) and thus disguise any reduction activity that may have taken place.

Unlike much of the previous work in which anaerobiosis did not begin until incubation, Lovley et al.'s studies utilized sediment samples that were collected and handled under either nitrogen or argon. Exposure to oxygen was significantly limited or completely avoided (Lovley and Phillips, 1986, 1987, 1988; Lovley et al., 1989; Lovley and Lonergan, 1990; Lonergan and Lovley, 1991; Coleman et al.,

1993). Media used for these experiments were prepared using strict anaerobic techniques, either roll tube method or modified Hungate technique described below. In their early work these workers relied on Fe(II) present in sediments to consume any traces of oxygen not removed by flushing with oxygen-free gas. In later studies sodium sulfide was added to further ensure anaerobic conditions. Using these anaerobic techniques. Lovley and co-workers (see list of papers in reference section, this chapter) were able to substantially expand our knowledge of Fe(III) and Mn(IV) reducing bacteria. They hypothesized that most of the Fe(III) and Mn(IV) reduction taking place in sedimentary environments is due to microbial activities and that these activities play a major role in the formation of iron and manganese minerals (Lovley, 1993a). These workers have also provided a rapidly increasing body of knowledge concerning the diversity of organic compounds that can be degraded via metal ion respirations. The low solubility of Fe(III) appears to be the limiting factor for iron respiration and can be alleviated by the use of chelating agents (Lovley et al., 1994b). Albrechtsen et al. (1994, 1995), utilizing strict anaerobic techniques for the collection and handling of samples, attempted to estimate the significance of microbial processes in the reduction of sediment-bound Fe(III) in a landfill leachate-polluted aquifer in Vejen, Denmark. Their results provided evidence that iron-reducing bacteria were present and that a substantial part of the iron reduction observed in the plume could be attributed to microbial activity.

Among the most interesting and significant findings of the recent work on metal ion respirations are the metabolic capabilities of the organisms at both ends of their redox chains. As mentioned earlier the old view that anaerobes possess limited catabolic ability has been giving way to the view that the catabolic capabilities of obligate anaerobes are likely to be comparable to those of aerobes. What is just as interesting, and possibly of greater significance, is that this versatility extends to both donors and acceptors that support growth, as well as to the modes of metabolism available to a single organism. *Geobacter metallireducens* (GS-15), as an example, can use as electron donors diverse compounds such as acetate, ethanol, a variety of other short-chain fatty acids and alcohols, toluene, benzoate, benzaldehyde, phenol, *p*-cresol, and a number of other aromatic compounds. As respiratory electron acceptors the organism can use Fe(III), Mn(IV), U(VI), and nitrate. Although *G. metallireducens* cannot utilize any form of sulfur as an electron sink, molecular biological analysis suggested that it is related to certain sulfur-reducing sulfidogens. Subsequently, it was shown that *Desulfuromonas acetooxidans*, an organism originally described as an acetate-oxidizing, elemental-sulfur-reducing sulfidogen can also grow by Fe(III) or Mn(IV) respiration (Roden and Lovley, 1993). In an acceptor-limited environment this versatility is clearly advantageous and is probably quite common. Isolates that grow by chromate (Cervantes, 1991), arsenate (Ahmann et al., 1994), and selenate respirations (Macy et al., 1993) have also been described, and it seems unlikely that these ions should be the environmentally significant primary respiratory acceptors for these organisms.

An even more interesting example of metabolic versatility among anaerobes is the genus *Pelobacter* (Dubourguier et al., 1986). Originally isolated as fermentative anaerobes that degrade the low-molecular-weight fermentation products of other bacteria, this genus was soon shown to be a group of versatile anaerobes capable of fermenting ethylene glycol, high-molecular-weight polyethylene glycol, trihydroxybenzenes, and an array of other compounds (Schink and Stieb, 1983; Samain et al. 1986; Dubourguier et al., 1986; Schink et al., 1987; Wagener and Schink, 1988; Frongs et al., 1992). Further work on members of this group showed that many were also capable of syntrophic growth when provided with a nonfermentable organic substrate and a sink organism such as a methanogen. More recent work has shown that members of the genus also possess Fe(III) and S^0 respiratory pathways similar to those of *Geobacter* and *Desulfuromas* (Lovley et al., 1995). Consequently members of this single genus can function as primary degraders of fermentable compounds, syntrophic oxidizers of non-fermentable compounds, or sink organisms. This versatility is likely a reflection of the highly competitive nature of anaerobic ecosystems. It is also a dramatic indication of how little we know of the true role of organisms within these ecosystems.

The high redox potential of Fe(III) reduction also means that iron reduction may be able to serve as an electron sink for the oxidation of compounds normally not considered susceptible to oxidative attack in anaerobic environments. In particular the sequential reductive dehalogenation of multihalogenated organic compounds frequently becomes very slow or no longer occurs when the number of halogen substituents has been decreased to one or two. Under these conditions it may be possible to exploit bioavailable Fe(III) as an electron acceptor and oxidatively attack the recalcitrant halogenated compounds.

Organic Electron Acceptors

Nonionic compounds such as dimethyl sulfoxide (Woods, 1981) and trimethylamine N-oxide (Strom et al., 1979) have long been known to serve as respiratory electron acceptors for a number of organisms. Efficient reduction of sterols (Eyssen et al., 1973) and saturation of fatty acids (Kemp et al., 1975) has also been demonstrated.

The literature contains many additional examples of reduction of oxidized sites within a wide variety of organic compounds. However, the area that is likely to be of greatest interest to the environmental scientist or engineer is reductive dehalogenation. Although the ability of the carbon–halogen bond to serve as an electron sink was recognized many years ago, the last decade has seen this area of research develop into a large and maturing area of environmental microbiology. Briefly summarized, reductive dehalogenation involves the removal of a halogen substituent of a molecule with simultaneous addition of electrons to the molecule. This method can be used to reduce alkyl or aryl halides. This process can initiate the biodegradation of environmentally important halogenated compounds such as hexachlorobenzene, tetrachloroethene, pentachlorophenol (PCP), and (PCBs) (see Chapter 10 for further discussion of PCP degradation and Chapter 11 for PCB catabolism). An interesting and unexpected finding in this area of research is the existence of at least one type of bacteria that can only grow if supplied with perchloroethylene as a respiratory electron acceptor and molecular hydrogen as electron donor. This isolate has been named *Dehalobacter restrictus* (Schumacher and Holliger, 1996).

Studies on the biodegradation of PCBs in the 1970s and 1980s focused on aerobic organisms, and several aerobic bacterial strains that are able to oxidize lower chlorinated PCBs were described. Aerobic biodegradation of PCBs has been demonstrated on congeners with five or fewer chlorines with at least two adjacent unsubstituted carbon atoms. More highly chlorinated PCBs are more recalcitrant and are of greater concern in the environment.

In 1984, Brown et al. reported dechlorination of Aroclor 1242 in the anaerobic sediments of the Hudson River south of Hudson Falls, New York. This particular congener is primarily composed of tri- and tetrabiphenyls with only 0.7% 2–chlorobiphenyl and 11.5% dichlorobiphenyl. Brown showed that the sediment extracts contained 10–43% 2-chlorobiphenyl and 21–50% dichlorobiphenyls. In 1987, Brown and co-workers showed evidence for the dechlorination of Aroclor 1260 in the anaerobic sediments of Silver Lake in Pittsfield, Massachusetts. The PCB extracts showed a 90–98% loss of the hexa- and heptachlorobiphenyl and the appearance of tri- and tetrachlorobiphenyls. These lower chlorinated homologs usually account for less than one percent of total PCBs in Aroclor 1260 but were now present as 57–82% of the total in the sediments. Dechlorination of such highly halogenated PCBs was never possible under aerobic conditions.

In the years since this work was published the range of aromatic and aliphatic compounds that can be reductively dehalogenated has grown to include virtually all compounds tested. Because this area has been extensively reviewed in recent years, interested readers are directed to these sources of detailed information (Mohn and Tiedje, 1990, 1992; Häggblom, 1992; Holliger et al., 1993; Ensley, 1991; Lowe et al., 1993c; Dolfing and Harrison, 1993; Tiedje et al., 1993; Fetzner and Lingens, 1994; Dolfing and Janssen, 1994; Utkin et al., 1994; Holliger and Schumacher, 1994; Kennes et al., 1996). Oxidative dehalogenation and mineralization of vinyl chloride coupled to Fe(III) reduction has also been demonstrated (Bradley and Chapelle, 1996). These findings have significant implications for in situ mineralization of vinyl chloride, which frequently proves to be recalcitrant.

Anaerobic dehalogenation can be employed in practical applications. Chlorinated solvents, pesticides, and many other useful compounds present major challenges for waste treatment processes as well as for environmental cleanup. Consequently, the exploitation of anaerobic bacteria has been a major goal of many research groups. This topic has been frequently reviewed in the literature. Among recent useful reviews on the application of bacterially mediated reductive dechlorination are: Holliger and Schraa (1994), Holliger (1995), Leisinger and Braus-Stromeyer (1995), and Leisinger (1996).

OTHER ANAEROBES

This review has focused entirely upon anaerobes that obtain their energy by chemical means, for example, biodegradation and biodegradation. No mention has been made of photosynthetic prokaryotes. Aside from the cyanobacteria, which are aerobes that possess the same chlorophyll *a* oxygenic photosystem found in eukaryotic algae and higher plants, all other photosynthetic prokaryotes are anaerobes. And there are many of them displaying a wide variety of characteristics. What they have in common is they lack the dual photosystems of oxygenic organisms and rely upon bacteriochlorphyll to mediate anoxygenic photosynthesis. This cyclic photophosphorylation system harvests energy, but unlike the dual system of the cyanobacteria it cannot supply electrons for biosynthesis. The nonoxygenic photosynthetic prokaryotes rely upon chemical oxidations to generated metabolic electrons. Organic and inorganic compounds are attacked. With respect to biodegradation this is a barely explored area of anaerobic microbiology. It would not be at all surprising if the biodegradative potential of photosynthetic prokaryotes is ultimately found to be comparable to the nonphotosynthetic organisms. This topic is beyond the competency of the current authors but is well worth study.

SIGNIFICANCE OF ANAEROBIC CULTURE TECHNIQUE

At the start of this chapter we noted that our understanding of the metabolic activities of anaerobic bacteria has greatly increased during the past few years and that a major cause of this was the application of appropriate cultivation techniques. We have described these techniques and their development in detail elsewhere (Kafkewitz and Togna, in press). In this chapter we will briefly summarize this material.

The study of the environmental activities of anaerobes is highly susceptible to artefactual results. The range of oxygen sensitivities of anaerobic prokaryotes— from indifferent to the presence of oxygen to exquisitely sensitive—is so great that an environmental sample cultured with virtually any type of anaerobic technique will yield growth. The question that must be asked is whether the organisms so obtained are environmentally relevant or merely most adaptable to the conditions of culture. Many of the organisms that are of greatest interest are trace components of their environments. It is often necessary to use enrichment cultures to increase their numbers with respect to the other organisms of the indigenous population. Unless strict anaerobic technique is utilized, minor but important constituents of the environmental sample may die before they have had a chance to establish themselves in the enrichment culture. It is for this reason that the evolution of methods for the growth and manipulation of extremely oxygen-sensitive bacteria has had such a major impact.

Anaerobic Media

The techniques necessary to cultivate very strictly anaerobic bacteria are more involved than those techniques generally taught in collegiate microbiology classes or those used in routine medical microbiology. These methods have been described in some detail in a relatively recent monograph (Levett, 1991).

The methods appropriate for the study of these environmentally important anaerobes were originally developed by Hungate (1969). The methods he developed were designed for the study of the metabolic activity of microbes found in the bovine and ovine rumen, the first stomach of cows, sheep, and other cloven-hoofed (ruminant) animals. Hungate's studies were among the first to utilize techniques that provided reasonable assurance that the significant organisms of an anaerobic ecosystem were not killed during manipulation and cultivation. Hungate deduced that media prepared in the conventional manner, with exposure to air during heating or use, were unsuitable for the study of strict anaerobes. Hungate also assumed that the conventional reliance on reducing agents to create an anaerobic environment add an excess of oxidized compounds to media in the form of the oxidized state of the reducing agent, thus creating too high a redox potential for the cultivation of strict anaerobes. Hungate addressed these problems by developing prereduced media techniques. The main principle of these techniques is that anaerobiosis must begin with media preparation and continue through sterilization to inoculation and incubation. Anaerobiosis in this cultivation system includes not only the exclusion of oxygen but also the maintenance of low redox potential. The original methods developed by Hungate to meet these requirements have been modified and simplified, but their principles are still the basis of anaerobic study today.

Techniques of Prereduced Media

Oxygen Exclusion   Oxygen has a very low solubility in water, and this decreases with increasing temperature. Therefore it is only necessary to boil water for a few minutes to drive out the oxygen, but it is important to keep an oxygen-free atmosphere during cooling to prevent its reentry. To accomplish this, water is brought to a boil under a stream of oxygen-free gas after dry media ingredients have been added. The boiling drives off the oxygen and the gas prevents the readmission of oxygen during cooling. The media are then sealed in culture vessels before being sterilized. This regimen establishes anaerobiosis before autoclaving and subsequent inoculation. This type of media preparation technique is referred to as PRAS: prereduced, anaerobically sterilized.

Reducing Agents and Indicators   Equally important as oxygen exclusion is the establishment and maintenance of a suitable redox potential. Although oxygen is the most abundant high redox compound found in most environments, creating the appropriate redox potential in media often requires more than oxygen exclusion. Other compounds and ions with high redox potential may also prevent the growth of strict anaerobes (Morris, 1975). Driving oxygen out of media may not result in a sufficiently low redox potential for the initiation of growth for many anaerobes. The presence of oxygen at saturation (1.48×10^{19} molecules of oxygen per liter of water) corresponds to a redox potential of about +800 mV. How low the redox levels has to be to permit the initiation of growth by anaerobes depends on the organisms of interest. To properly handle environmental samples, conditions compatible with the initiation of growth of methanogens from small inocula is an approach usually taken by investigators. Methanogens have been found to require a redox potential of approximately −300 mV in order to be able to initiate growth. As calculated by

Hungate (1969), this would correspond to 1.48×10^{-56} molecules of oxygen per liter of water. Of course, oxygen or any other compound does not exist in negative concentrations, but this number illustrates that provision of appropriate redox potential extends beyond oxygen exclusion. Therefore, in conjunction with oxygen exclusion the addition of a chemical reducing agent may be necessary in order to bring the redox potential of the media into this acceptable range.

The most commonly used reducing agents in conventional microbiological practice are sodium thioglycolate and cysteine–HCl. These are thiol compounds (mercaptans) with sulfhydryl groups that act as the reductant. These compounds are not strong enough reducing agents to convert oxygen into water. They instead produce hydrogen peroxide, which can be toxic to microbes. Sodium sulfide, at a final concentration of 0.02–0.05%, is a commonly used reducing agent for PRAS media. It lowers the redox potential of the media to about $-571\,mV$. Higher concentrations of sulfide may be toxic and may induce trace element deficiency by precipitation of divalent and trivalent metal ions.

In addition to reducing agents, recipes for PRAS media often include an indicator dye, such as resazurin with a final concentration of about 0.01%. The purpose of the dye is to confirm that the redox level is being maintained in the media below -40 to $-50\,mV$. This redox level is far too high to cultivate sensitive anaerobes but does provide visual assurance that no oxygen has leaked into the media.

General Method for Media Preparation

Media, trace element, and vitamin solution recipes for anaerobes have been published by many researchers (Balch et al., 1979; Ljungdahl and Wiegel, 1986; Hunter-Cervera et al., 1986). Typically, media recipes for biodegradation work include: water, bicarbonate buffer, macro mineral elements, trace elements, vitamin mix, resazurin, reducing agent, and the organic compound to be degraded. Yeast extract, trypticase, or clarified rumen fluid are sometimes included in media to fulfill general nutrient requirements that may be undefined for particular cultures.

Media are generally boiled for approximately 15 min using a hot plate with a magnetic stir-bar. The media are then further boiled and degassed for an additional 15 min under an oxygen-free stream of gas such as nitrogen or a mixture of 20% CO_2 and 80% nitrogen. This degassing step is accomplished by means of a simple gassing manifold, construction of which is outlined by Balch and Wolf (1976). The manifold allows degassing, evacuation of culture of vessels, and purging of culture vessels by means of tubing fitted with either needles or a Hungate-style gassing cannula. After the media are degassed, they are dispensed into gas-purged culture vessels using a gas-purged pipette or peristaltic pump with gas-purged tubing. Alternately, the flask of media can be stoppered and transferred into an anaerobic glovebox. The media can then be dispensed into culture vessels (tubes or serum bottles) within the glovebox under the nitrogen/hydrogen headspace in the chamber. In either case, the culture vessels are sealed with stoppers. Typically, test tubes and serum bottles are sealed with butyl rubber stoppers and crimped with an aluminum seal to keep the stopper in place. Certain organic compounds can be absorbed by butyl rubber stoppers, falsely creating the appearance of biodegradation. In these cases Teflon-lined stoppers or disks must be used to seal the culture vessels. After the vessels have been properly sealed, they are autoclaved, resulting in PRAS media.

All transfers into and out of PRAS media are made using gas-purged syringes, as demonstrated by the modified Hungate techniques of Miller and Wolin (1974). All manipulations of the cultures must be made using aseptic technique, therefore it is necessary to purged sterile needles and syringes with sterile gas.

To summarize the basic principles of anaerobic methodology as applied to environmental studies: It is necessary to exclude oxygen from media, establish a sufficiently low redox potential, and maintain anaerobiosis throughout media preparation, sterilization, and subsequent inoculation and handling of cultures. However complex these methods may sound, they are essential for the study and exploitation of anaerobes.

Acknowledgments

The preparation of this chapter was supported, in part, by grants from the Hazardous Substance Management Research Center, New Jersey Institute of Technology, and by the Department of Biological Sciences, Rutgers University.

REFERENCES

Aeckersberg, F., F. Bak, and F. Widdel (1991). Anaerobic Oxidation of Saturated Hydrocarbons to CO_2 by a New Type of Sulfate-reducing Bacterium, *Arch. Microbiol.*, **156**:5–14.

Ahmann, D., A. L. Roberts, L. R. Krumholz, and F. M. Morel (1994). Microbe Grows by Reducing Arsenic, *Nature*, **371**:750.

Albrechtsen, H. J. and T. H. Christensen (1994). Evidence for Microbial Iron Reduction in a Landfill Leachate-polluted Aquifer (Vejen, Denmark), *Appl. Environ. Microbiol.*, **60**:3920–3925.

Albrechtsen, H. J., G. Heron, and T. H. Christensen (1995). Limiting Factors for Microbial Fe(III)-Reduction in a Landfill Leachate Polluted Aquifer (Vejen, Denmark), *FEMS Microbiol. Ecol.*, **16**:233–248.

Angelidaki, I and B. K. Ahring (1995). Establishment and Characterization of an Anaerobic Thermophilic (55°C) Enrichment Culture Degrading Long-chain Fatty Acids, *Appl. Environ. Microbiol.*, **61**:2442–2445.

Auburger, G. and J. Winter (1996). Activation and Degradation of Benzoate, 3-Phenyl-propionate, and Crotonate by *Syntrophus buswellii* Strain GA. Evidence for Electron-Transport Phosphorylation during Crotonate Respiration, *Appl. Microbiol. Biotechnol.*, **44**:807–815.

Balch, W. E. and R. S. Wolfe (1976). New Approach to the Cultivation of Methanogenic Bacteria: 2-Mercaptoethanesufonic Acid (HS-CoM)-Dependent Growth of *Methanobacterium ruminatium* in a Pressurized Atmosphere, *Appl. Environ. Microbiol.*, **32**:781–789.

Balch, W. E., G. E. Fox,. L. J. Margrum, C. R. Woese, and R. S. Wolfe (1979). Methanogens: Reevaluation of a Unique Biological Group, *Microbiol. Rev.*, **43**:260–296.

Barker, H. A. (1981). Amino Acid Degradation by Anaerobic Bacteria, *Ann. Rev. Biochem.*, **50**:23–40.

Bélaich, J. P. and M. Bruschi, Eds. (1990). *Microbiology and Biochemistry of Strict Anaerobes Involved in Interspecies Hydrogen Transfer*, Plenum Press, New York.

Belkin, S., C. O. Wirsen, and H. W. Jannasch (1985). Biological and Abiological Sulfur Reduction at High Temperatures, *Appl. Environ. Microbiol.*, **49**:1057–1061.

Bradley, P. M. and F. H. Chapelle (1996). Anaerobic Mineralization of Vinyl Chloride in Fe(III)-Reducing Aquifer Sediments, *Env. Sci. Tech.*, **30**:2084–2086.

Brown J. F., R. E. Wagner, B. L. Bedard, M. J. Brennan, J. C. Carnahan, and R. J. May (1984). Transformations in Upper Hudson Sediments, *Northeast Environ. Sci.*, **3**:167–179.

Brown, J. F., B. L. Bedard, M. J. Brennan, J. C. Carnahan, H. Feng, and R. E. Wagner (1987). Polychlorinated Biphenyl Dechlorination in Aquatic Sediments, *Science*, **236**:709–712.

Caldwell, D. R. (1995). *Microbial Physiology and Metabolism*, Wm. C. Brown, Dubuque, Iowa.

Cervantes, C. (1991). Bacterial Interactions with Chromate, *Antonie van Leeuwenhoek*, **59**:229–233.

Coates, J. D., R. T. Anderson, and D. R. Lovley (1996). Oxidation of Polycyclic Aromatic Hydrocarbons under Sulfate Reducing Conditions, *Appl. Environ. Microbiol.*, **62**:1099–1101.

Coleman, M. L., D. B. Hedrick, D. R. Lovley, D. C. White, and K. Pye (1993). Reduction of Fe(III) in Sediments by Sulphate-Reducing Bacteria, *Nature*, **361**:436–438.

Colleran, E., F. Finnegan, and P. Lens (1995). Anaerobic Treatment of Sulphate-Containing Wastes Streams, *Antonie van Leeuwenhoek*, **67**:29–46.

Dalsgaard, T. and F. Bak (1994). Nitrate Reduction in a Sulfate-Reducing Bacterium *Desulfovibrio desulfuricans*, Isolated from Rice Paddy Soil: Sulfide Inhibition, Kinetics, and Regulation, *Appl. Environ. Microbiol.*, **60**:291–297.

DeFilippi, L. J. (1994). Bioremediation of Hexavalent Chromium in Water, Soil and Slag Using Sulfate-Reducing Bacteria, in *Remediation of Hazardous Waste Contaminated Soils*, D,. L. Wise and D. J. Trantolo, Eds., Marcel Dekker, New York, pp. 437–457.

Dolfing, J. (1988). Acetogenesis, in *Biology of Anaerobic Microorganisms*, A. J. B. Zehnder, Ed., J. Wiley, New York, pp. 417–468.

Dolfing, J. and B. K. Harrison (1993). Redox and Reduction Potentials as Parameters to Predict the Degradation Pathway of Chlorinated Benzenes in Anaerobic Environments, *FEMS Microbiol. Ecol.*, **13**:23–30.

Dolfing, J. and D. B. Janssen (1994). Estimates of Gibbs Free Energies of Formation of Chlorinated Aliphatic Compounds, *Biodegradation*, **5**:21–28.

Dolfing, J., J. Zeyer, P. Binder-Eicher, and R. P. Schwarzenbach (1990). Isolation and Characterization of a Bacterium That Mineralizes Toluene in the Absence of Molecular Oxygen, *Arch. Microbiol.*, **154**:336–341.

Dong, X., G. Cheng, and A. J. M. Stams (1994). Butyrate Oxidation by *Syntrophospora bryantii* in Co-culture with Different Methanogens and in Pure Culture with Pentenoate as Electron Acceptor, *Appl. Microbiol. Biotechnol.*, **42**:647–652.

Drake, H. L. (1994). *Acetogenesis*, Chapman and Hall, New York.

Dubourguier, H. C., E. Samain, G. Prensier, and G. Albagnac (1986). Characterization of Two Strains of *Pelobacter carbinolicus Isolated from Anaerobic Digesters, Arch. Micrbiol.*, **145**:248–253.

Edwards, E. A. and D. Grbic-Galic (1992). Complete Mineralization of Benzene by Aquifer Microorganisms under Strictly Anaerobic Conditions, *Appl. Environ. Microbiol.*, **58**:2663–2666.

Ensley, B. D. (1991). Biochemical Diversity of Trichloroethylene Metabolism, *Ann. Rev. Microbiol.*, **45**:283–299.

Evans, P. J., T. M. Dzung, K. S. Kim, and L. Y. Young (1991). Anaerobic Degradation of Toluene by a Denitrifying Bacterium, *Appl. Environ. Microbiol.*, **57**:1139–1145.

Eyssen, H. J., G. G. Parmentier, F. C. Compernolle, G. De Pauw, and M. Piessens-Denef (1973). Biohydrogenation of Sterols by *Eubacterium* ATCC 21, 408—Nova Species, *Eur. J. Biochem.*, **36**:411–421.

Ferguson, S. J. (1994). Denitrification and Its Control, *Antonie van Leeuwenhoek*, **66**:89–110.

Fetzner, S. and F. Lingens (1994). Bacterial Dehalogenases: Biochemistry, Genetics, and Biotechnological Applications, *Microbiol. Rev.*, **58**:641–685.

Frongs, J., E. Schramm, and B. Schink (1992). Enzymes Involved in Anaerobic Polyethylene, Glycol Degradation by *Pelobacter Venetianus* and *Bacteroides* Strain PG1, *Appl. Environ. Microbiol.*, **58**:2164–2167.

Ghiorse, W. C. (1988). Microbial Reduction of Manganese and Iron, in *Biology of Anaerobic Microorganisms*, A. J. B. Zehnder, Ed., J Wiley, New York, pp. 305–332.

Gorny, N., G. Wahl, A. Brune, and B. Schink (1992). A Strictly Anaerobic Nitrate-Reducing Bacterium Growing with Resorcinol and Other Aromatic Compounds, *Arch. Microbiol.*, **158**:48–53.

Gottschalk, G. (1985). *Bacterial Metabolism*, 2nd ed., Springer-Verlag, New York,

Häggblom, M. M. (1992). Microbial Breakdown of Halogenated Aromatic Pesticides and Related Compounds, *FEMS Microbiol. Rev.*, **103**:29–72.

Hansen, T. A. (1994). Metabolism of Sulfate-Reducing Prokaryotes, *Antonie Van Leeuwenhoek*, **66**:165–185.

Holliger, C. (1995). The Anaerobic Microbiology and Biotreatment of Chlorinated Ethenes, *Cur. Opin. Biotech.*, **6**:347–351.

Holliger, C. and G. Schraa (1994). Physiological Meaning and Potential for Application of Reductive Dechlorination by Anaerobic Bacteria, *FEMS Microbiol. Rev.*, **15**:297–305.

Holliger, C. and W. Schumacher (1994). Reductive Dehalogenation as a Respiratory Process, *Antonie Van Leeuwenhoek*, **66**:239–246.

Holliger, C., G. Schraa, A. J. M. Stams, and A. J. B. Zehnder (1993). A Highly Purified Enrichment Culture Couples the Reductive Dechlorination of Tetrachlorethylene to Growth, *Appl. Environ. Microbiol.*, **59**:2991–2997.

Hungate R. E. (1950). The Anaerobic Mesophilic Celluloytic Bacteria, *Bacteriol. Rev.*, **14**:1–49.

Hungate, R. E. (1966). *The Rumen and Its Microbes*, Academic Press, New York.

Hungate, R. E. (1969). A Role Tube Method for Cultivation of Strict Anaerobes, in *Methods in Microbiology*, Vol. 3b, J. R. Norris and D. Ribbons, Eds., Academic Press, New York, pp. 117–132.

Hunter-Cervera, J. C., M. E. Fonda, and A. Belt (1986). Isolation of Cultures, in *Manual of Industrial Microbiology and Biotechnology*, A. L. Demain and N. A. Solomon, Eds., American Society for Microbiology, Washington, D C, pp. 1–23.

Jannasch, H. W., C. O. Wirsen, S. J. Molyneaux, and T. A. Langworthy (1988). Extremely Thermophilic Fermentative Archaebacteria of the Genus *Desulfurococcus* from Deep-Sea Hydrothermal Vents, *Appl. Environ. Microbiol.*, **54**:1203–1209.

Jones, J. G., S. Gardener, and B. M. Simon (1983). Bacterial Reduction of Ferric Iron in a Stratified Eutrophic Lake, *J. Gen. Microbiol.*, **129**:131–139.

Jones, J. G., S. Gardener, and B. M. Simon (1984). Reduction of Ferric Iron by Heterotrophic Bacteria in Lake Sediments, *J. Gen. Microbiol.*, **130**:45–51.

Kafkewitz, D., P. M. Armenante, G. Hinshalwood, and G. San Agustin (1994). Immobilization of Heavy Metals in Incinerator Ash by the Activity of *Desulfovibrio desulfuricans, Hazard. Waste Hazard. Mat.*, **11**:519–527.

Kafkewitz, D. and M. T. Togna (submitted). The methods of Anaerobic Microbiology.

Kelly, R. M. and M. W. W. Adams (1994). Metabolism in Hyperthermophilic Microorganisms, *Antonie van Leeuwenhoek*, **66**:247–270.

Kemp, P., R. W. White, and D. J. Lander (1975). The Hydrogenation of Unsaturated Fatty Acids by Five Bacterial Isolates from the Sheep Rumen, Including a New Species, *J. Gen. Microbiol.*, **90**:100–114.

Kennes, C., W. M. Wu, L. Bhatnagar, and J. G. Zeikus (1996). Anaerobic Dechlorination and Mineralization of Pentachlorophenol and 2,4,6, Trichlorophenol by Methanogenic Pentachlorphenol-Degrading Granules, *Appl. Microbiol. Biotech.*, **44**:801–806.

Kuhn, E. P., J. Zeyer, P. Eicher, and R. P. Schwarzenbach (1988). Anaerobic Degradation of Alkylated Benzenes in Denitrifying Laboratory Aquifer Columns, *Appl. Environ. Microbiol.*, **54**:490–496.

Leisinger, T. (1996). Biodegradation of Chlorinated Aliphatic Compounds, *Curr. Opin. Biotech.*, **7**:295–300.

Leisinger, T. and S. A. Braus-Stromeyer (1995). Bacterial Growth with Chlorinated Methanes, *Env. Hlth. Perspec.*, **103**(suppl. 5):33–36.

Levett, P. N., Ed. (1991). *Anaerobic Microbiology: A Practical Approach*, Oxford University Press, New York.

Ljungdahl, L. G. and J. Wiegel (1986). Working with Anaerobic Bacteria, in *Manual of Industrial Microbiology and Biotechnology*, A. L. Demain and N. A. Solomon, Eds., American Society for Microbiology, Washington, D C, pp. 84–96.

Lonergan, D. J. and D. R. Lovley (1991). Microbial Oxidation of Natural and Anthropogenic Aromatic Compounds Coupled to Fe(III) Reduction, in *Organic Substances and Sediments in Water*, (Vol. 1), R. A. Baker, Ed., Lewis Publishers, Chelsea, MI, pp. 357–338.

Lovley, D. R. (1991). Dissimilatory Fe(III) and Mn(IV) Reduction, *Microbiol. Rev.*, **55**:259–287.

Lovley, D. R. (1993a). Anaerobic Bacteria into Heavy Metal: Dissimilatory Metal Reduction in Anoxic Environments, *Trends Ecol. Evol.*, **8**:213–217.

Lovley, D. R. (1993b). Dissimilatory Metal Reduction, *Ann. Rev. Microbiol.*, **47**:263–290.

Lovley, D. R. and D. J. Lonergan (1990). Anaerobic Oxidation of Toluene, Phenol, and *p*-cresol by the Dissimilatory Iron-Reducing Organism, GS-15, *Appl. Environ. Microbiol.*, **56**:1858–1864.

Lovley, D. R. and E. J. P. Phillips (1986). Organic Matter Mineralization with Reduction of Ferric Iron in Anaerobic Sediments, *Appl. Environ. Microbiol.*, **51**:683–689.

Lovley, D. R. and E. J. P. Phillips (1987). Rapid Assay for Microbially Reducible Ferric Iron in Aquatic Sediments, *Appl. Environ. Microbiol.*, **53**:1536–1540.

Lovley, D. R. and E. J. P. Phillips (1988). Novel Mode of Microbial Energy Metabolism: Organic Carbon Oxidation Coupled to Dissimilatory Reduction of Iron or Manganese, *Appl. Environ. Microbiol.*, **54**:1472–1480.

Lovley, D. R., E. J. P. Phillips, and D. J. Lonergan (1989). Hydrogen and Formate Oxidation Coupled to Dissimilatory Reduction of Iron and Manganese by *Alteromonas putrefaciens*, *Appl. Environ. Microbiol.*, **55**:700–706.

Lovley, D. R., E. R. Roden, E. J. P. Phillips, and J. C. Woodward (1993a). Enzymatic Iron and Uranium Reduction by Sulfate-Reducing Bacteria, *Marine Geol.*, **113**:41–53.

Lovley, D. R., S. J. Giovannoni, D. C. White, J. E. Champine, E. J. P. Phillips, Y. A. Gorby, and S. Goodwin (1993b). *Geobacter metallireducens* gen. nov. sp. nov., a Microorganism Capable of Coupling the Complete Oxidation of Organic Compounds to the Reduction of Iron and Other Metals, *Arch. Microbiol.*, **159**:336–344.

Lovley, D. R., J. D. Coates, J. C. Woodward, and E. J. P. Phillips (1994a). Benzene Oxidation Coupled to Sulfate Reduction, *Appl. Environ. Microbiol.*, **61**:953–958.

Lovley, D. R., J. C. Woodward, and F. H. Chapelle (1994b). Stimulated anoxic Biodegradation of Aromatic Hydrocarbons Using Fe(III) Ligands, *Nature*, **370**:128–131.

Lovley, D. R., E. J. P. Phillips, D. J. Lonergan, and P. K. Widman (1995). Fe(III) and $S°$ Reduction by *Pelobacter carbinolicus, Appl. Environ. Microbiol.*, **61**:2132–2138.

Lowe, S. E., M. K. Jain, and J. G. Zeikus (1993c). Biology, Ecology, and Biotechnological Applications of Anaerobic Bacteria Adapted to Environmental Stresses in Temperature, pH, Salinity, and Substrates, *Microbiol. Rev.*, **57**:453–509.

Macy, J. M., S. Rech, G. Auling, M. Dorsch, E. Stackebrandt, and L. I. Sly (1993). *Thauera selenatis gen. nov., sp. nov.*, a Member of the Beta Subclass of Proteobacteria with a Novel Type of Anaerobic Respiration, *Int. J. Sys. Bacteriol.*, **43**:135–142.

McFarland, M. J. and R. C. Sims (1991). Thermodynamic Framework for Evaluating PAH Degradation in the Subsurface, *Groundwater*, **29**:885–896.

Mclnerney, M. J. (1986). Transient and Persistent Associations Among Prokaryotes, in *Bacteria in Nature*, Vol. 2. *Methods and Special Applications in Bacterial Ecology*, J. S. Pondexter and E. R. Leadbetter, Eds., Plenum Press, New York, pp. 293–338.

Mihelcic, J. R. and R. G. Luthy (1988a). Degradation of Polycyclic Aromatic Hydrocarbon Compounds under Various Redox Conditions in Soil-Water Systems, *Appl. Environ. Microbiol.*, **54**:1182–1187.

Mihelcic, J. R. and R. G. Luthy (1988b). Microbial Degradation of Acenaphthene and Napthalene under Denitrification Conditions in Soil-Water Systems, *Appl. Environ. Microbiol.*, **54**:1188–1198.

Miller, T. L. and M. J. Wolin (1974). A Serum Bottle Modification of the Hungate, Technique for Cultivating Obligate Anaerobes, *Appl. Microbiol.*, **27**:985–987.

Mohn, W. M. and J. M. Tiedje (1990). Strain DCB-1 Conserves Energy for Growth from Reductive Dehalogenation Coupled to Formate Oxidation, *Arch. Microbiol.*, **153**:267–271.

Mohn, W. M. and J. M. Tiedje (1992). Microbial Reductive Dehalogenation, *Microbiol. Rev.*, **56**(3):482–507.

Mormile, M. R., K. R. Gurijala, J. A. Robinson, M. J. McInerney, and J. M. Suflita (1996). The Importance of Hydrogen in Landfill Fermentations, *Appl. Environ. Microbiol.*, **62**:1583–1588.

Morris, J. G. (1975). The Physiology of Obligate Anaerobiosis, in *Advances in Microbial Physiology*, Vol. 12, A. H. Rose and D. W. Tempest, Eds., Academic Press, New York, pp. 169–233.

Mulder, A., A. A. Van de Graaf, L. A. Robertson, and J. G. Kuenen (1995). Anaerobic Ammonium Oxidation Discovered in a Denitrifying Fluidized Bed Reactor, *FEMS Microbiol. Ecol.*, **16**:177–184.

Odom, J. M. and R. Singleton (1993). The Sulfate-Reducing Bacteria: Contemporary Perspectives. Springer-Verlag, New York.

Oremland, R. S. (1988). Biogeochemistry of Methanogenic Bacteria, in *Biology of Anaerobic Microorganisms*, A. J. B. Zehnder, Ed., Wiley, New York, pp. 641–662.

Ottow, J. C. G. (1971). Iron Reduction and Gley Formation by Nitrogen Fixing Clostridia, *Oecologia (Berl.)*, **6**:164–175.

Ottow, J. C. G. and H. Glathe (1971). Isolation and Identification of Iron-Reducing Bacteria from Gley Soils, *Soil. Biol. Biochem.*, **3**:43–55.

Pfennig, N. (1989). Metabolic Diversity Among the Dissimilatory Sulfate-Reducing Bacteria, *Antonie van Leeuwenhoek*, **56**:127–138.

Pfennig, N. and F. Widdel (1982). The Bacteria of the Sulphur Cycle, *Phil. Trans. R. Soc. Lond. B.*, **298**:433–441.

Postgate, J. R. (1992). *Microbes and Man*, 3rd ed., Cambridge University Press, Cambridge.

Postgate, J. R. (1984). *The Sulphate-Reducing Bacteria*, 2nd ed., Cambridge University Press, Cambridge.

Rabus, R., R. Nordhaus, W. Ludwig, and F. Widdel (1993). Complete Oxidation of Toluene under Strictly Anoxic Conditions by a New Sulfate-Reducing Bacterium, *Appl. Environ. Microbiol.*, **59**:1444–1451.

Reeve, J. N. (1992). Molecular Biology of Methanogens, *Ann. Rev. Microbiol.*, **46**:165–191.

Roden, E. R. and D. R. Lovley (1993). Dissimilatory Fe(III) Reduction by the Marine Microorganism *Desulfuromonas acetoxidans, Appl. Environ. Microbiol.*, **59**:734–742.

Rueter, P., R. Rabus, H. Wilkes, F. Aeckersberg, F. A. Rainey, H. W. Jannasch and F. Widdel (1994). Anaerobic Oxidation of Hydrocarbons in Crude Oil by New Types of Sulphate-Reducing Bacteria, *Nature*, **372**:455–458.

Sørensen, J. (1982). Reduction of Ferric Iron in Anaerobic, Marine Sediment and Interaction with Reduction of Nitrate and Sulfate, *Appl. Environ. Microbiol.*, **43**:319–324.

Samain, E., G. Albagnac, and H. C. Dubourguier (1986). Initial Steps of the Catabolism of Trihydroxybenzenes in *Pelobacter acidigallici, Arch. Micrbiol.*, **144**:242–244.

Schauder, R. and A. Kröger (1993). Bacterial Sulphur Respiration, *Arch. Microbiol.*, **159**:491–497.

Schink, B. (1988). Principles and Limits of Anaerobic Degradation: Environmental and Technological Aspects, in *Biology of Anaerobic Microorganisms*, A. J. B. Zehnder, Ed. Wiley, New York, pp. 771–793.

Schink, B. and M. Stieb (1983). Fermentative Degradation of Polyethylene Glycol by a Strictly Anaerobic, Gram-Negative, Nonsporeforming Bacterium *Pelobacter venetianus* sp. Nov., *Appl. Envron. Microbiol.*, **45**:1905–1913.

Schink, B., D. R. Kremer, and T. A. Hansen (1987). Pathway of Propionate Formation from Ethanol in *Pelobacter propionicus, Arch. Micrbiol.*, **147**:321–327.

Schnürer, A., B. Schink, and B. O. Svensson (1996). *Clostridium ultuinense* sp. Nov.. A Mesophilic Bacterium Oxidizing Acetate in Syntrophic Association with a Hydrgenotrophic Methanogenic Bacterium, *Int. J. Sys. Bacteriol.*, **46**:1145–1152.

Schocher, R. J., B. Seyfried, F. Vazquez, and J. Zeyer (1991). Anaerobic Degradation of Toluene by Pure Cultures of Denitrifying Bacteria, *Arch. Microbiol.*, **157**:7–12.

Schumacher, W. and C. Holliger (1996). The Proton/Electron Ratio of Menaquinone-Dependent Electron Transport from Dihydrogen to Tetrachlorethylene, *J. Bacteriol.*, **178**:2328–2333.

Sokatch, J. R. (1969). *Bacterial Physiology and Metabolism*, Academic Press, New York.

Stams, A. J. M. (1994). Metabolic Interactions between Anaerobic Bacteria in Methanogenic Environments, *Antonie van Leeuwenhoek*, **66**:271–294.

Straub, K. L., M. Benz, B. Schink, and F. Widdel (1996). Anaerobic, Nitrate Dependent Microbial Oxidation of Ferrous Iron, *Appl. Environ. Micrbiol.*, **62**:1458–1460.

Strom, A. R., J. A. Olafsen, and H. Larsen (1979). Trimethylamine Oxide: A Terminal Electron Acceptor in Anaerobic Respiration of Bacteria, *J. Gen. Microbiol.*, **112**:315–320.

Thauer, R. K., K. Jungermann, and K. Decker (1977). Energy Conservation in Chemotrophic Anaerobic Bacteria, *Microbiol. Rev.*, **41**:100–180.

Tiedje, J. M. (1988). Ecology of Denitrification and Dissimilatory Nitrate Reduction to Ammonium, in *Biology of Anaerobic Microorganisms*, A. J. B. Zehnder, Ed., Wiley, New York, pp. 179–244.

Tiedje, J. M., J. F. Quensen, J. Chee-Sanford, J. P. Schimel, and S. A. Boyd (1993). Microbial Reductive Dechlorination of PCBs, *Biodegradation*, **4**:231–240.

Utkin, I., C. Woese, and J. Wiegel (1994). Isolation and Characterization of *Desulfitobacterium dehalogens* gen. *nov.*, spec. *nov.*, an Anaerobic Bacterium Which Reductively Dehalogenates Chlorophenolic Compounds, *Int. J. Syst. Bacteriol.*, **44**:612–619.

Van de Graaf, A. A., A. Mulder, P. deBruijn, M. S. M. Jetten, L. A. Robertson, and J. G. Kuenen (1995). Anaerobic Oxidation of Ammonium is a Biologically Medated Process, *Appl. Environ. Microbiol.*, **61**:1246–1251.

Van de Graaf, A. A., P. deBruijn, L. A. Robertson, M. S. M. Jetten, and J. G. Kuenen (1996). Autotrophic Growth of Anaerobic Ammonium-Oxidizing Microorganisms in Fluidized Bed Reactor, *Microbiology*, **142**:2187–2196.

Vogel, T. and D. Grbic-Galic (1986). Incorporation of Oxygen from Water into Toluene and Benzene During Anaerobic Fermentative Transformation, *Appl. Environ. Microbiol.*, **52**:200–202.

Vogels, G. D., J. T. Keltjens, and C. van der Drift (1988). Biochemistry of Methane Production, in *Biology of Anaerobic Microorganisms*, A. J. B. Zehnder, Ed., Wiley, New York, pp. 707–778.

Wagener, S. and B. Schink (1988). Fermentative Degradation of Nonionic Surfactants and Polyethylene Glycol by Enrichment Cultures and by Pure Cultures of Homoacetogenic and Propionate-Forming Bacteria, *Appl. Environ. Microbiol.*, **54**:561–565.

White, D. (1995). *The Physiology and Biochemistry of Procaryotes*, Oxford University Press, New York.

White, C. and G. M. Gadd (1996). Mixed Sulphate-Reducing Bacterial Cultures for Bioprecipitation of Toxic Metals: Factorial and Response-Surface Analysis of the Effects of dilution Rate, Sulphate and Substrate Concentration, *Microbiology*, **142**:2197–2205.

Widdel, F. (1988). Microbiology and Ecology of Sulfateand Sulfur Reducing bacteria, in *Biology of Anaerobic Microorganisms*, A. J. B. Zehnder, Ed., Wiley, New York, pp. 469–586.

Wolin, M. J. (1976). Interactions between H_2-Producing and Methane Producing Species, in *Microbial Formation and Utilization of Gases (H_2, CH_4, CO_2)*, H. G., Schlegel, G. Gottschalk, and N. Pfennig, Eds., Goltze, KG, Göttingen, pp. 141–149.

Wood, P. M. (1981). The Redox Potential for Dimethyl Sulphoxide Reduction to Dimethyl Sulphide, *FEMS Lett.*, **124**:11–14.

Woolfolk, C. A. and H. R. Whitely (1962). Reduction of Inorganic Compounds with Molecular Hydrogen by *Micrococcus lactylticus, J. Bacteriol*, **84**:647–658.

Zehnder, A. J. B. and B. H. Svensson (1986). Life Without Oxygen: What Can and What Cannot? *Experientia*, **42**:1197–1205.

Zeyer, J., E. P. Kuhn, and R. P. Schwarzenbach (1986). Rapid Microbial Mineralization of Toluene and 1,3-Dimethylbenzene in the Absence of Molecular Oxygen, *Appl. Environ. Microbiol.*, **52**:944–947.

Zitomer, D. H. and R. E. Speece (1993). Sequential Environments for Enhanced Biotransformations of Aqueous Contaminants *Environ. Sci. Technol.*, **27**:227–244.

CHAPTER 13

Composting

John A. Hogan

Department of Environmental Sciences, Cook College, Rutgers University, New Brunswick, New Jersey 08903

Physiochemical methods typically utilized for the treatment of soils contaminated with hazardous wastes include removal and burial in a secured landfill, vitrification, thermal desorption, incineration, chemical extraction, chemical fixation, stabilization and solidification, and asphalt incorporation (Lindsey and Jakobson, 1989; Preslo et al., 1989). Except for incineration, these methods usually do not result in the destruction of the wastes but rather incorporate the hazardous compounds into a stable matrix or move them from one medium or site to another. In contrast biological treatment can mineralize many hazardous compounds or transform the parent compound into acceptable metabolites (Skladany and Metting, 1993; see Chapter 2 for earlier references). Composting, a self-heating, substrate-dense, managed microbial ecosystem, is one solid-phase biological treatment technology suitable to the treatment of large amounts of contaminated solid materials (such as soil).

Traditional composting practice involves the informal processing of "clean" organic wastes such as leaves, animal manures, and agricultural wastes to produce an end product (compost) usable as a soil amendment. More problematic wastes, including domestic and industrial sewage sludges and municipal solid waste (MSW), call for facilities to conduct composting in a highly controlled manner in order to maximize the rate of organic matter decomposition. Benefits of a high rate of decomposition include decreases in required facility size, amount of on-site material being processed, and potential for malodor production (Finstein et al., 1986). The indications are, however, that although many facilities possess competent material-handling systems, biological process control aspects are often inadequate (Finstein et al., 1987a–d).

Recently, the possibility of treating hazardous wastes through composting has undergone limited investigation. As compared to nonhazardous operation, the

Biological Treatment of Hazardous Wastes, Edited by Gordon A. Lewandowski and Louis J. DeFilippi
ISBN 0-471-04861-5   ©1998 John Wiley & Sons, Inc.

composting of hazardous wastes is subject to additional operational demands including the need to reliably attain a clear treatment end point, as set through strict contaminant-level regulations. Moreover, many hazardous compounds are resistant to microbial degradation due to chemical structure attributes, associated microbial toxicity, and compound concentrations that are too low to induce enzyme production or support growth. Additionally, microbial limitations with regard to moisture, pH, inorganic nutrients, and particle size might be more severe. Also, because the composting of hazardous wastes typically involves the bioremediation of contaminated, substrate-sparse soils, support of microbial self-heating will generally necessitate the incorporation of the proper type and amounts of supplemental, readily available organic matter. Finally, process emissions could pose a significant health risk, as opposed to being merely a malodorous nuisance. For these reasons, the successful composting of hazardous wastes requires strict focus on both biological and physical process control.

Hazardous compounds reported to disappear through composting include aliphatic and aromatic hydrocarbons (Hogan et al., 1989a; Hunter et al., 1981; Martens, 1982; Muller and Korte, 1975; Taddeo et al., 1989), pesticides (Lemmon and Pylypiw, 1992; Petruska et al., 1985; Rose and Mercer, 1968), explosives (Williams et al., 1992; Williams and Myler, 1990), and certain halogenated compounds (Hogan et al., 1989a; Hunter et al., 1981; also refer to Chapter 10 for discussions on PCP degradation through composting). It must be noted that a decrease in the measured concentration of the parent compound may not correspond to its transformation to inorganic constituents (mineralization). In some cases transformation may be incomplete, leading to an accumulation of intermediate products. Although the intermediates are often of a benign nature, they also can be more toxic than the parent compound (Sutherland, 1992). Other possible routes contributing to disappearance include volatilization, assimilation (incorporation into cellular biomass), adsorption, polymerization, and leaching (Fogarty and Touvinen, 1991). Ultimately, a complete accounting of the fate of the target hazardous compounds is necessary to determine whether composting is a suitable treatment technology.

COMPOSTING COMPARED TO OTHER SOLID-WASTE TREATMENT PROCESSES

To best define what composting is, as well as what it is not, requires comparison to other biological, solid-phase treatment systems. Systems referred to as *land treatment* or *landfarming* involve the treatment of contaminated soil in a surface soil layer. Treatment mostly occurs in the upper 15–30 cm of soil, though additional immobilization and degradation might occur in the underlying soils up to a depth of 150 cm (Lynch and Genes, 1989). Contamination of the soil may have been unintentional (i.e., spills) or through deliberate waste application for the purposes of treatment. Biological management practices often consist of initial and periodic adjustment of water and inorganic nutrients (N, P, K), and tilling to mix and supply oxygen. Substrate density is dependent on waste loading rates, but it is typically low. Common applications of land treatment include the processing of municipal sewage sludges and petroleum wastes, including both solids and aqueous emulsions (Bartha and Bossert, 1984). Since land treatment systems are typically conducted in open fields, wastes are able to migrate through the mechanisms of leaching, runoff, and volatilization. Moreover, slow biodegradable or nonbiodegradable materials can accumulate to unacceptable levels with repeated application. Because of these hazards, open land treatment of certain wastes is now either prohibited or strictly regulated (USEPA, 1992).

From a regulatory standpoint, operating land treatmentlike systems in a closed configuration is preferred. This requires that the treatment area be constructed with liners and covered so as to permit the collection of process leachate and gaseous emissions for treatment. This method has been referred to as solid-phase bioremediation (Skladany and Metting, 1993). Although open (and to a lesser extent closed) land treatment can be an inexpensive treatment method, the temperature of the soil is subject to ambient fluctuations, and winter operation is impossible for some locations. Additionally, land treatment systems can become oxygen limited, depending on factors such as waste loading rates, depth of waste application, and frequency of tilling (Devinny and Islander, 1989).

Oxygen limitation can be averted by actively forcing air through the soil via the controlled use of blowers (Finstein et al., 1983). Typical configuration includes the formation of large piles with an underlying ductwork for aeration (Fig. 13.1). This approach has been referred to as *engineered soil piles* and *forced aeration treatment* (Skladany and Metting, 1993). Again, and for the same reasons, closed configuration is preferred to open. Agitation, if employed, is for mixing purposes, not controlled aeration. In this approach, the substrate density plays a major role in determining the nature of the microbial dynamics. A theoretical construct for this low substrate density system is developed below and will serve as a reference to elucidate the nature of composting (high substrate density).

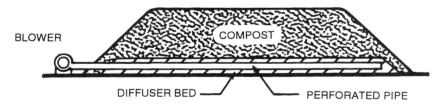

Figure 13.1 Common physical configuration of an open composting/forced-aeration treatment system (cutaway side view). Ventilative air makes a single pass through the material. Airflow direction is dependent on whether the blower is operated in a positive or negative mode. Transformation into a closed configuration requires the use of an appropriate cover or container (see Fig. 13.4 for an example of closed configuration).

Imagine a large pile of soil contaminated with low levels of hazardous waste. The material is at ambient temperature, thoroughly colonized with indigenous microbial populations, and contains sufficient water and inorganic nutrients for microbial metabolism. To promote aerobic conditions, high interstitial oxygen concentrations are maintained through the use of timer-actuated blowers (Fig. 13.2a). Aside from limited amounts of stabilized soil organic matter, the

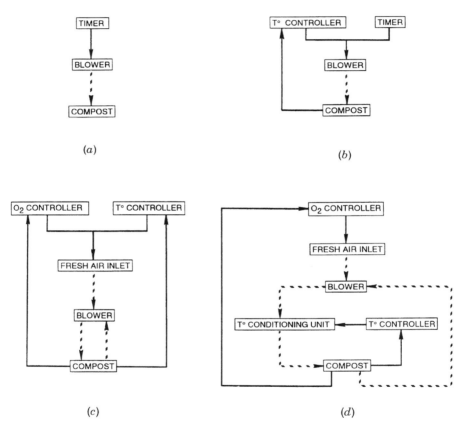

(a)

(b)

(c)

(d)

Figure 13.2 Four possible composting ventilation process control strategies are depicted diagramatically as follows: (a) Timed introduction of ventilation for the purpose of oxygenation (no feedback loop). (b) Automatic temperature regulation (feedback loop). Timer provides oxygenation during the ascent to and descent from temperature setpoint control period. (c) Automatic control of both oxygen and temperature using a multiple-pass ventilation system. This strategy requires containment of composting material to allow recirculation of ventilative gas and proper oxygen sampling. (d) System is identical to (c) with the exception of an added automatic temperature conditioning unit (usually cooling) that regulates recirculating gas temperature. This allows separation of the oxygenation and cooling functions of recirculating ventilative gas. Note: Solid lines indicate control and monitoring signals. Dashed lines indicate gas flow.

hazardous waste will be the principal carbon and energy source. Since only small amounts of organic matter are present, metabolic heat generation is limited. Although the pile is sufficiently large to self-insulate, the accumulation of heat is insufficient to induce an appreciable temperature elevation. As with land treatment, the temperature of the material is essentially dictated by ambient conditions.

Conversion of the above arrangement into a composting system can be achieved by establishing a high substrate density through the addition of a readily metabolizable organic amendment. Alternatively, high substrate densities can exist due to the presence of high concentrations of a biodegradable hazardous waste. Initially, mesophilic microbes (tolerance range = 15–45°C) generate heat at the expense of the organic material, and the accumulation of waste metabolic heat is now sufficient to increase the temperature of the mass. Population structures are altered and rates of metabolism typically increase with temperature. If substrate or other factors are not limiting, mesophilic populations will reach their upper temperature threshold, and thermal debilitation will result. In contrast, it is this temperature elevation that allows thermophilic organisms (45°C+) to become established and continue to elevate the temperature. Thermophilic populations typically reach significant thermal inhibition when temperatures exceed 55–60°C (Bach et al., 1987; Finstein, et al., 1983; Kuter et al., 1985; MacGregor et al., 1981; McKinley and Vestal, 1984). This can be avoided, however, through deliberate heat removal in reference to a selectable temperature setpoint (Fig. 13.2*b*). Eventually conditions such as substrate, water, or temperature limitation allow for heat loss to exceed generation, resulting in a temperature descent.

Although an objective of nonhazardous waste composting is the rapid degradation of putrescibles, the primary focus of hazardous waste composting is to maximize the rate and extent of degradation of the hazardous target compounds. Like other biological treatment systems, alleviating microbiological limitations regarding oxygen, moisture, inorganic nutrients, and pH is necessary. Additionally, the two distinct features of composting, high substrate density and self-heating, might be exploited so as to further enhance degradation of target compounds. Optimizing composting process design and control requires the addressing of these and other ecological factors in a systematic manner.

Because composting is conducted at elevated temperatures and requires the use of large volumes of air for oxygen supply and cooling, there is a potential for the volatilization of hazardous compounds (Hogan et al., 1989a; Petruska et al., 1985). The capture and treatment of process emissions is therefore likely to be a regulatory requirement for most operations. This necessitates that the composting material be housed in a closed container to allow exhaust process gas to be conducted to a treatment unit. Additionally, a high substrate density will require controlled aeration and cooling of the composting material. The use of agitation, passive oxygen diffusion, or conductive heat removal mechanisms for these purposes is contraindicated (Finstein et al., 1986). The superior and only practicable method is the use of forced ventilation, through the use of blowers.

PREPARATION OF WASTE FOR COMPOSTING

Waste Suitability

As composting is a solid-phase biological treatment process, target compounds must be either a solid or a liquid associated with a solid matrix. They also must be able to be biologically mineralized or transformed into acceptable intermediate metabolites. An extensive array of hazardous compounds are known to be biologically transformed (Dragun, 1988), and theoretically most of these compounds and conceivably some yet to be evaluated can be treated through composting. Co-contamination with certain hazardous inorganics such as heavy metals can yield unacceptable residues despite the destruction of hazardous organics and might require a secondary inorganics removal process. Wastes that are overtly

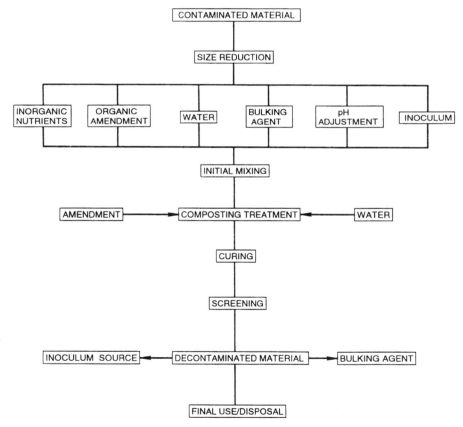

Figure 13.3 Generalized overall process scheme involved in designing a composting treatment sequence. See text for details of each step/process. Not all steps depicted may be necessary for every treatment scenario.

toxic to the microbial community, yet otherwise capable of behaving as biodegradable target compounds, might require dilution with a "clean" material, such as uncontaminated or previously treated soil, before treatment begins.

Once the waste material is judged suitable for treatment through composting, it must be prepared so as to maximize the biological treatment potential. This requires the adjustment of various physical, chemical, and biological factors, as discussed below (Fig. 13.3).

Contaminant Availability

As compared to aqueous processes, where contact is facilitated by solubilization of target compounds and enhanced microbial translocation, solid-phase processes pose significant physical and ecological barriers to through microbial colonization. Solid particles can physically shield target compounds trapped within from microbial contact and limit the penetration of oxygen, nutrient and water (particularly if hydrophobic). Furthermore, because the rates of dissolution can regulate when a compound becomes biologically accessible, water-insoluble compounds might exhibit restricted availability (Thomas et al., 1986). Additionally, some wastes might become unavailable due to their incorporation into stable organic compounds (i.e., humic acids) or adsorption onto soil particles.

Initial preparation of the composting material should include measures aimed at increasing contaminant availability. An effective physical method involves increasing the exposed surface area of the target compound by mechanically reducing particle size, thereby increasing the opportunity for both microbial contact and target compound dissolution. Useful size reduction is limited by a decrease in material porosity and a subsequent increase in resistance to ventilation. The utility of enhancing availability through chemical methods, such as the use of surfactants, is unclear (see Chapters 2 and 3 for further discussion and references concerning surfactants).

Substrate Density

In contaminated soils, the hazardous target compounds and indigenous soil organic matter are the principal microbial growth substrates. If target compounds are able to serve as a carbon and energy source and are in sufficient concentrations, and if the antagonistic interactions generated by microbes utilizing the soil organic matter are not restrictive, the growth and enzyme induction necessary for the degradation of the hazardous target compounds can occur. In this case, alleviating microbially-limiting conditions with respect to water, oxygen, inorganic nutrients, and pH might suffice for satisfactory treatment (i.e., forced aeration treatment). In contrast, target compounds may be present in very low concentrations, yet still exceed regulatory limits. These low levels might not support competent microbial populations or induce the required enzyme systems as needed to degrade target compounds. Alleviating this condition by increasing the concentration of the target compound is not a viable option. An alternative is to add analog, or nonanalog, nonhazardous carbon sources known to stimulate the induction and production of broad-specificity enzymes able to transform both the carbon amendment and target compound (Keck et al., 1989). While this practice might not alter compound-specific metabolic rates, the increased biomass can enhance overall degradation rates (Lindstrom and Brown, 1989).

Another possible condition is the presence of compounds that are susceptible to transformation by enzymatic attack but do not serve as a carbon or energy source. This indicates a need to stimulate cometabolic activity. Cometabolism involves the transformation of compounds that are not able to serve as an organism's energy or carbon source (e.g., certain hazardous wastes) at the expense of another growth substrate (e.g., organic amendment) (Slater and Lovatt, 1984). In noncomposting systems, cometabolic mechanisms have been shown to enable or enhance the degradation of certain recalcitrant compounds in waste mixtures (Beam and Perry, 1974; Keck et al., 1989; Raymond et al., 1967; Wilson and Wilson, 1985). Although specific investigation is lacking, it is probable that these mechanisms would be operative in a composting system. Stimulation of cometabolism can be achieved through the addition of carefully selected carbon sources that support the nonspecific enzyme activity responsible for the target compound transformation (Lindstrom and Brown, 1989).

The amounts and type of carbon present in the composting matrix will serve as a strong determinant of the number, types, and metabolic characteristics of microbes present. Supplementation with certain nonspecific carbon sources might bring to dominance microbial populations that are incapable of degrading target compounds. This will increase the potential for competition and catabolite repression, possibly suppressing populations capable of target compound transformation. The choice and proportioning of organic amendment is therefore crucial and, when possible, should be carefully conducted so as to support microbial populations proven effective at degrading target compounds.

Most hazardous waste composting experimentation has given little attention to carbon source selection. Rather, conveniently obtained, nonhazardous, bulk organic supplements that are perceived as wastes themselves are commonly chosen for the sole purpose of temperature elevation. Results obtained from this approach are anecdotal and might falsely restrict the application of hazardous waste composting. Further development is essential and might require the devising of specific supplementation "formulas" of organic materials that create suitable conditions for target compound degradation.

Depending on site-specific conditions, addition of small amounts of the proper types of supplemental carbon might be adequate for target compound degradation. In some circumstances, however, a high substrate density might elicit greater rates of degradative activity. In proper physical configuration, a high substrate density also permits the biological heat generation and storage conditions required to elevate matrix temperatures (composting). Elevated temperatures have the potential to enhance rates of degradation by increasing metabolic and growth rates (Atlas and Bartha, 1981), microbial population selection (Strom, 1985), and dissolution rates of the target compounds.

A possible, though not verified, degradation-enhancing mechanism would exploit the temperature elevation generated in the degradation of nonspecific supplemental carbon by organisms not competent in degrading target compounds. In effect, the carbon source is used strictly as a "fuel" to elevate composting material temperatures. If competitive and catabolite repression effects are not significant, temperatures could be maintained at levels known to enhance the growth and metabolism of microbes capable of degrading target compounds.

Considering in isolation the possible benefits exhibited by temperature elevation, the need is for a prolonged period of elevated temperatures. Supplementation with rapidly depleted materials, such as sugars or other simple carbohydrates will elicit only short periods of temperature elevation, while resistant organics such as sawdust might not produce noticeable temperature elevations. Sustained temperature elevation of contaminated soils can be maintained with moderately complex wastes, such as animal manures (Finstein and Hogan, 1993a; Williams et al., 1992).

Another potential benefit of a substrate-dense system includes the alteration of substrate characteristics during the course of processing, reflecting a dynamic progression of physical, chemical, and biological conditions with time. This can be associated with an ecological succession, thereby exposing the contaminants to a greater diversity of microbial populations that in turn potentially increases the opportunity for degradation. This may be of particular importance for the sequential degradation of various components of mixed wastes. Organic amendments can also serve to stabilize microbial populations in inhibitory environments (Lindstrom and Brown, 1989), increase porosity and water-holding capacity, and provide necessary inorganic nutrients.

In contrast to potential amendment-associated benefits, the presence of organic material can place an increased demand on the need for oxygen, water, and inorganic nutrients. Amendment also increases the amount of material that requires treatment and handling, thereby leading to larger treatment facilities, increased amounts of residue, and increased costs due to amendment acquisition and storage. Resultant temperature elevations can increase the rates of contaminant volatilization and ventilation requirements for temperature control, possibly resulting in the volatilization of otherwise biodegradable contaminants and increased exhaust treatment facility and operation costs. For these reasons, organic amendment should be utilized only when shown to be advantageous, and in the smallest amounts consistent with effectiveness.

Water Content

The availability of water plays a complex role in regulating microbial metabolism. Water functions as a medium for bacterial movement, to transport solubilized substrates and nutrients to the cell, and facilitate the removal of metabolic waste products. In principle, the greater the amount of water available, the higher the potential for microbial growth. Conversely, the diffusion of oxygen through water is about 10,000 times slower than through free air (Miller, 1989). Therefore the proper moisture content is one that provides sufficient water for microbial needs without excessive restriction of oxygen diffusion.

Depending on the moisture status of the contaminated material, bulking agent, and organic amendment, initial wetting of the composting mixture might be required. Uniform water addition is essential for proper ventilation, as air will preferentially seek areas of higher porosity (lower moisture), leading to channeling. After wetting, mixing of the entire batch or loading the composting unit uniformly with successive thin layers will help provide even moisture distribution.

Typical initial gravimetric moisture levels for the composting of nonhazardous organic materials range from about 50–70% (wet weight basis). In contrast, many hazardous waste composting scenarios might require a lower gravimetric water content to establish proper water status. This is due to the incorporation of large amounts of nonabsorbent materials, such as soil that typically exhibits free drainage at lower moisture concentrations than most organic-rich materials. It is therefore crucial to establish the moisture levels in reference to a microbially favorable level, rather than a "universal" gravimetric moisture content. Standardization of this level might be facilitated through the determination of the matrix water potential, which is a measure of the water tension within the composting material (Miller, 1989).

Although water is produced during the mineralization of organic matter, actively aerated composting matrices typically exhibit a net loss of water during the course of treatment (Finstein et al., 1983; Finstein et al., 1986). With time, water can become microbially limiting. Depending on initial moisture levels, and rates of heat evolution and ventilation, midcourse rewetting of the matrix may be necessary (see later).

Inorganic Nutrients

Although metabolic limitation can theoretically be induced by any one nutrient, nitrogen (and to a lesser extent, phosphorus) levels are the primary limiting inorganic nutrients in nonhazardous waste composting. Optimal N and P levels are generally a function of the amount of available carbon in the composting material. While mixed populations (bacteria and fungi) function most effectively at narrow C:N (10:1) and C:P (100:1) ratios (Atlas and Bartha, 1981), bacteria often possess a competitive advantage (Rosenzweig and Stotzky, 1980). In contrast, fungi are typically able to function at much wider C:N ratios, some exceeding 1600:1 (Levi and Cowling, 1969). Therefore the C:N ratio can act as a strong population selection factor.

While the role of nutrient status in hazardous waste composting has received little study, it is likely to be similar to other solid-phase microbial ecosystems. During landfarming, the addition of N and P has been reported to enhance microbial degradation of oily wastes (Dibble and Bartha, 1979). It was observed that a C:N ratio of 60:1 was optimal in the range 15:1–300:1. A C:P ratio of 800:1 was optimal in the range 200:1–4000:1. This would indicate that an excess, as well as a limitation, of nutrients may reduce biodegradation of target compounds. Deviance from the expected C:N:P ratio of 100:10:1 might be explained by the incomplete assimilation of C and turnover of nutrients within the populations (Dibble and Bartha, 1979).

There are certain circumstances when nutrient limitation may evoke alternate metabolic pathways beneficial to waste treatment objectives. For example, it has been observed that under conditions of N limitation, the fungus *Phanerochaete chrysosporium* produces extracellular lignin peroxidases capable of degrading persistent compounds such as benzo (*a*) pyrene (Bumpus, 1989) and 2,4,6-trinitro-toluene (Fernando et al., 1990). Therefore, the role of nutrient availability as applied to hazardous waste composting is a complex matter requiring further broad-scale study.

The addition of certain organic amendments might provide enough inorganic nutrients for the degradation of both the hazardous contaminants and nonhazardous organics. If not, supplemental nutrients (in a readily available form) can be dissolved and added with the water used for initial moisture adjustment.

pH

While the effect of pH on the rates of hazardous waste degradation during composting has not been evaluated, relevant experimentation has been performed. During the simulated land treatment of oily wastes, it was reported that a pH of 7.8 (experimental range = 5.0–7.8) promoted the highest rates of degradation (Dibble and Bartha, 1979). Similarly, it was found that octadecene and naphthalene mineralization rates increased when pH was increased from 5.0 to 8.0 (Hambrick et al., 1980) in an acidic estuarine sediment. In contrast, lignin degradation by a pure culture of *P. chrysosporium* was observed to be most rapid at a pH of 3.0 (experimental range = 3.0–6.5) (Boyle et al., 1992). These data suggest that optimal pH levels can be species site and waste specific.

The pH of nonhazardous wastes encountered in composting rarely exceeds the thresholds for microbial activity, but hazardous wastes may present pH-related limitations more frequently. Therefore pH adjustment of the composting material might be required. The addition of organic amendment may buffer the overall mixture; however, additional pH control (e.g., liming for a more alkaline pH) may be required. Novel treatment approaches might require adjustment to strongly acidic or basic conditions, in which case proper pH altering agents must be identified. Further examination is required to elucidate both proper levels and the mechanisms by which pH affects the degradation of target compounds.

Bioaugmentation

The initial microbial lag phase for nonhazardous waste composting materials is ordinarily brief because the composting material is thoroughly colonized with indigenous populations. While this might also be true of many hazardous-waste-contaminated materials, the nature or concentrations of some wastes can be so toxic as to be microbially inhibitive. Also, when exposure to hazardous wastes is recent (e.g., a single action, such as a spill rather than low-level chronic exposure), there might be a substantial period before suitable populations are developed. Inoculation of waste mixtures with an acclimatized consortia can immediately establish large populations of competent organisms. Initial sources of an acclimatized consortia can include soils or sediments from areas of long-term contamination, while subsequent sources would be best derived from previously composted material. Inoculum addition should minimize both the time required to establish a competent community and the amount of fresh waste it displaces. When thoroughly mixed, adding 5–10% (v/v) of inoculum material is likely more than sufficient (Hogan et al., 1989a).

The use of commercial inocula in general, and genetically engineered microorganisms (GEMs) in particular, to enhance hazardous waste degradation remains a controversial issue and appears unwarranted except in special cases. Significant impediments exist when attempting to utilize specialized inocula in competitive, mixed culture ecosystems (Lajoie et al., 1992), and evidence for their beneficial use in composting has not been established (see Chapter 2 for related discussions on this topic).

Bulking Agents

Contaminated materials (especially soils) that have a very fine particle size and/or high moisture levels can result in minimal porosity, making ventilation nearly impossible. Although organic amendments can simultaneously serve as a bulking agent under certain circumstances, it may be necessary to intentionally incorporate bulking agents of a suitable particle size and water-absorbing capacity to impart adequate pore volume and continuity. Bulking agents should be relatively inert so as not to interfere with the intended purpose of the organic amendment, added in minimal quantities (consistent with effectiveness), and recycled (if recovery is possible) to minimize associated aquisition, storage, and operational costs. Although nonhazardous composting applications commonly utilize woodchips and recycled compost as bulking agents, it is unclear whether these would be useful for hazardous waste applications. Further research is necessary to determine appropriate bulking agents.

Initial Mixing

A crucial and final step in material preparation is the thorough mixing of the composting mixture. This serves to evenly distribute the contaminated material and associated target compounds, organic amendment, water, nutrients, inoculum, and bulking agent (if required). Mixing should be performed so as to uniformly enhance porosity, while avoiding unnecessary compaction. Initial mixing can be as uncomplicated as turning batches with front-end loaders or consist of a series of conveyors that simultaneously feed materials to a mechanical mixer. Mixing might simultaneously reduce particle size, though special machinery may be required.

PROCESS DESIGN AND OPERATION

Once the material has been prepared for composting, provisions must be taken to establish and maintain near-optimal environmental conditions so as to maximize the rate of hazardous waste degradation. This entails designing composting treatment facilities to exercise strict control over a variety of physical, chemical, and biological factors, as discussed below.

Ventilation Configuration

Ventilation serves to supply oxygen, while simultaneously removing heat through the mechanisms of evaporation of water (major mechanism) and sensible heating of the air (minor mechanism) (MacGregor et al., 1981). Ventilation is typically executed through the use of blowers in either a forced-pressure or vacuum-induced mode. Theoretical (Goyal, 1983) and experimental (Miller et al., 1982) evidence demonstrate that the forced-pressure mode provides greater energy efficiency and superior oxygen and temperature control. Forced-pressure ventilation typically uses blowers to pressurize an underlying plenum, forcing air through a "false" (perforated) floor into the overlying composting material (Fig. 13.4). Reversing the direction of airflow (top to bottom) might be useful when utilizing containers in specific applications (see later). Blower selection is determined by the total amount of air to be delivered for cooling purposes and by the backpressure caused by the false floor and the composting material (a function of pile height and porosity). Blowers must be able to supply enough air, at operational pressures, to comfortably exceed the volumes required at peak usage.

An important consideration regards the method of air usage. Air can be passed through the composting matrix only once (single pass) (Fig. 13.1) or it can be

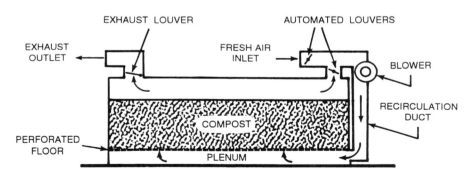

Figure 13.4 Physical configuration of a closed (contained) composting system with forced ventilation (cutaway side view). This particular example depicts a generalized multiple-pass ventilation system, often referred to as a tunnel (arrows depict possible gas flow routes). To reduce the volume of exhaust gas, a temperature conditioning unit could be installed in the recirculation duct (see text and Fig. 13.2*d*). Removal of the recirculation hardware would allow system to operate in a single-pass ventilation mode.

constantly recirculated (multiple pass) through the matrix (Fig. 13.4) (Harper et al., 1992; Hogan and Finstein, 1991). As seen in Figure 13.2, implementation of single-pass ventilation is typically straightforward, whereas the utilization of multiple pass requires a more sophisticated control system.

Oxygen Control

The optimal oxygen status for the microbial degradation of hazardous wastes will be compound specific. For example, anaerobic conditions were required for the reductive dehalogenation and mineralization of a variety of halobenzoates (Suflita et al., 1982). In the majority of cases, however, microbes requiring oxygen as a terminal electron acceptor play a dominant role in the degradation of both hazardous and nonhazardous compounds. While operating a treatment system to produce anaerobic conditions is simply a matter of avoiding oxygenation, aerobic operation requires supplying oxygen at a rate that meets or exceeds the maximum rate of utilization.

Ventilation serves to actively force fresh air through the interstices of the composting matrix, thereby replenishing oxygen and removing carbon dioxide and other gaseous products. While oxygen can easily be supplied in excess, this can lead to excessive heat loss (suboptimal temperatures) and energy expenditure. Oxygen supply therefore must be regulated.

Maintenance of desired interstitial oxygen concentrations can be accomplished through automatic oxygen control by delivering air in reference to an oxygen setpoint, as governed by an oxygen sensor and controller. Due to the technical difficulties of obtaining a representative sample from open systems, as a practical matter automatic oxygen control is restricted to contained systems where the headspace process gas can be sampled before mixing with ambient air (Fig. 13.2 and 13.4). An economical alternative to automatic oxygen control is the use of "baseline" ventilation (Fig. 13.2a). This method delivers ventilation at scheduled intervals through the timer actuation of blowers (e.g., 2 min every 10 min) (Finstein et al., 1983; MacGregor et al., 1981). While this does not take into account the time-variable nature of oxygen consumption encountered in composting, when combined with ventilative temperature control (discussed later) baseline ventilation is necessary only during the short period of temperature ascent and during the quiescent temperature descent stage (Fig. 13.2b) (Finstein and Hogan, 1993b). If desired, periodic adjustment of the delivery schedule can be made in reference to manual oxygen measurements.

In actual practice, the maintenance of high interstitial oxygen concentrations does not ensure uniform oxygenation. The inner region of particles can become oxygen limited due to slow diffusion rates, especially when wet (Haug, 1980). Therefore anaerobic conditions can exist, possibly to a significant extent, within a material exhibiting high interstitial oxygen concentrations. While reducing particle size decreases the potential for anaerobic conditions by decreasing the length of the gas transport pathway to the center of the particle, it can also inhibit uniform air distribution by reducing porosity. Thorough oxygen distribution therefore requires reconciliation of potentially conflicting factors such as particle size, porosity, and moisture content.

Maintaining aerobic conditions increases the potential for high rates of microbial metabolism and subsequent heat storage. In the absence of controlled heat removal, microbially inhibitive temperatures can be attained. Therefore the maintenance of aerobic conditions is necessary but not sufficient for effective treatment. Provisions for temperature regulation are also required.

Temperature Control

Regulation of composting material temperatures is essential to promote the establishment of desired microbial populations and maximal metabolic rates. This requires the control of heat flow to provide for both heat storage (when temperatures are ascending to or descending from desired levels) and heat loss (when temperatures exceed the optimum). Promoting heat storage entails minimizing heat loss by delivering ventilation solely for oxygenation purposes. Inducing heat loss is best accomplished through the use of the ventilation in reference to a temperature setpoint (Finstein et al., 1983; MacGregor et al., 1981).

A consequence of using ventilation for the control of temperature is that oxygen is supplied. The relationship between the amount of air required for cooling and that required for supplying oxygen has been termed the *air function ratio* (Finstein et al., 1986). At temperatures encountered in composting, more air is required to remove a given amount of heat than is necessary to supply the oxygen necessary for its generation. The actual value of the ratio is determined by the thermodynamic conditions of the inlet air and outlet gas. Given inlet conditions of 20°C and 50% humidity and outlet conditions of 60°C and 100% humidity (saturated air), with a metabolic conversion rate of 14,000 kJ/kg (6020 BTU/lb) oxygen, approximately 9 times as much air is required for cooling as for oxygen resupply. If inlet conditions remain as above, but outlet temperatures are decreased (i.e. a lower control temperature), the ratio widens. At saturated outlet conditions of 35°C, the corresponding air function ratio would be 29:1 (Finstein and Hogan, 1993b).

Implementation of ventilative temperature control includes an automatic controller that continuously monitors matrix temperatures using probes (preferably the average of several strategically positioned within the material). As the temperature exceeds a selected setpoint (shown to optimize degradation of the target compounds), ventilation is supplied to force the temperature back down to setpoint. During periods of ventilative temperature control, sufficient oxygenation is assured due to the air function ratio. In contrast, when temperatures are below setpoint (the ascent to and descent from setpoint), automatic oxygen regulation or adequate baseline ventilation is required to maintain oxygenated conditions.

Temperature Setpoint Selection

The temperature setpoint selected should enhance particular treatment objectives. An obvious goal for nonhazardous waste composting is the maximization of heat generation, which is equivalent to the degradation of putrescibles. Hazardous waste composting differs since the goal is the degradation of the target contaminants. This may or may not translate to temperatures that maximize organic amendment destruction.

Because microbes exhibit substrate specificity and possess individual temperature optima, temperature may strongly affect the rate and extent of degradation of the target compounds. Similarly, temperature can influence the availability (solubility) and volatilization of the target compounds. It is therefore conceivable that each target compound possesses a temperature optimum with regard to treatment effectiveness. For example, mesophilic setpoints resulted in higher rates of disappearance during the composting of certain polycyclic aromatic hydrocarbons (PAHs) (Finstein et al., 1990; Hogan et al., 1989a), while for certain explosives' wastes thermophilic setpoints proved to be superior (Williams et al., 1992). For certain wastes containing mixtures of hazardous compounds, a compromise temperature or the use of sequential temperature stages may be indicated.

Additionally, relatively small variations in temperature can significantly alter treatment rates. For example, we reported that although the treatment of various PAHs was effective at setpoints of 35, 40, 45, and 50°C, the setpoints of 55 and 60°C severely decreased treatment effectiveness (Finstein et al., 1990). Given the apparent sensitivity to temperature, further research is necessary to determine optimum temperature setpoints through both empirical and mechanistic methods.

Temperature Gradients

The flow of cooling air into the composting mass establishes a temperature gradient in the direction of air flow, with the coolest areas being at the point of air entry. The steepness of the gradient is a function of the rate of air mass delivery, which in turn is controlled by rates of metabolic heat generation. Therefore periods of intense microbial activity will induce steep gradients and can establish both mesophilic and thermophilic zones within the same mass (Finstein et al., 1983; Hogan et al., 1989b). Additionally, the composting mass is subject to conductive heat loss, creating a gradient between outer surfaces (cooler) and the interior (warmer).

Although temperature gradients might induce a greater diversity of microbial populations, it would do so in a stratified manner. Additionally, if a particular temperature has been determined to bring about the most efficient treatment of the compounds of interest, even small gradients could substantially compromise treatment. It is therefore useful under certain circumstances to minimize temperature gradients. One way this could be accomplished is by limiting the amount of organic amendment. Limited supplementation might be contraindicated, however, since it also can restrict degradative activity and duration of temperature elevation. Conduction-induced temperature gradients can be decreased by insulating the container or by common-wall construction. Periodic mechanical agitation will temporarily restore temperature uniformity and expose material to a greater diversity of microbial populations. Constant agitation can be used to essentially eliminate temperature gradients, though costs might be high.

Exhaust Gas Minimization and Treatment

The composting of hazardous wastes involves high rates of ventilation and relatively high temperatures, possibly leading to the volatilization of organic contaminants. Additionally, exhaust process gases are typically saturated with water, indicating the potential for co-distillation of target compounds. Therefore collection and treatment of exhaust process gases will be required for most hazardous waste composting operations.

To decrease the required size and operational costs of emissions treatment units, the generation of exhaust process gas should be minimized consistent with effective treatment. This is perhaps most effectively accomplished through separation of the ventilative functions of heat removal and oxygenation by employing automatically cooled multiple-pass ventilation (Fig. 13.4) (Hogan and Finstein, 1991). This requires the automatic introduction of fresh air only for purposes of oxygenation. Although this will remove some heat, the remaining heat removal can be achieved by automatically cooling the recirculating process gas in an external heat exchanger. If properly executed, nearly all heat removal can be accomplished without the use of fresh air, thereby reducing the volume of exhaust to that required for oxygenation.

When employing single-use ventilation, or multiple-pass systems that lack automatic cooling of the recirculated process gas, the utilization of high-temperature setpoints (i.e., 60°C) will narrow the air function ratio. While this decreases the amount of ventilation required for heat removal, higher temperatures may not be optimal for degradation. The use of less organic amendment will also reduce ventilation requirements (and overall residue) but again might restrict degradative activity and the period of temperature elevation. Ultimately, the minimization of exhaust process gas is a complex, site- and process-specific problem requiring the analysis of various performance and economic considerations.

Conveying the exhaust gases to the treatment area requires that the material be in a closed container. Design of the exhaust treatment unit is dependent on factors such as the chemical nature and concentrations of compounds it must treat or capture, the need for moisture removal, and the volume of gas requiring treatment. Available treatment options include biological filters that employ a biologically active, solid matrix controlled so as to degrade volatile organic compounds (VOCs) (Williams and Miller, 1992a); liquid spray scrubbers with subsequent biological or chemical treatment of the liquids; trickling filters (see Chapter 5); and incineration (Ottengraf, 1986). The presence of chemically dissimilar compounds in the exhaust process gases might require using some combination of the above treatment processes.

Water Addition

Because the use of ventilation results in the removal of water, moisture might become biologically limiting during the course of treatment. For this reason, composting systems ideally should have provisions to add water in a tightly controlled and uniform manner as needed during the composting cycle.

Since the mixing of the hazardous waste composting material in the open is contraindicated, water addition should occur within the container. Although applying water to the composting material and allowing gravitational drainage is the simplest method for midcourse water addition, this practice will tend to fill pore spaces with excess water, resulting in poor water and ventilation distribution. Furthermore, upward ventilative airflow preferentially dries the material at the bottom due to the undersaturation of the incoming air (Finstein et al., 1983). Therefore agitation in combination with, or after, water addition will be necessary for uniform water distribution. A possible method is the utilization of top-to-bottom ventilation in combination with upper surface water addition. This practice would supply water at the site of excessive drying and possibly lessen the need for water distribution-associated agitation.

Mechanical Agitation

While there is a paucity of data concerning the effects of agitation, its capacity to abrade material may prove to be an essential feature of hazardous waste composting, since aggregates can physically shield inner contaminants from microbial degradation. Additional benefits include the uniform distribution of water, substrate, and microbes, and the restoration of porosity to compacted materials. Agitation cannot usually serve as the sole means of oxygenation due to the limited, uncontrolled and intermittent nature of gas exchange (Finstein et al., 1986).

Agitation systems typically employed in large-scale operations utilize machines that move through the matrix, including suspended augers and inclined mixers/translocators. Due to the dense, heterogeneous, and abrasive nature of many wastes, these systems are typically maintenance intensive. Fixed horizontal rotating drum systems offer superior mixing abilities, as they can turn the entire batch simultaneously. This would also facilitate controlled midcourse rewetting. In contrast, the intrinsic properties of this physical configuration usually requires ventilative air to be introduced from the end of the drum and only pass over the material rather than through it. As the ventilative air passes through the drum, its capacity to supply oxygen and remove heat are increasingly diminished, making uniform aeration and temperature control problematic.

Extended Treatment (Curing)

The time required to complete processing will depend on the nature of the target compounds, the associated solids and organic amendment, and treatment method. Primary treatment duration could conceivably range from several days to several months. Until more data become available, pilot-scale operation is required to determine both efficacy and processing period.

To minimize the use of costly primary treatment units, partially processed material can be removed from the primary units for further treatment in open or roofed areas with provisions for leachate collection. Process control requirements and duration of extended treatment will be a function of target compound concentrations and the extent of organic amendment stabilization. Subsequent to an effective primary stage, periodic agitation might suffice for oxygen regulation in a secondary stage.

CASE STUDIES

Laboratory Studies

Laboratory experimentation concerning the composting of hazardous compounds often employs unrealistic conditions including the use of artificial waste preparations and single, as opposed to mixed, wastes. Additionally, while composting test systems permit elevated temperatures, most fall short of the complex process of composting simulation. Many impose fixed temperatures or nonrepresentative temperature programs that control material temperature based on an experimental regimen, rather than simulating actual conditions. Similarly, field-scale effects such as compaction, waste heterogeneity, and daily and seasonal temperature fluctuations are absent. Nevertheless, laboratory studies can be of value, since they allow for contaminant accountability (mass balance), environmentally controlled conditions, convenient test quantities, and are not subject to a complex regulatory permitting process. Several examples follow, with emphasis on work completed in this laboratory.

To investigate the fate of common hazardous contaminants in sewage sludge during composting, ^{14}C-labeled carbaryl (a pesticide) and phenanthrene (a polynuclear aromatic hydrocarbon or PAH) were added in low concentrations (1.3–2.2 ppm) to a sewage sludge–woodchip mixture (Racke and Frink, 1989). After 18–20 days in a laboratory composting apparatus, between 1.6 and 4.9% of the carbaryl was recovered as ^{14}CO$_2$ with nearly all of the remainder bound to the soil organic matter. Between 89 and 93% of initial phenanthrene remained unchanged at the termination of the composting period. Due to a lack of deliberate temperature regulation, temperatures peaked at 69°C or higher in all but one of the trials. These temperatures might have exceeded tolerance levels of competent populations, possibly accounting for the low levels of degradation.

Using a composting physical model developed in our laboratory (Hogan et al., 1989b), a mixture of sewage sludge, sludge compost, and acclimated inoculum was contaminated with various hazardous compounds and composted for a period of 20 days (Hogan et al., 1989a). Temperature setpoints representing mesophilic (35°C) and thermophilic (50°C) conditions were tested. The target compounds and their percent of disappearance (including volatilization and partial degradation) at 35 and 50°C, respectively, were 1-octadecene (98 and 95%), 2,6,10,15,19,23-hexamethyltetracosane (75 and 51%), phenanthrene (99.7 and 97%), fluoranthene (94 and 90%), and pyrene (93 and 87%). This indicates that a predominance of mesophilic temperatures might provide a slight advantage for the treatment of these compounds. In the same study (setpoint = 50°C), 81% of PCB Aroclor 1232 disappeared. As these studies only determined disappearance of the parent compound, and no radioisotope determination was performed, the extent of transformation, and irreversible partition to solid and vapor phases, are unknown.

A subsequent study (Finstein et al., 1990) using this physical model was conducted to determine the effect of temperature on the rate of disappearance of 13 PAHs contained in coal tar oil (creosote). A mixture of whole ground rice hulls and rice flour was dosed with 0.3% creosote and composted at each of the setpoints (35, 40, 45, 50, 55, and 60°C) for a 20-day period. As an example, the proportion of initial pyrene remaining in the composting mixture after 20 days was 8.3, 8.4, 4.1, 8.5, 74.6, 85.3% for those setpoints, respectively. These results reflect the behavior of the majority of the PAHs, in that overall disappearance was typically greatest at the 45°C setpoint, while the 55 and 60°C setpoints significantly inhibited PAH degradation.

A follow-up investigation studied the effect of organic amendment on soil spiked with creosote during simulation of land treatment, forced aeration treatment, and composting (Finstein and Hogan, 1993a). Three treatments included no amendment (land and forced aeration treatment), and amendment with fresh cow manure (land treatment and composting). Each treatment received two different levels of creosote addition (0.1 and 0.3%). Although the sawdust-amended soil was subjected to composting simulation, there was no temperature elevation, and it is therefore considered to be forced aeration treatment. The manure-amended composting treatment was the only treatment to self-heat and displayed the greatest amount of creosote volatilization. Composting treatments showed that organic amendment inhibited PAH degradation at the 0.3% creosote dose, while the 0.1% dose was best treated through manure amendment. The simulation of land treatment showed that disappearance was greatest without organic amendment at both creosote concentrations, with the 0.3% treatment demonstrating the greatest difference between amended and unamended treatments. These results suggest that organic amendment might play a positive role at the lower creosote concentrations but may spare PAH degradation at higher creosote concentrations. Therefore a recommended protocol might be to exclude organic amendment and alleviate oxygen, inorganic nutrient, and water limitation in soils contaminated with higher concentrations of creosote. Conversely, soils contaminated with low concentrations might be best treated through composting (organic amendment added). Further research is required to determine the efficacy of this approach and its application to other types of hazardous compounds.

Field Studies

Because large amounts of actual wastes are employed, field-scale experimentation is subject to substantial economic and regulatory constraints. Additionally, performing tightly controlled studies with target compound accountability is problematic. Despite these obstacles, a limited number of field studies involving hazardous waste composting have been documented.

An area receiving significant attention with respect to the utilization of composting for hazardous waste treatment is the remediation of explosives-contaminated soil (Isbister et al., 1984; Kaplan and Kaplan, 1982; Osman and Andrews, 1978; Williams et al., 1992; Williams and Myler, 1990). One field demonstration utilized an open, forced-aeration, temperature-controlled composting system for the treatment of TNT (2,4,6-trinitrotoluene), HMX (octahydro-1,3,5,7-tetraazocine), and RDX (hexahydro-1,3,5-trinitro-1,3,5-triazine) contaminated soils and sediments using both thermophilic (55°C) and mesophilic (35°C) temperature setpoints (Williams et al., 1992). The treatment matrix consisted of the contaminated soils/sediments and a mixture of alfalfa, straw, manure, horse feed, and wood chips. Half-lives of 11.9 and 21.9 days for TNT, 17.3 and 30.1 days for RDX, and 22.8 and 42.0 days for HMX were observed under thermophilic and mesophilic conditions-respectively. Although the route of disappearance is unknown (degradation, volatilization, polymerization, etc.), these data suggest that thermophilic conditions are more suitable for these particular wastes. Commonly formed nitroaromatic transformation products were detected during the first several weeks of the experiment but later diminished to low levels.

The efficacy of composting grass clippings from a lawn deliberately treated (all treatments < 40 ppm) with the commonly used pesticides diazonon, chloropyrifos, isofenphos, and pendimethalin was examined (Lemmon and Pylypiw, 1992). Lawn clippings were composted in small, open bins ($1.2 \times 1.2 \times 1.2$ m), without forced aeration so as to mimic informal composting methods typically practiced by homeowners. Although temperatures were not actively controlled, observed composting material temperatures never exceeded 52°C. Within 17 weeks, all four pesticides had reached undetectable levels.

Another study measured the disappearance of PAHs from a coal-tar-contaminated soil (6330 ppm total hydrocarbons) treated using a forced-aeration system (Taddeo et al., 1989). Soil preparation included incorporation of an unspecified "inert" bulking agent and commercial inorganic nutrient mixture. Composting mass temperatures were manually maintained within the range of 18–29°C. It is not clear whether weak self-heating or deliberate temperature control was the cause for the minimally elevated temperatures. If the bulking agent was truly inert, the majority of heat production would likely have stemmed from the metabolism of the coal tar. Regardless, at the end of a 78-day test period, 94% of the total PAHs were not able to be recovered. The routes of disappearance were not investigated.

FUTURE APPLICATION

As demonstrated by the limited amount of operational experience, it is evident that the field of hazardous waste composting remains in its infancy. Studies are commonly anecdotal and unrepresentative of realistic conditions. Laboratory-scale studies typically employ questionable simulation devices, and field-scale experiments lack the infrastructure (adequate, uniform, experimental processing units) required to conduct composting in the manner of actual treatment situations. This lack of data and operational experience hinders "real-world" application.

Extensive field-scale (and to some degree laboratory-scale) research using actual contaminated solids, in properly designed composting systems, is necessary to establish standard treatment criteria. Principal research areas requiring systematic investigation include determining the types of hazardous wastes treatable through composting, fate of target compounds, optimization of organic amendment selection and quantity, temperature setpoint, and oxygen, moisture, inorganic nutrient, and pH levels.

Proprietary systems suitable for research quality field-scale experimentation (and full-scale treatment) include, but are not limited to, composting "tunnels" initially developed in the mushroom growing industry (Fig. 13.4). While traditionally used for agricultural purposes, tunnels are now being used for the treatment of solid wastes (Finstein, 1993). These contained units provide multiple-pass ventilation and sophisticated temperature and oxygen control. Such units could be designed to incorporate the automatic cooling of recirculating process gas so as to minimize the generation of exhaust gases, as described earlier. Contained systems employing single-use ventilation that possess competent temperature and oxygen control are also adequate. Because both unit types are contained, exhaust gases could be led to an exhaust treatment facility. Provisions for midcourse agitation should be included whenever possible.

Composting applications requiring organic amendment to promote self-heating should undergo comparison against unamended systems (forced-aeration treatment), since the benefits of high substrate density, such as increased rates and extent of degradation through temperature, cometabolic, and other effects, are possible in some circumstances, but not necessarily in all. As evidenced by laboratory experimentation, it is possible that the presence of organic amendments may actually retard treatment. Amendments also increase operational requirements and cost, including increased process residue, facility size, power requirements, exhaust gas, and amendment acquisition and storage costs. This indicates that default treatment should utilize nonamended systems and employ composting only when performance and economic results outweigh the additional demands.

Portable Systems

Since most hazardous waste composting treatment scenarios would involve the remediation of substantial volumes of soil, on-site treatment is preferable as it reduces materials handling and transportation costs. Additionally, field-scale research and pilot-scale operation might be best performed on-site. One possible conceptualization is that of a portable, modular container system, possibly constructed through conversion of a tractor trailer. Each portable trailer could possess its own set of ventilation and process controls, blowers, and possibly an electric generator for remote operation. Exhaust gas treatment, loading/unloading, and agitation equipment could be shared between many units. Upon successful treatment of the waste, residues might be returned to their place of origin.

CONCLUSIONS

Although composting has a long-standing history as an organic waste treatment method, its use for hazardous waste remediation is a recent development. It requires not only stricter attention to biological process control but also a shift in approach, since supplemental organic material will often be added to the waste. These amendments must be selected and environmental conditions controlled so as to facilitate the rapid and complete biodegradation of the target compounds, which are often a small percentage of the overall organics present. Composting can be conducted with a high degree of control over a wide range of complex, interactive microbial growth conditions, substrate types and concentration, temperature, oxygen, and moisture content. Further laboratory- and field-scale research needs to be redirected in order to avoid perpetuating flawed, yet frequently employed practices (such as those without adequate temperature control), and to provide systematic investigation, using properly designed facilities, into the underlying principles governing the composting of hazardous wastes.

Acknowledgments

The author deeply thanks Drs. Melvin S. Finstein and Peter F. Strom for their helpful discussions and review of this material.

REFERENCES

Atlas, R. M. and R. Bartha (1981). *Microbial Ecology Fundamentals and Applications*, Addison-Wesley. Reading, MA.

Bach, P. D., K. Nakasaki, M. Shoda, and H. Kubota (1987). Thermal Balance in Composting Operations, *J. Ferment. Tech. (Jpn.)*, **65**:199–209.

Bartha, R. and I. D. Bossert (1984). The Treatment and Disposal of Petroleum Refinery Wastes, in *Petroleum Microbiology*, R. M. Atlas, Ed., MacMillan, New York, pp. 553–577.

Beam, H. W. and J. J. Perry (1974). Microbial Degradation of Cycloparaffinic Hydrocarbons via Co-metabolism and Commensalism *J. Gen. Microbiol.*, **82**:163–169.

Boyle, C. D., B. R. Kropp, and I. D. Reid (1992). Solubilization and Mineralization of Lignin by White Rot Fungi, *Appl. Environ. Microbiol.*, **58**(10):3217–3224.

Bumpus, J. A. (1989). Biodegradation of Polycyclic Aromatic Hydrocarbons by *Phanerochaete chrysosporium*, *Appl. Environ. Microbiol.*, **55**(1):154–158.

Devinny, J. S. and R. L. Islander (1989). Oxygen Limitation in Land Treatment of Concentrated Wastes, *Haz. Waste Haz. Mat.*, **6**(4):421–433.

Dibble, J. T. and R. B. Bartha (1979). Effect of Environmental Parameters on the Biodegradation of Oil Sludge, *Appl. Environ. Microbiol.*, **37**(4):729–739.

Dragun, J. (1988). *The Soil Chemistry of Hazardous Materials*, Hazardous Materials Control Research Institute, Silver Spring, MD.

Fernando, T., J. A. Bumpus, and S. D. Aust (1990). Biodegradation of TNT (2,4,6-Trinitrotoluene) by *Phanerochaete chrysosporium*, *Appl. Environ. Microbiol.*, **56**(6):1666–1671.

Finstein, M. S. (1993). Guide to Matching Composting Technology to Circumstance, *Composting Frontiers*, Winter, pp. 9–14, 19.

Finstein, M. S. and J. A. Hogan (1993a). *Optimization of Biological Soil Remediation*, Report to the Hazardous Substance Management Research Center, New Jersey Institute of Technology, Newark, NJ.

Finstein, M. S. and J. A. Hogan (1993b). Integration of Composting Microbiology, Facility Structure and Decision Making, in *Science and Engineering of Composting: Design, Environmental, Microbiological and Utilization Aspects*, (H. Hoitink and H. M. Keener, Eds.), Renaissance, Worthington, OH, pp. 1–23.

Finstein, M. S., F. C. Miller, P. F. Strom, S. T. MacGregor, and K. M. Psarianos (1983). Composting Ecosystem Management for Waste Treatment, *Bio/Tech.*, **1**:347–353.

Finstein, M. S., F. C. Miller, and P. F. Strom (1986). Waste Treatment Composting as a Controlled System, in *Biotechnology, Vol. 8, Microbial Biodegradations*, (W. Schonborn, ed.), VCH Verlagsgesellschaft, Weinheim, FRG, pp. 363–398.

Finstein, M. S., F. C. Miller, J. A. Hogan, and P. F. Strom (1987a). Analysis of EPA Guidance on Composting Sludge. Part I: Biological Heat Generation and Temperature, *Biocycle*, **28**(1):20–26.

Finstein, M. S., F. C. Miller, J. A. Hogan, and P. F. Strom (1987b). Analysis of EPA Guidance on Composting Sludge. Part II: Biological Process Control, *Biocycle*, **28**(2):42–47.

Finstein, M. S., F. C. Miller, J. A. Hogan, and P. F. Strom (1987c) Analysis of EPA Guidance on Composting Sludge. Part III: Oxygen, Moisture, Odor and Pathogens, *Biocycle*, **28**(3):13–19.

Finstein, M. S., F. C. Miller, J. A. Hogan, and P. F. Strom (1987d). Analysis of EPA Guidance on Composting Sludge. Part IV: Decomposition Rate, and Facility Design and Operation, *Biocycle*, **28**(4):20–25.

Finstein, M. S., J. V. Hunter, and J. A. Hogan (1990). *Treatment of Industrial Solid Wastes Through Composting: Process Optimization at Mesophilic and Thermophilic Temperatures*, Final Report, Hazardous Substance Management Research Center, New Jersey Institute of Technology, Newark, NJ.

Fogarty, A. M. and O. H. Touvinen (1991). Microbiological Degradation of Pesticides in Yard Waste Composting, *Microbiol. Rev.*, **55**(2):225–233.

Goyal, S. (1983). Choosing Between Induced and Forced-Draft Fans, *Chem. Eng.*, 7 February, pp. 92–93.

Hambrick, G. A., III, R. D. DeLuane, and W. H. Patrick, Jr. (1980). Effect of Estuarine Sediment pH and Oxidation-Reduction Potential on Microbial Hydrocarbon Degradation, *Appl. Environ. Microbiol.*, **40**(2):365–369.

Harper, E., F. C. Miller, and B. J. Macauley (1992). Physical Management and Interpretation of an Environmentally Controlled Composting Ecosystem, *Austral. J. Exper. Agric.*, **32**:657–667.

Haug, R. T. (1980). *Compost Engineering Principles and Practice*, Ann Arbor Publishers, Ann Arbor, MI.

Hogan, J. A. and M. S. Finstein (1991). Composting of Solid Waste During Extended Human Travel and Habitation in Space, *Waste Mgmt. Res.*, **9**:453–463.

Hogan, J. A., G. R. Toffoli, F. C. Miller, J. V. Hunter, and M. S. Finstein (1989a). Composting Physical Model Demonstration: Mass Balance of Hydrocarbons and PCBs, in *Proc. Intern. Conf. Physiochemical Biological Detoxification Hazardous Wastes,*Yeun, C. Wu, Ed. Technomic, Lancaster, PA, pp. 742–758.

Hogan, J. A., F. C. Miller, and M. S. Finstein (1989b). Physical Modeling of the Composting Ecosystem, *Appl. Environ. Microbiol.,* **55**(5):1082–1092.

Hunter, J. V., M. S. Finstein, D. J. Suler, and R. R. Bobal (1981). *Fate of Concentrated Industrial Wastes During Laboratory Scale Composting of Sewage Sludge*, Report of New Jersey Agricultural Experiment Station, New Brunswick, NJ.

Isbister, J. D., G. L. Anspach, J. F. Kitchens, and R. C. Doyle (1984). Composting for Decontamination of Soils Containing Hazardous Wastes, *Microbiologies,* **7**:47–73.

Kaplan, D. L. and A. M. Kaplan (1982). Thermophilic Biotransformations of 2,4,6-Trinitro-toluene under Simulated Composting Conditions, *Appl. Environ. Microbiol.,* **44**(3):757–760.

Keck, J., R. C. Sims, M. Coover, K. Park, and B. Symons (1989). Evidence for Cooxidation of Polynuclear Aromatic Hydrocarbons in Soil, *Wat. Res.,* **23**(12):1467–1476.

Kuter, G. A., H. A. J. Hoitink, and L. A. Rossman (1985). Effects of Aeration and Temperature on Composting of Municipal Sludge in a Full-Scale Vessel System, *J. Wat. Poll. Cont. Assoc.,* **57**(4):309–315.

Lajoie, C. A., S. -Y. Chen, K. -C. Oh, and P. F. Strom (1992). Development and Use of Field Application Vectors to Express Nonadaptive Foreign Genes in Competitive Environments, *Appl. Environ. Microbiol.,* **58**(2):655–663.

Lemmon, C. R. and H. M. Pylypiw, Jr. (1992). Degradation of Diazinon, Chloropyrifos, Isofenphos, and Pendimethalin in Grass and Compost, *Bull. Environ. Contam. Toxicol.,* **48**:409–415.

Levi, M. P. and E. B. Cowling (1969). Role of Nitrogen in Wood Deterioration. VII. Physiological Adaptation of Wood-Destroying and Other Fungi to Substrates Deficient in Nitrogen, *Phytopathology,* **59**:460–468.

Lindsey, A. W. and K. Jakobson (1989). Demonstrating Technologies for the Treatment of Hazardous Wastes at EPA, in *Proc. Intern. Conf. Physiochemical Biological Detoxification Hazardous Wastes*, (Yeun, C. Wu, Ed.). Technomic, Lancaster, PA, pp. 17–34.

Lindstrom, J. E. and E. J. Brown (1989). Supplemental Carbon Use by Microorganisms Degrading Toxic Organic Compounds and the Concept of Specific Toxicity, *Haz. Waste Haz. Mater.,* **6**(2):195–200.

Lynch, J. and B. R. Genes (1989). Land Treatment of Hydrocarbon Contaminated Soils, in *Petroleum Contaminated Soils*, Vol. 1. P. T. Kostecki and E. J. Calabrese, Eds., Lewis Publishers, Chelsea, MI, pp. 163–174.

MacGregor, S. T., F. C. Miller, K. M. Psarianos, and M. S. Finstein (1981). Composting Process Control Based on Interaction Between Microbial Heat Output and Temperature, *Appl. Environ. Microbiol.,* **41**:1321–1330.

Martens, R. (1982). Concentrations and Microbial Mineralization of Four to Six Ring Polycyclic Aromatic Hydrocarbons in Composted Municipal Waste, *Chemosphere,* **11**(8):761–770.

McKinley, V. L. and J. R. Vestal (1984). Biokinetic Analysis of Adaptation and Succession: Microbial Activity in Composting Municipal Sludge, *Appl. Environ. Microbiol.,* **47**(5):933–941.

Miller, F. C. (1989). Matric Water Potential as an Ecological Determinant in Compost, a Substrate Dense System, *Microb. Ecol.*, **18**:59–71.

Miller, F. C., S. T. MacGregor, K. M. Psarianos, J. Cirello, and M. S. Finstein (1982). Direction of Ventilation in Composting Wastewater Sludge, *J. Wat. Poll. Cont. Fed.*, **54**(1):111–113.

Muller, W. P. and F. Korte (1975). Microbial Degradation of Benzo[a]pyrene, Monolinuron, and Deildrin in Waste Composting, *Chemosphere*, **3**:195–198.

Osman J. L. and C. C. Andrews (1978). *The Biodegradation of TNT in Enhanced Soil and Compost Systems*, U. S. Army Armament Research and Development Command Report ARLD-TR-77032, Dover, NJ.

Ottengraf, S. P. P. (1986). Exhaust Gas Purification, in *Biotechnology Vol. 8, Microbial Biodegradations*, (W. Schonborn, Ed.), VCH Verlagsgesellschaft, Weinheim, FRG, pp. 425–452.

Petruska, J. A., D. E. Mullins, R. W. Young, and E. R. Collins, Jr. (1985). A Benchtop System for Evaluation of Pesticide Disposal by Composting, *Nucl. Chem. Waste Mgmt.*, **5**:177–182.

Preslo, L., M. Miller, W. Suyama, M. McLearn, P. Kostecki, and E. Fleischer (1989). Available Remedial Technologies for Petroleum Contaminated Soils, in *Petroleum Contaminated Soils*, Vol. 1, P. T. Kostecki and E. J. Calabrese, Eds., Lewis Publishing Chelsed, MI pp. 115–136.

Racke, K. D. and C. R. Frink (1989). Fate of Organic Contaminants During Sewage Sludge Composting, *Bull. Environ. Contam. Toxicol.*, **42**:526–533.

Raymond, R. L., V. W. Jamison, and J. O. Hudson (1967). Microbial Hydrocarbon Co-oxidation. I. Oxidation of Mono- and Dicyclic Hydrocarbons by Soil Isolates of the Genus Nocardia, *Appl. Microbiol.*, **15**(4):857–865.

Rose, W. W. and W. A. Mercer (1968). *Fate of Insecticides in Composted Agricultural Wastes*, National Canners Association, Washington, DC.

Rosenzweig, W. D. and G. Stotzky (1980). Influence of Environmental Factors on Antagonism of Fungi and Bacterium in Soil: Nutrient Levels, *Appl. Environ. Microbiol.*, **39**(2):354–360.

Skladany, G. J. and F. B. Metting, Jr. (1993). Bioremediation of Contaminated Soil, in *Soil Microbial Ecology*, F. B. Metting, Jr., Ed., Marcel Dekker, New York, pp. 483–513.

Slater, J. H. and D. Lovatt (1984). Biodegradation and the Significance of Microbial Communities, in *Microbial Degradation of Organic Compounds*, D. T. Gibson, Ed., Marcel Dekker, New York, pp. 439–485.

Strom, P. F. (1985). Effect of Temperature on Bacterial Species Diversity in Thermophilic Solid-Waste Composting, *Appl. Environ. Microbiol.*, **50**:899–905.

Suflita, J. M., A. Horowitz, D. R. Shelton, and J. M. Tiedje (1982). Dehalogenation: A Novel Pathway for the Biodegradation of Haloaromatic Compounds, *Science*, **218**(10):1115–1116.

Sutherland, J. B. (1992). Detoxification of Polycyclic Aromatic Hydrocarbons by Fungi, *J. Indust. Microbiol.*, **9**:53–62.

Taddeo, A., M. Findlay, M. Danna, and S. Fogel, (1989). Field Demonstration of a Forced Aeration Composting Treatment for Coal Tar, in *Proc. of the 2nd National Conference*, The Hazardous Materials Control Research Institute, Applied Biotreatment Association Case History Compendium, Nov. 27–29, Washington D C, pp. 57–62.

Thomas, J. M., J. R. Yordy, J. A. Amador, and M. Alexander (1986). Rates of Dissolution and Biodegradation of Water-Insoluble Organic Compounds, *Appl. Environ. Microbiol.*, **52**(2):290–296.

U. S. Environmental Protection Agency (1992). *Code of Federal Regulations, No. 40, Parts 264 and 268*, Office of the Federal Register National Archives and Records Administration, U. S. Government Printing Office, Washington, DC.

Williams, T. O. and F. C. Miller (1992a). Odor Control Using Biofilters, *Biocycle*, **33**(10):72–77.

Williams, T. O. and F. C. Miller (1992b). Biofilter and Facility Operations, *Biocycle*, **33**(11):75–79.

Williams, R. T. and C. A. Myler (1990). Bioremediation Using Composting, *Biocycle*, **31**(11):78–82.

Williams, R. T., P. S. Ziegenfuss, and W. E. Sisk (1992). Composting of Explosives and Propellant Contaminated Soils under Thermophilic and Mesophilic Conditions, *J. Indust. Microbiol.*, **9**:137–144.

Wilson, J. T. and B. H. Wilson (1985). Biotransformation of Trichloroethylene in Soil, *Appl. Environ. Microbiol.*, **49**:242–243.

Index

Acclimated, developed, and enriched cultures
(microorganisms), 37, 87, 122–125, 219, 243, 255,
278–279, 283–284, 288–290, 331, 346, 368, 375
Acenaphthene
biodegradation of, 57–58, 89
solubility, volatility, 89
Acetic acid and acetate, 122, 250, 273, 284, 286, 336,
338, 340–341, 343
Acetogens, anaerobic processes, 340–341
Acid mine drainage, 340
Actinomycetes, 87
Activated carbon, see Carbon and charcoal, activated
Activated-sludge process, see also Sludge
biomass (sludge) age, comparisons, formation,
recycle, and use, 42–44, 46, 52, 54–55, 62, 70,
111, 187, 280, 285, 288, 290–292
biomass separators (clarifiers), 19–20
principle of operation, design, and performance, 6–7,
16, 19–20, 46, 52, 80, 92
Adenosine triphosphate (ATP) as measure of
biodegradative potential, 244, 333–334, 340
Adsorption/desorption, see also Carbon, activated
biofiltration of VOC vapors, 122, 129–130, 133–134,
137, 143, 146
bioremediation and composting, limitations and
design
nutrient availability, 254
rate of contaminant desorption and availability,
255–262, 363
definition, 74–75
fixed film reactor, 50, 53
hydrogeologic factors, transport parameters, 201–203
permeability, effect on, 217
soils and other particles in hydrogeological, bioslurry
reactor, CSTR, and composting applications,
[permeability and desorption enhancement
(transfer to aqueous phase)] 38, 69, 88, 201,
233, 289, 358, 363
adsorption partition coefficient (K_d), 75
apparent distribution ratio for ionogenic substances
(K), 76
enhancement of desorption, 76–79
equilibrium curve (adsorption isotherm), 75–76
interfaces, effect of, 74
normalized partition coefficient (K_{oc}), 75–76
surfactants, effect of, 76–79, 262–263
Advanced fixed-film reactor, 41–47, see also Fixed-film
reactors
Advection, see Flow processes
Aerated lagoons, bioslurry reactor, 71–72
Aeration, see also Membrane biofilm reactor; Oxygen;
Flow processes, soil–air
airlift (gaslift), 17
aerobic vs. anaerobic considerations, 328
basins, 7, 16–17
bubble (diffused), 16–17, 103
composting, 359–361, 363, 370–371, 374, 376–378
diffusers, 48–49
hydrogeological, 208
jet reactors, 18
membrane bioreactor, 103, 115, 294
packed-bed and fixed-film reactors, 22–23, 50–51,
53, 61–63, 288
reactor classification, gas sparging, and off-gas
collection requirements, 2, 7, 12, 15–16, 18
rotating disk, 26–27
sequencing batch, 93
sequential anaerobic–aerobic, 22
slurries, 72–73, 85–86, 93
sparging
aerobic requirement, 2
systems, 13
stirred, 7, 12
surface, 15–16
systems, 13–15
Aerobic processes, see also Oxygen; individual
chemicals
affect of depth in fixed film, 43–45
biomass formation from, 55
efficiency vs. anaerobic processes, 56, 328–329,
332–333
naphthalene, 248

385

RETURN
TO ➡